Progress in Mathematics
Volume 111

Matthias Schwarz

Morse Homology

Birkhäuser Verlag
Basel · Boston · Berlin

Author:

Matthias Schwarz
ETH Zentrum
Mathematik
8092 Zürich
Switzerland

Library of Congress Cataloging-in-Publication Data

Schwarz, Matthias, 1967–
Morse homology / Matthias Schwarz.
p. cm. – (Progress in mathematics ; vol. 111)
Includes bibliographical references and index.
ISBN 0-8176-2904-1
1. Morse theory. 2. Homology theory. I. Title. II. Series:
Progress in mathematics (Boston, Mass.) ; vol. 111.
QA331.S43 1993
515'.73 – dc20

Deutsche Bibliothek Cataloging-in-Publication Data

Schwarz, Matthias:
Morse homology / Matthias Schwarz. – Basel ; Boston ; Berlin
: Birkhäuser, 1993
(Progress in mathematics ; Vol. 111)
ISBN 3-7643-2904-1 (Basel ...)
ISBN 0-8176-2904-1 (Boston)
NE: GT

Camera-ready copy prepared by the author
Printed on acid-free paper produced of chlorine-free pulp
Printed in Germany
ISBN 3-7643-2904-1
ISBN 0-8176-2904-1

9 8 7 6 5 4 3 2 1

Contents

List of symbols

$\overline{\mathbb{R}}$	compactification of $\mathbb{R}$ to $\mathbb{R} \cup \{\pm\infty\}$	22
$\mathcal{L}(X;Y)$	space of continuous linear Banach space operators	29
$\mathcal{F}(X;Y)$	space of Fredholm operators between Banach spaces	29
$\mathcal{K}(X;Y)$	space of compact operators	30
$\mathcal{S}$	set of non-degenerate, conjugated self-adjoint operators on $\mathbb{R}^n$	29
$\mathcal{A}$	matrix-valued curves with ends within $\mathcal{S}$	29
$\Sigma_{\xi,\nabla}$	set of special Fredholm operators	35
Θ_{F_A}	equivalence class of operators from Σ_{triv} with fixed ends	35
Det	determinant space of a Fredholm operator	105
$\widehat{K}_\psi$	surjectively extended Fredholm operator	109
$H^{1,2}(\mathbb{R},\mathbb{R}^n)$	Sobolev space of curves square-integrable together with the first derivative	22
$H^{1,2}_{\overline{\mathbb{R}}}$	Banach space of sections in a vector bundle on $\overline{\mathbb{R}}$ with $H^{1,2}$-quality	24
$L^2_{\overline{\mathbb{R}}}$	Banach space of sections with L^2-quality	25
C^∞_{loc}	space of smooth mappings with C^∞-compact-open topology	61
C^∞_0	smooth mappings with compact support	41
$C^\infty_{x,y}$	$\overline{\mathbb{R}}$-smooth curves with fixed ends x, y	22
$\mathcal{P}^{1,2}_{x,y}$	Banach manifold of $H^{1,2}$-curves with fixed ends	24
$\mathcal{M}^f_{x,y}$	trajectory space associated to the time-independent gradient flow of a Morse function f	21
$\mathcal{M}^{h^{\alpha\beta}}_{x_\alpha,x_\beta}$	trajectory space associated to the time-dependent gradient flow of a Morse homotopy	21
$\mathcal{M}^{H^{\alpha\beta}}_{x_\alpha,y_\beta}$	λ-parametrized trajectory space associated to a homotopy of Morse homotopies	21
$h^{\alpha\beta}$	Morse homotopy	21
$\widehat{\mathcal{M}}^f_{x,y}$	unparametrized trajectory space associated to time-independent gradient flow	54
$\mathcal{G}_{g_0}$	Banach space of variations of the Riemannian metric	45
D_u	differential $DF_u(0)$ of the Fredholm map $\left(\frac{\partial}{\partial t} + \nabla f\right)_{\mathrm{loc}}$ in specific local coordinates on $\mathcal{P}^{1,2}_{x,y}$	75
$\gamma \bullet \tau = \gamma_\tau$	time-shifting of a $\overline{\mathbb{R}}$-curve	52
$\#_\rho$	gluing operation for trajectories	68
$\#^o_\rho$	pre-gluing of trajectories, etc.	71
$\widehat{\#}_\rho$	gluing operation for unparametrized trajectories	89
χ	short notation for the triple of a broken trajectory and a gluing parameter	73

β^+, β^-	frequently used cut-off functions	71
$\Lambda^{\max}$	space of volume forms of a finite dimensional vector space	104
$\sigma[u, K]$	coherent orientation of the Fredholm operator K along the curve u	129
$o_{\text{geom}}[u_{xy}]$	geometric orientation of a trajectory	222
Λ	set of equivalence classes of Fredholm operators along curves	129
Γ	transformation group for coherent orientations	129
$\mathcal{C}_\Lambda$	set of coherent orientations	129
Ω	special natural isomorphism between orientations	127
$\text{Crit}_k f$	set of critical points of a Morse function f with fixed Morse index k	135
$C_k(f)$	k-th chain group generated by the elements of $\text{Crit}_k f$	135
$C_k(f, f_A)$	k-th relative chain group associated to an admissible pair of Morse functions	184
$H_k(f, f_A)$	k-th relative homology group of the relative Morse complex associated to (f, f_A)	184
$\mathcal{D}$	injectivity domain within the tangent bundle with respect to the exponential map	211
h^*TM	pull-back bundle with respect to the smooth curve h	211
$h^*\mathcal{D}$	pull-back of the injectivity domain with respect to h	211
$\nabla_2 \exp$	fibre linearization of the exponential map with respect to a Levi-Civita connection	215
Θ	special bundle map on the tangent bundle associated to the Levi-Civita connection	215
$\nabla_t \xi$	covariant time-derivative of a vector field along a curve	215

Chapter 1

Introduction

1.1 Background

The subject of this book is Morse homology as a combination of relative Morse theory and Conley's continuation principle. The latter will be used as an instrument to express the homology encoded in a Morse complex associated to a fixed Morse function independent of this function. Originally, this type of Morse-theoretical tool was developed by Andreas Floer in order to find a proof of the famous Arnold conjecture, whereas classical Morse theory turned out to fail in the infinite-dimensional setting. In this framework, the homological variant of Morse theory is also known as Floer homology. This kind of homology theory is the central topic of this book. But first, it seems worthwhile to outline the standard Morse theory.

1.1.1 Classical Morse Theory

The fact that Morse theory can be formulated in a homological way is by no means a new idea. The reader is referred to the excellent survey paper by Raoul Bott [Bo].

The classical theory in its formulation by Marston Morse relates the analytic information by means of the critical set of the so-called Morse function to the global topology of the underlying manifold (see [M1]). The local information about the critical points of a Morse function $f \in C^\infty(M, \mathbb{R})$ ("Morse" means $df \pitchfork 0_M \subset T^*M$, i.e. at each $x \in \operatorname{Crit} f$ the Hessian $H^2 f(x)$ is non-degenerate) is reflected by the Morse index,

$$\mu(x) = \dim \operatorname{Eig}^- H^2 f(x)$$

as the number of negative eigenvalues of the Hessian operator at $x \in \operatorname{Crit} f$ counted with multiplicity. This relation between analysis and topology is expressed in the form of the so-called strong Morse inequalities: We denote the Betti numbers and numbers of critical points of index k by

$$\begin{aligned} &\beta_k = \dim H_k(M, \mathbb{R}) \,,\ k = 0, \ldots, n = \dim M \ , \\ &C_k = \# \operatorname{Crit}_k f \,,\ \operatorname{Crit}_k f = \{\, x \in \operatorname{Crit} f \mid \mu(x) = k \,\} \,,\ k = 0, \ldots, n \ . \end{aligned}$$

Then the following holds:

$$\begin{aligned} c_k - c_{k-1} + \ldots \pm c_0 &\geqslant \beta_k - \beta_{k-1} + \ldots \pm \beta_0 \text{ for } k = 0, \ldots, n-1 \text{ and} \\ c_n - c_{n-1} + \ldots \pm c_0 &= \chi(M) \ . \end{aligned}$$

The key observation which finally provides this result is the change of the homotopy type of the sublevel sets

$$M^a = \{\, x \in M \mid f(x) \leqslant a \,\} \ ,$$

when a crosses a critical value. For each critical point x with $f(x) = a$, a cell of dimension $\mu(x)$ is attached, when a crosses the value $f(x)$,

$$M^{a+\epsilon} \simeq M^{a-\epsilon} \cup \bigcup_{x \in \operatorname{Crit} f \cap f^{-1}(c)} e_x^{\mu(x)} \ .$$

Starting from the absolute minimum[1], one obtains a cell decomposition of M up to homotopy equivalence,

$$M \simeq \bigcup_{x \in \operatorname{Crit} f} e_x^{\mu(x)} \ .$$

The first steps toward an intrinsically defined Morse homology were taken successively by Thom, Smale and Milnor (40's–60's) by not only taking into account the number and types of cells in the cell decomposition indexed by the critical points but also the attaching maps. Let us henceforth stick to the convention of considering the negative gradient flow

$$\begin{aligned} &\psi : \mathbb{R} \times M \to M \\ &\frac{\partial}{\partial t}\psi(t,x) = -\nabla f\big(\psi(t,x)\big) \,,\ \psi(0,\cdot) = \mathrm{id}_M \end{aligned} \tag{1.1}$$

associated to a Morse function f on a compact, n-dimensional manifold M. Given this dynamical system, one is led to the stable and unstable manifolds

$$\begin{aligned} W^u(x) &= \{\, p \in M \mid \lim_{t \to -\infty} \psi(t,p) = x \,\} \ , \\ W^s(x) &= \{\, p \in M \mid \lim_{t \to +\infty} \psi(t,p) = x \,\} \end{aligned}$$

1This exists if M is assumed to be compact or f to be coercive, i.e. all M^a compact.

associated to the isolated[2] critical points $x \in \operatorname{Crit} f$. As early as the 1940's, R. Thom recognized that the disjoint decomposition of the compact manifold M,

$$M = \bigcup_{x \in \operatorname{Crit} f} W^u(x) \ ,$$

describes naturally the homotopically equivalent cell decomposition analysed in Morse theory. In fact, the unstable and stable manifolds are cells of dimension $\mu(x)$ and $n - \mu(x)$, respectively, with tangent spaces

$$T_x W^u(x) = \operatorname{Eig}^- H^2 f(x) \qquad \text{and} \qquad T_x W^s(x) = \operatorname{Eig}^+ H^2 f(x) \ .$$

Unfortunately, this immediate cell decomposition does not represent directly a CW-complex, because in general the dynamical system might misbehave in such a way that the closure condition is not properly fulfilled. Nevertheless, in the 50's, S. Smale found the crucial condition for this gradient dynamical system[3], which allows us to describe all ingredients of a Morse theoretical homology, namely, the critical points as generators and the attaching maps of the associated Morse complex as boundary relations in terms of the unstable and stable manifolds. He discovered that the requirement of transversal intersections

$$W^u(x) \pitchfork W^s(y) \qquad \text{for all } x, y \in \operatorname{Crit} f$$

appears as a generic condition for the dynamical system. For instance, one can always find a generic Riemannian metric on M. This is the so-called Morse-Smale condition. Given this condition, it is immediate that $W^u(x) \cap W^s(y)$ is a $(\mu(x) - \mu(y))$-dimensional manifold which intersects the hypersurface $f^{-1}(a)$ for any regular value $a \in (f(y), f(x))$ transversally. Seeing that $f^{-1}(a)$ carries a natural orientation by $-\nabla f$, one is able to define a number $n(x, y)$ for pairs $(x, y) \in \operatorname{Crit} f \times \operatorname{Crit} f$ with relative Morse index $\mu(x) - \mu(y) = 1$. $n(x, y)$ is defined as the intersection number of $W^u(x)$ and $W^s(y)$ restricted to the niveau surface $f^{-1}(a)$. It is well-defined if we are provided with arbitrary orientations of the unstable manifolds $W^u(x)$, $x \in \operatorname{Crit} f$ which imply orientations of the complementary manifolds $W^s(x)$ given that M is oriented. In other terms, the short exact sequence

$$0 \to T_p\big(W^u(x) \cap W^s(y)\big) \to T_p W^u(x) \oplus T_p W^s(y) \to T_p M \to 0 \ ,$$

associated to the transversal intersection

$$p \in W^u(x) \cap W^s(y) \cap f^{-1}(a) \ ,$$

yields the characteristic intersection number $n(x, y)$. These numbers happen to describe exactly the attaching maps for the cells within the Morse complex.

2by the above assumption of non-degeneracy
3generalized to gradient-like vectorfields

This idea was taken up by J. Franks in [Fr]. In his paper he dealt with these attaching maps in terms of the Thom-Pontryagin construction.

Thus one is able to define the following chain complex. The chain groups, graded by means of the Morse index, are defined as the free Abelian groups

$$C_k(f) = \mathbb{Z}^{\mathrm{Crit}_k f}\ , \quad k = 0, \ldots, n\ ,$$

generated by the critical points $x \in \mathrm{Crit}_k f$. The boundary operator $\partial_k : C_k(f) \to C_{k-1}(f)$ is defined by

$$\partial_k x = \sum_{y \in \mathrm{Crit}_{k-1} f} n(x,y)\, y\ , \quad x \in \mathrm{Crit}_k f\ . \tag{1.2}$$

The key observation is that the associated homology $H_k\big(C_*(f), \partial\big)$, $k = 0, \ldots, n$, recovers the standard homology of the underlying compact manifold. For a fruitful application of this interrelation, the reader is referred to the famous h-cobordism theorem, see [M2]. This isomorphism theorem for homology theories immediately implies the strong Morse inequalities. In fact, this is the essence of classical Morse theory, and the homological formulation is assigned to Thom, Smale and Milnor in equal parts.

In 1982, E. Witten (see [W]) rediscovered this way of recovering the (co-)homology of the underlying manifold by solely taking into account the critical points of a Morse function, which are graded by their Morse index, and the isolated connecting orbits between those points of relative index 1, provided that the Morse-Smale condition holds. With this observation, Andreas Floer is said to have been led to the idea of considering Morse theory in thoroughly relative terms. Note that the above homology groups are constructed merely on the basis of non-degenerate critical points of a functional, which are graded by some Morse index, together with the information about the sets $\mathcal{M}^f_{x,y}$ of connecting orbits with relative index 1. Up to this moment the latter has been described in terms of the entire global gradient flow,

$$\mathcal{M}^f_{x,y} \equiv W^u(x) \pitchfork W^s(y)\ .$$

Also the proof of the fundamental isomorphism result,

$$H_*\big(C_*(f), \partial\big) \cong H_*(M)\ ,$$

relies on the analysis of the cell decomposition of the manifold based on the global gradient flow where M is assumed to be compact. In fact, it is largely due to Andreas Floer that the essential ingredients of this Morse homology can be extracted from the spaces of connecting orbits, the so-called trajectory spaces, in a completely relative way. Counting the isolated trajectories between the critical points x and y yields the intersection number $n(x,y)$ within (1.1).

1.1.2 Relative Morse Theory

The prime motivation for the development of a purely relative concept of Morse theory is rooted in the existence problem of periodic solutions for a time-dependent Hamiltonian system considered under the non-degeneracy assumption. For a detailed presentation of this subject leading to the so-called Floer homology, the reader is referred to [McD], [S], [S-Z1] and [S-Z2]. The variational approach consists of restating the existence problem of solutions of a differential equation as an existence problem of critical points of an appropriate functional. This functional ought to be related to the topology of the underlying spaces in a more obvious way. It is well known from the classical theory that Marston Morse applied it to the existence problem of closed geodesics as critical points of the energy functional

$$\left(\gamma : S^1 \to M\right) \mapsto E(\gamma) = \int_{S^1} \left|\dot{\gamma}(t)\right|^2 dt \ ,$$

defined on an appropriate space of loops in the manifold M. Hence, one is naturally led to try to use these methods equally for the action functional $\mathcal{A}$ associated to the above Hamiltonian system,

$$\mathcal{A}(\gamma) = \int\limits_{D^2} \tilde{\gamma}^* \omega - \int_0^1 H\big(t, \gamma(t)\big)\, dt$$

with respect to the symplectic form ω and

$$\tilde{\gamma} : D^2 \to M \,, \quad \tilde{\gamma}_{|\partial D^2} = \gamma : S^1 \to M \ .$$

One introduces an L^2-scalar product on the vector spaces playing the role of tangent spaces to the loop space on which the functional $\mathcal{A}$ is defined. Then the zeroes of the gradient of $\mathcal{A}$ due to this scalar product can immediately be identified with the solutions of the Hamiltonian equation via

$$\operatorname{grad} \mathcal{A}(\gamma) = J(\dot{\gamma} - X_H \circ \gamma) \tag{1.3}$$

where

- X_H is the Hamiltonian vector field depending explicitly and periodically on the time t with period 1, and
- J is an almost complex structure associated to the Hamiltonian system with the symplectic structure ω.

The standard approach, however, which aims at the application of classical Morse theory in the infinite-dimensional setting, fails for the following reasons.

First, unlike the energy functional, the action is unbounded below and above[4], such that there is no absolute minimum with which to start a Morse complex in the sense of a cellular decomposition. Second, and even more striking, is the lack of a finite Morse index. The subspaces on which the Hessian operator is maximally positive and negative definite are both of infinite dimensions. Finally, the most natural candidate for a gradient of this action functional, $\operatorname{grad}\mathcal{A}$ as mentioned above, does not even give rise to a well-defined flow on the loop space in question. This gradient behaves in a so-called unregularized way.

Yet there is still hope for Morse theory. Namely, Andreas Floer recognized that in this situation all essential conditions for a Morse theory reduced to a purely relative version are fulfilled. This means a restriction to the relative gradient flow, i.e. a flow between fixed critical points x and y as limit sets, and to a relative Morse index which, roughly speaking, measures something like a codimension of the 'unstable manifold' of y with respect to the 'unstable manifold' of x. Here, we take the classical Morse theory considered above as a finite-dimensional model. This relative Morse index turns out to be finite. Setting up the nonlinear partial differential equation

$$\begin{aligned} &u:\mathbb{R}\times S^1\to M\,, \qquad \lim_{s\to\pm\infty} u(s,\cdot)=x(\cdot)\,,\ y(\cdot)\ \text{respectively,} \\ &\tfrac{\partial}{\partial s}u(s,t)+J\tfrac{\partial}{\partial t}u(s,t)-JX_H\big(t,u(s,t)\big)\ =0\ , \end{aligned} \tag{1.4}$$

Floer was able to handle finite-dimensional solution sets $\mathcal{M}_{x,y}$, which in Morse theoretical terms describe exactly the 'trajectory spaces' of connecting orbits between fixed critical points x and y. In fact, equation (1.4) might be restated formally using (1.3) as

$$\frac{\partial}{\partial s}u+\operatorname{grad}\mathcal{A}\circ u\ =0\ .$$

Floer identified the local dimension of these trajectory spaces $\mathcal{M}_{x,y}$ as the appropriate relative Morse index within this setting. The $\mathcal{M}_{x,y}$ prove to be manifolds provided that one is given a generic J. Using this relative grading on the set of critical points for the action functional and the sets of isolated connecting orbits[5], he found all ingredients for a homological concept known today as Floer homology.

4Note that the first term in the formula for $\mathcal{A}$ describes the area enclosed by γ measured with sign.

5The local dimensions of the trajectory manifolds lead to such a well-defined relative grading only if certain conditions on the topology of the underlying manifold are fulfilled. Therefore, the Arnold conjecture still remains unproven in strict generality.

1.1.3 The Continuation Principle

The second fundamental part of the homology theory, besides the idea of relativity in Morse theory, is the continuation principle. This aspect has been investigated by Charles Conley in order to describe the topological type of invariant sets of flows in a way which is invariant under continuous deformations of the whole system. Concerning this topic the reader is referred to [C] and [C-Z]. The continuation principle appears as an essential idea within the theory of the Conley index. Due to this principle Floer was able to relate his homology theory based on the periodic solutions of a fixed Hamiltonian system to the topology of the underlying manifold. The idea proved very transparent.

First, one has to choose a time-independent Morse function H^o on the given compact manifold small enough in the C^2-norm, so that there are no 1-periodic solutions of the Hamilton equation

$$\dot{\gamma}(t) = J\nabla H^o\big(\gamma(t)\big)\,, \;\big(\text{ note that } X_{H^o}(t,p) = J\nabla H^o(t,p)\;\big)$$

apart from the constant ones, namely the critical points of H^o. Actually, the Morse homology associated to H^o is naturally isomorphic to the Floer homology of H^o by this identification (for a generic J). Conley's continuation principle is now integrated by the observation that any generic homotopy between H^o and an arbitrary fixed Hamiltonian induces a unique isomorphism for the associated Floer homology groups in a canonical way, i.e. independent of the concrete choice of the homotopy. Additionally, one obtains the following functorial behaviour for these isomorphisms $\Phi_{\beta\alpha} : HF_*(H_\alpha) \xrightarrow{\cong} HF_*(H_\beta)$ between the homology groups for the Hamiltonians H_α and H_β:

$$\Phi_{\gamma\beta}\circ\Phi_{\beta\alpha} = \Phi_{\gamma\alpha} \quad\text{and}\quad \Phi_{\alpha\alpha} = \mathrm{id}_{HF_*(H_\alpha)}\,. \tag{1.5}$$

Therefore, the family

$$\big\{\, HF_*(H_\alpha)\,\big|\, H_\alpha \text{ an admissible Hamiltonian for Floer homology}\,\big\}$$

together with these canonical isomorphisms forms what Conley named a connected simple system. By means of these isomorphisms one can carry out an inverse limit process and obtain the Floer homology $HF_*(M)$ explicitly independent of the concrete Hamiltonian. This is the continuation principle which provides homology groups intrinsically independent of a special Hamiltonian. Recalling that the time-independent Morse function H^o considered above establishes the normal homology of M, provided that M is compact, it is now obvious that by Conley's continuation principle the Floer homology reproduces the ordinary homological information about M.

Actually, this very inventive and fruitful theory by Andreas Floer not only concerns the Arnold conjecture on symplectic manifolds but was also transferred successfully to 3-dimensional topology. Specifically this abstract method based on relative Morse theory and the continuation principle by Conley can be applied to the Chern-Simons functional in 3-dimensional gauge field theory thus leading to the instanton homology (see [F4]), which is often denoted by Floer homology, also.

1.2 Overview

The main significance of this study, entitled 'Morse homology', is that it gives a presentation of a homological concept of Morse theory in strict analogy with Floer homology. Therefore it may serve as a finite-dimensional model. The results are based principally on the same methods and techniques and the prime emphasis lies on the detailed presentation of the underlying analysis. A survey of the analytic programme will be given in the next section on 'Remarks on the Methods'. Since the organization of this work is primarily governed by this analytic programme, the following overview is intended to convey an idea of the logical approach to the Morse homology rather than give a mere outline of the contents. The latter is postponed to the end of the introduction.

Notwithstanding the first remark, the subject of this study is not only a finite-dimensional modelling of Floer homology. It also provides a construction of Morse homology as an axiomatic homology theory. This means an axiomatic theory in the sense of Eilenberg and Steenrod (see [E-S]). As a consequence of this axiomatic approach one immediately obtains the uniqueness result for Morse homology, that is a natural equivalence to any other axiomatic homology theory which is defined on the same suitably large category of topological spaces, like for instance singular theory. The advantage of this result based on the axiomatic approach is that, in contrast to the classical investigations by Thom, Smale and Milnor, it is valid also for non-compact manifolds and for all Morse functions simultaneously. The latter in particular is due to Conley's continuation principle.

1.2.1 The Construction of the Morse Homology

Before we present the actual axiomatic approach to Morse homology we shall first outline the development of the Morse homology groups themselves.

The first step is to define the canonical chain complex associated to any fixed Morse function. This complex, which has been related to the strong Morse inequalities in the previous section, is called the Morse complex $C_*(f)$ of f. In constrast to the treatment by Thom, Smale, Milnor and Witten, we shall prove the fundamental complex property, $\partial^2 = 0$, intrinsically. In other words, we shall deduce the identity

$$\partial^2 x = \sum_{\mu(z)=\mu(y)+1} \sum_{\mu(y)=\mu(x)+1} \langle \partial z\,, y\rangle \langle \partial y\,, x\rangle z = 0$$

directly from the analysis of the trajectory spaces for arbitrary smooth manifolds, without assuming orientability, and for general coefficients. The fact that Witten's idea of counting the isolated trajectories of the negative gradient flow (1.1) leads to a ∂-operator of descending degree 1, relies on the $\mathbb{R}$-symmetry of the trajectory spaces with respect to time-shifting. $\gamma(\cdot + \tau)$ yields a new trajectory with the same endpoints, provided that they are different. Hence, one obtains a set of isolated trajectories (modulo time-shifting) for relative Morse index 1.

In the second step we shall introduce Conley's continuation principle. Motivated by the classical observation that the Morse complex for any Morse function on a compact manifold recovers the standard homology of the manifold, one is naturally inclined to search for an isomorphism between the graded homology groups associated to two different Morse functions. Moreover, similar to the canonical ∂-operator, the isomorphism should arise from a given pair of Morse functions in a canonical way. In fact, this can be verified if we consider the isolated trajectories for the time-dependent gradient field

$$\dot{\gamma} = -\nabla h_t \circ \gamma \ , \tag{1.6}$$

where

$$f^\alpha \overset{h_t}{\simeq} f^\beta$$

is a (nearly arbitrary) homotopy between the two Morse functions. Since, due to explicit dependence on the time t, the solution curves of (1.6) are no longer endowed with time-shifting invariance, we exactly obtain isolated trajectories under generic conditions for critical endpoints $x_\alpha \in \operatorname{Crit} f^\alpha$ and $x_\beta \in \operatorname{Crit} f^\beta$ with relative Morse index

$$\mu(x_\alpha) - \mu(x_\beta) = 0 \ .$$

In formal analogy with the definition of the ∂-operator we can thus define the homomorphism

$$\Phi^{\beta\alpha} x_\alpha = \sum_{\mu(x_\beta)=\mu(x_\alpha)} n(x_\alpha, x_\beta)\, x_\beta \tag{1.7}$$

by counting the isolated connecting orbits in an appropriate way. This is the candidate for a morphism between the chain complexes $C_*(f^\alpha)$ and $C_*(f^\beta)$, which complies with the grading. Hence, there are two items to be analysed. First, one has to show that this homomorphism commutes with the ∂-operators associated to f^α and f^β. Second, one has to verify that the induced homomorphism $\Phi_*^{\beta\alpha}$ between the associated graded homology groups does not depend on the choice of the homotopy h_t.

Quite interesting and revealing may be the fact that both aspects are treated by means of exactly the same techniques which are used for the proof of the $(\partial^2 = 0)$-relation. In fact, given two different homotopies h_t^0 and h_t^1 between f^α and f^β, we choose a suitable homotopy

$$h_t^0 \overset{H_t^\lambda}{\simeq} h_t^1$$

introducing an additional parameter $\lambda \in [0,1]$ for the analysis. If we consider the trajectories of the associated λ-parametrized and explicitly time-dependent gradient flow we shall obtain isolated trajectories for the relative Morse index -1. Actually, referring to the analogous definition of a homomorphism like in (1.2) and (1.7) we shall be led to a chain homotopy operator, from which we can conclude that both homotopies, h_t^0 and h_t^1, induce the same morphism on the level of the homology groups.

Altogether, the induced morphism $\Phi_*^{\beta\alpha} : H_*(f^\alpha) \to H_*(f^\beta)$ fits in with the concept of Conley's continuation principle as presented by (1.5) in the previous section. The equivalence relation

$$\{x_\alpha\} \sim \{x_\beta\} \overset{\text{def}}{\Longleftrightarrow} \Phi_*^{\beta\alpha}\{x_\alpha\} = \{x_\beta\}$$

on the family

$$\big(H_*(f^\alpha)\big)_{\{f^\alpha \text{ Morse function on } M\}}$$

gives rise to the definition of Morse homology groups independent of a fixed Morse function,

(1.8) $$H_*^{\text{Morse}}(M) = \big(H_*(f^\alpha)\big)_{\{f^\alpha\}} \Big/ \sim_{\Phi_*^{\beta\alpha}} .$$

We recall that that these homology groups are well-defined on arbitrary smooth manifolds provided that we consider Morse functions to be smooth functions $f \in C^\infty(M,\mathbb{R})$ with non-degenerate critical points and compact sublevel sets

$$M^a = \{\, x \in M \mid f(x) \leqslant a \,\}, \quad a \in \mathbb{R} \ ;$$

that is in short:

(1.9) $$f \in C^\infty(M,\mathbb{R}), \ \ df \pitchfork 0_M \subset T^*M \ \text{ and } \ f \text{ coercive} .$$

1.2.2 The Axiomatic Approach

Let us now sketch a portrait of the functorial and axiomatic treatment of Morse homology. The fundamental idea behind this approach is based upon the concentration of the Morse homology by means of the continuation principle, thus providing homology groups which are explicitly independent of a certain

Morse function. By reverse arguing, this concentration, organized as a limit process, always allows us to choose specific Morse functions, which fit the given situation best, and to control any change of the function in a unique manner without deviating from a fixed equivalence class $\big[\{x_\alpha\}\big]$. The following Morse functions, which are provided in a natural way on products and bundles with the local structure of a product, are of capital importance: If we are given manifolds $(M\,,f)$ and $(N\,,g)$ endowed with arbitrary Morse functions, then the operation

$$(f \oplus g)(m,n) = f(m) + g(n)$$

yields a Morse function on $M \times N$ and analogously in the case of a local product structure. The particular Morse function $f \oplus g$ obviously implies the identification

$$\mathrm{Crit}_n(f \oplus g) = \bigcup_{0 \leqslant k \leqslant n} \mathrm{Crit}_k f \times \mathrm{Crit}_{(n-k)} g \ . \tag{1.10}$$

Actually, this fundamental operation for Morse functions not only gives rise as a key feature to the entire functorial concept, it also forms the base for product operations within the (co-)homology theory which we wish to develop. The latter statement refers to the identity for the tensor chain complex

$$C_*(f) \otimes C_*(g) = C_*(f \oplus g) \ ,$$

which is concluded from (1.10).

Relative Morse Homology and the Excision Axiom

In order to develop an axiomatic homology theory we first have to begin with relative homology associated to a pair of manifolds, (M, A), where A forms a submanifold of M. Regarding the homotopy invariance restated below as the homotopy axiom we may restrict our treatment to open submanifolds, as any closed submanifold is equipped with an open tubular neighbourhood. We shall remain in analogy with the construction of the singular relative homology groups by means of suitable quotient complexes. Hence, we need an appropriate representation of subcomplexes in terms of a Morse complex. Essentially, this is provided by a restriction of the given Morse function to a subset of M, which contains the critical points of the subcomplex in question and from which no trajectories of the negative gradient flow emanate. Hence, we take functions as so-called relative Morse functions on admissible pairs of manifolds, such that the gradient field restricted to ∂A forms an outward pointing normal field. From this we shall derive graded relative homology groups $H_*(M, A)$ together with the axiomatically required long exact homology sequence. Since for the definition of $H_*(M, A)$ we only need to know the Morse function of M on a neighbourhood of $M \backslash A$ together with the transversality of the gradient with respect to ∂A, the excision axiom is verified immediately according to the construction.

The Functorial Property

The functorial behaviour of the Morse homology, however, cannot be deduced directly from the definition. It is clear that arbitrary smooth maps between manifolds do not preserve the properties of a Morse function as specified in (1.9). Actually, it is at this stage that we have to refer to the facilities provided by the independence from the actual Morse function as mentioned above. In the first place, let us restrict ourselves to closed embeddings,

$$\varphi : M \hookrightarrow N \ .$$

As a consequence, we obtain a chain homomorphism $\varphi_\bullet$ between suitable Morse complexes in the following way. Let f be an arbitrary Morse function on M and let ν be the normal bundle of the closed submanifold $\varphi(M)$. Then the embedding φ and the Morse function

$$q_{\varphi(m)}(x) = \langle x, x \rangle_{\varphi(m)}$$

defined by means of any fiberwise Riemannian metric on the bundle ν give rise to a Morse function

$$\varphi_*^o f\big(\varphi(m), x\big) = f(m) + q_{\varphi(m)}(x) \tag{1.11}$$

on the normal bundle ν in the way that has been described above, and we obtain a canonical identification

$$\mathrm{Crit}_*(\varphi_*^o f) = \mathrm{Crit}_* f \ .$$

Now considering a tubular neighbourhood of $\varphi(M)$, we are supplied with a function which can be extended to a Morse function on the entire manifold N, such that the Morse complex of f forms a subcomplex of $C_*(\varphi_* f)$ up to identification. Hence, every closed embedding induces a chain homomorphism for Morse homology.

Referring to similar investigations by means of the product property (1.10) this functorial concept may be extended to projections

$$pr_2 : M \times N \to N \ .$$

Then, finally, the homology functor for arbitrary smooth maps $\varphi : M \to N$ in general is obtained from the factorization

$$\varphi = (\mathrm{id}\,, \varphi) \ \circ \ pr_2 \tag{1.12}$$

into the closed embedding of the graph manifold and the projection.

The Homotopy Axiom and the Dimension Axiom

Finally, we finish this outline of the axiomatic approach with some remarks on the remaining two axioms. Regarding the homotopy axiom, this is immediately deduced from the continuation principle. Actually, the entire construction of the Morse homology is based upon the invariance under homotopy via the identification isomorphisms $\Phi_*^{\beta\alpha}$. Thus it is sufficient to consider homotopies of closed embeddings which give rise to homotopies of Morse functions.

Perhaps more novel and interesting is the investigation of the dimension axiom. Let us first review the framework for the Morse complex as it was analysed by Smale and Witten. They assumed the underlying manifold to be compact and orientable. As a consequence they obtained a complex both for the Morse function f and for $-f$, that is to say that they were able to consider the ascending and the descending boundary operator on equal terms. In fact, based upon these assumptions for M one naturally recovers the symmetry known as Poincaré duality. This symmetry is abolished immediately if one generalizes for non-compact manifolds or coefficients in $\mathbb{Z}$ in the non-orientable case. We observe that in general the right behaviour as homology theory, i.e. concerning the descending boundary operator for the negative gradient flow, is related to the coercivity of f as stated in (1.9). Obviously, any continuous function on a compact manifold is coercive. In summary it can be stated that coercivity is the crucial feature which underlies the dimension axiom, that is to say,

$$H_0^{\text{Morse}}(M,\mathbb{Z}) \cong \mathbb{Z} \ ,$$

if M is pathwise connected.

Products and Further Extensions

As we have observed above, the natural operation $f \oplus g$ for Morse functions in some local framework plays a crucial role within this kind of Morse homology theory. It is thus quite revealing to see that this feature gives rise to a rich variety of extensions to further items of algebraic topology, for example to product operations. Making use of the well-known dualization concept by means of the hom-functor, which provides a cohomology theory starting from chain complexes, together with the corresponding functorial behaviour with respect to the diagonal map

$$\triangle : M \longrightarrow M \times M$$

we recover the construction of the cup-product.

At the end of this chapter we shall give a brief outlook on the possibilities of the treatment of features like the Thom class or the Euler class associated to a vector bundle within the framework of the Morse (co-)homology theory. This possibility is due to the concept of relative Morse functions. If we are given any vector bundle $\pi : E \to M$ endowed with a Riemannian metric, the latter gives rise to the fiberwise quadratic, decreasing function

$$-q_m(x) = -\langle x, x\rangle_m\,,\quad \pi(x) = m \in M\ .$$

Combining this function with any arbitrary Morse function f on the base M we obtain a relative Morse function $f \oplus (-q)$ for the pair $\big(E\,, E\backslash\{0\}\big)$. Then the natural identification

$$\mathrm{Crit}_k f \;\cong\; \mathrm{Crit}_{n+k} f \oplus (-q)$$

should yield the starting point from which one can develop further elements of algebraic topology concerned with vector bundles.

1.3 Remarks on the Methods

As we have already mentioned, the essence of this treatise on Morse homology is the methodical relation to Floer homology. The main techniques which will appear within the analysis refer to Banach calculus on Banach spaces and Banach manifolds, for example the fundamental contraction mapping principle, as well as to Fredholm theory, which means either the theory of linear Fredholm operators or of nonlinear Fredholm maps. The framework for this calculus is typically infinite-dimensional. The purpose is to give a largely self-contained presentation. We also add a detailed deduction for the infinite-dimensional Banach manifolds which arise during the discussion and for which it is particularly troublesome to give an appropriate reference.

Before we come to a listing of the analytic programme which ought to serve as Ariadne's thread through the labyrinth of infinite-dimensional analysis, we first give an outline of the prime investigations.

The main interest of this approach to Morse homology in contrast to earlier treatments is the analysis of the trajectory spaces of negative gradient flow. Instead of analysing such spaces geometrically as transversal intersection manifolds $W^u(x) \pitchfork W^s(y)$, we shall treat them as zero sets of a nonlinear operator

$$F = \frac{\partial}{\partial t} + \nabla f : \gamma \mapsto \dot\gamma + \nabla f \circ \gamma \tag{1.13}$$

on a Banach manifold $\mathcal{P}^{1,2}_{x,y}$ of curves. This happens in strict analogy to Floer's manifolds of instantons. In our case $\mathcal{P}^{1,2}_{x,y}$ denotes the Hilbert manifold of $H^{1,2}$-curves[6]

$$\gamma : \mathbb{R} \cup \{\pm\infty\} \longrightarrow M$$

6These are square integrable together with the first weak derivative.

with fixed endpoints

$$\gamma(-\infty) = x\,, \;\; \gamma(+\infty) = y\;\;.$$

It forms the infinite-dimensional fundamental space for our analysis.

Now, exactly like in the case of the nonlinear Cauchy-Riemann operator with the additional Hamiltonian term in Floer's approach, the linearization of the elliptic operator F turns out to be of Fredholm type. This is largely due to the asymptotic ellipticity provided by the non-degeneracy of the Hessians at the critical endpoints. The crucial observation which gives rise to relative Morse theory is the statement that the local index of F is given by the relative Morse index $\mu(x) - \mu(y)$. This follows from the analysis of the spectral flow of the linearization of ∇f as the curve between the Hessians $H^2 f(x)$ and $H^2 f(y)$ with values in the set of endomorphisms in the pull-back bundle of TM along the trajectory γ.

The next step is to identify the classical Morse-Smale condition as the regularity of the zero section as a value for the non-linear operator F. Thus, provided a generic Riemannian metric, we can conclude the manifold property of the trajectory space $\mathcal{M}^f_{x,y} = F^{-1}(0)$ from the implicit function theorem. The convincing advantage of this Banach calculus is expressed as the fact that the analogous results on the trajectory spaces for the time-dependent gradient flow follow directly from a simple and immediate generalization of F to

$$\frac{\partial}{\partial t} + \nabla h_t\;\;.$$

We will recognize that these analytical methods allow us to unify the proofs of

- the chain complex identity $\partial^2 = 0$
- the chain morphism identity $\partial^\beta \Phi^{\beta\alpha} = \Phi^{\beta\alpha}\partial^\alpha$
- and the chain homotopy identity $\partial^\beta \Psi^{\beta\alpha} - \Psi^{\beta\alpha}\partial^\alpha = \Phi^{\beta\alpha}_1 - \Phi^{\beta\alpha}_0$

to a large extent. The intrinsic deduction of these three fundamental identities is a consequence of the study of the respective trajectory spaces. That study will involve the questions of "compactness" "gluing" and "orientation" for the totality of the trajectory spaces

$$\left\{\, \mathcal{M}^f_{x,y} \,\middle|\, (x,y) \in \operatorname{Crit} f \times \operatorname{Crit} f \,\right\}\;\;.$$

We shall obtain the above three identities from discussions of coherent orientations on all these trajectory spaces and of cobordism relations with respect to these orientations. In fact "compactness" and "gluing" are complementary concepts which yield the result that the trajectory manifolds without boundary[7]

7with respect to the $H^{1,2}$-topology

$\mathcal{M}^f_{x,y}$ are compact up to the splitting-up of trajectories, that is, up to the existence of sequences which converge to so-called broken trajectories in the C^∞_{loc}-topology. These are tuples $(u_1, \ldots, u_\nu)$ of trajectories from lower-dimensional trajectory spaces with coinciding critical endpoints,

$$\begin{aligned} & u_i(+\infty) = u_{i+1}(-\infty) \text{ for } i = 1, \ldots, \nu - 1 \\ & \text{and } u_1(-\infty) = x\,,\ u_\nu(+\infty) = y\ . \end{aligned}$$

"Gluing" amounts to exactly the reverse process, namely to mapping such broken trajectories equipped with an additional suitable parametrization into the appropriate higher-dimensional trajectory space,

$$\mathcal{M}_{x,y} \times \mathcal{M}_{y,z} \ni (u, v) \mapsto u\#_\rho v \in \mathcal{M}_{x,z}, \ \ \rho \geqslant \rho_0\ ,$$

such that the sequences $(u\#_{\rho_n} v)_{n\in\mathbb{N}}$ with $\rho_n \to \infty$ correspond exactly to the aforementioned obstructions to compactness. This complementarity of gluing and compactness engenders a cobordism relation between the trajectory manifolds in the sense of the C^∞_{loc}-topology. Especially for isolated trajectories, compactness guarantees that there are only finitely many trajectories to count.

Finally, 'coherent orientation' describes an orientation concept for these trajectory spaces which is respected by those cobordisms. This concept is based upon analysis of the so-called determinant bundle, which is a line bundle on the topological space of Fredholm operators for given fixed Hilbert spaces. The idea is to generalize the classical notion of the orientation of a finite-dimensional manifold within this special framework of Fredholm maps. Since the tangent spaces of the trajectory manifolds may be identified with the kernels of the linearized Fredholm sections F, this new concept also yields an orientation for an appropriate composite of kernel and cokernel of a non-surjective Fredholm operator. Thus, orientations of trajectory spaces are well-defined even if no regularity (i.e. the property of being a manifold) is guaranteed. This turns out to be exactly the right concept for an oriented cobordism analysis.

We thus obtain the proofs for the above three fundamental identities by means of a calculus of oriented cobordisms for the trajectory spaces associated to the negative gradient flow of a Morse function. Moreover, it is worth mentioning that the class of Fredholm operators which after trivialization are of the type

$$\begin{aligned} & \frac{\partial}{\partial t} + A : H^{1,2}(\mathbb{R}, \mathbb{R}^n) \to L^2(\mathbb{R}, \mathbb{R}^n)\,, \\ & A(\pm\infty) \text{ has real non-zero eigenvalues}\ , \end{aligned}$$

forms the key feature for the entire methodological foundation of this approach to Morse homology.

Discussions of the analogous analysis of the Floer type instanton spaces, which correspond to the trajectory spaces, can be found in [F1], [F2],

[F3], [F5], [F6], [McD], [S], [S-Z1] and [S-Z2]. Whereas in these treatments only Floer homology groups with coefficients in $\mathbb{Z}_2$ are considered, a new joint paper [F-H], by Floer and Hofer, presents the method of the so-called coherent orientation of certain Fredholm operators, which allows one to generalize the theory for coefficients in $\mathbb{Z}$.

After these remarks on the methods we now present the fundamental analytic programme which is exactly the same as for Floer homology:

1. analytical setup, definition of the trajectory spaces
2. analysis of the index problem
3. transversality: regularity results
4. compactness
5. gluing
6. coherent orientation.

1.4 Table of Contents

As a guide to the reader, we once again briefly sum up the contents of the monograph:

Chapter 2 comprises the entire analytical work on the trajectory spaces except for the orientation problem. The organization is firmly related to the above analytic programme. The purpose is to develop the complementarity concept of compactness and gluing as described above. Section 2.1 contains the construction of the trajectory spaces as zero sets of appropriate sections in Banach bundles on Banach manifolds of curves. Section 2.2 analyses the Fredholm property of these sections and identifies the Fredholm index as the relative Morse index by means of the spectral flow. Section 2.3 is devoted to the transversality problem. The regularity result is obtained from a parameter version of the theorem of Sard-Smale. Section 2.4 gives an account of the effect of splitting up of trajectories as an obstruction to the strong compactness of the trajectory spaces. This gives the first half of the crucial complementarity of compactness and gluing. Section 2.5 establishes gluing as the reverse operation by means of the Banach contraction mapping principle and finally yields the first basic results on the cobordism relations among the trajectory spaces.

Chapter ,3 on the subject of orientation, is separated from Chapter 2 on trajectory spaces because of its methodical complexity. In this chapter we shall develop an orientation concept for Fredholm operators which generalizes the notion of the orientation of a finite-dimensional manifold, so that together with the outcome of Chapter 2 we obtain a theory of oriented cobordisms for our trajectory manifolds.

Chapter 4 finally contains the main proofs for the construction of Morse homology. Section 4.1 is devoted to the three fundamental identities stated above, which are proved by means of the oriented cobordisms calculus. In Section 4.2 we go through the axioms of Eilenberg and Steenrod and finally come to the uniqueness result in Section 4.3, where the link to any other standard homology theory is established.

Chapter 5 presents an account of the possibilities for recovering the product operations and the naturally present Poincaré duality. Finally, in this chapter we give an outlook on possible further developments toward algebraic topology on vector bundles.

Chapter A of the appendix gives all the technical details which are necessary to construct and handle the Banach manifold of curves $\mathcal{P}^{1,2}_{x,y}$ together with some Banach bundles on $\mathcal{P}^{1,2}_{x,y}$ which turn up within the analysis of trajectory spaces.

Chapter B, finally, is intended to yield the missing link between the geometrical definition of the boundary operator according to Thom, Smale, Milnor and Witten and the definition within this present work. There we verify the equivalence of the coherent orientation concept with the classical concept of transversal intersection of the unstable and stable manifolds for the gradient flow $W^u(x) \pitchfork W^s(y)$, as far as the trajectory manifolds are concerned.

1.5 Acknowledgments

The author was partially supported by DAAD-Procope and the Graduiertenkolleg 'Geometrie und Mathematische Physik' at the Ruhr-Universität Bochum.

This monograph is an outgrowth of my diploma thesis completed in January 1992 under the direction of Prof. H. Hofer at the Ruhr-Universität Bochum. I am very much indebted to my teachers Helmut Hofer and Ralph Stöcker, from whom I learned much of this material and whose stimulating courses have generated great enthusiasm for the subject. I would like to express my deep gratitude to Professors Hofer and E. Zehnder for convincing me to publish this work and for their constant encouragement and invaluable support. I am also very grateful to Martin Reimann at the University of Bern for his warm hospitality and his stimulating interest. I profited from innumerable discussions with friends and colleagues. I am particularly indebted to Dietmar Salamon, Graeme Segal, Jürgen Jost and Norbert A'Campo for their advice and encouragement. Many people generously devoted considerable time and effort to reading the manuscript and helping me improve the exposition. Among them I am especially grateful to Stefan Handzsuj, Claudia Putz, Prof. Gordon Wassermann, Tilmann Wurzbacher and to Birkhäuser Verlag.

Zürich, April 1993 Matthias Schwarz

Chapter 2

The Trajectory Spaces

2.1 The Construction of the Trajectory Spaces

In this chapter we shall construct the trajectory manifolds needed for the definition of Morse homology. These trajectory spaces associated to the negative gradient flow correspond to the moduli spaces and instanton manifolds in Floer homology theory. In the most general form appearing in the following analysis, these trajectory spaces are composed of the solutions of the non-autonomous ordinary differential equation

$$\dot{\gamma} = -\frac{\nabla h_t}{\sqrt{1 + |\dot{h}_t|^2 |\nabla h_t|^2}} \circ \gamma \, . \tag{2.1}$$

The scaling factor in the denominator has been chosen in order to guarantee compactness in the case of time dependency (see the section on 'compactness').

To give a preview, the following trajectory spaces will be investigated:

- The time-independent case $h_t \equiv f$ yields the negative gradient flow of f. It serves to define the ∂-operator in the Witten complex. The time-independent trajectory spaces are denoted by $\mathcal{M}^f_{x,y}$.
- The time-dependent case of a Morse homotopy (see Definition 2.40)

  $$f^\alpha \overset{h_t^{\alpha\beta}}{\simeq} f^\beta$$

  with $\frac{\partial}{\partial t} h_t^{\alpha\beta} = 0$ for $|t| \geqslant R$ will be studied in order to construct the transformation homomorphisms $\Phi^{\beta\alpha}$ which provide the means for identifying homology groups associated to different Morse functions. Analogous to the time-independent case, we denote these time-dependent trajectory spaces by $\mathcal{M}^{h^{\alpha\beta}}_{x_\alpha,x_\beta}$.

- Finally, we shall analyze the parametrized case of the so-called λ-parametrized trajectory spaces used to define the chain homotopy operator, i.e.

$$\mathcal{M}^{H^{\alpha\beta}}_{x_\alpha,y_\beta} = \left\{ (\lambda,\gamma) \mid \gamma \in \mathcal{M}^{H^{\alpha\beta}_\lambda}_{x_\alpha,x_\beta},\ \lambda \in [0,1] \right\} ,$$

where $H^{\alpha\beta} : [0,1] \times \mathbb{R} \times M \to \mathbb{R}$ is a generic homotopy of Morse homotopies.

By close analogy to Floer homology we shall employ methods for the analysis of the required properties which basically rely on the calculus of Fredholm operators and their orientation.

In order to obtain the results about the trajectory spaces in an analytical way they will be equipped with a local Banach space structure. In fact, we will represent them as sub-manifolds of suitable Banach manifolds. On the one hand, with respect to the Fredholm operators mentioned above, it seems worthwhile to study a local $H^{1,2}(\mathbb{R},\mathbb{R}^n)$-structure. On the other hand, we are interested in compact $\mathbb{R}$-curves, that is to say in mappings

$$\gamma : \mathbb{R} \to M \qquad \text{with} \lim_{t\to\pm\infty} \gamma(t) = x \text{ and resp. } y \in M \ .$$

Therefore, it is with respect to endpoints $x, y \in M$ chosen in a specific way that we shall construct the Banach manifolds $\mathcal{P}^{1,2}_{x,y}$ with $H^{1,2}(\mathbb{R},\mathbb{R}^n)$-structure on which the entire calculus is founded. It should be emphasized that due to this differentiable structure these manifolds of maps cannot be identified with subspaces of the Hilbert manifold $H^{1,2}([0,1],M)$ of which the reader familiar with Klingenberg's variational approach to the geodesic problem might be reminded.

Definition 2.1 *We compactify $\mathbb{R}$ as $\overline{\mathbb{R}} = \mathbb{R}\cup\{\pm\infty\}$ equipped with the structure of a bounded manifold by the requirement that*

$$h : \left\{ \begin{array}{ccc} \overline{\mathbb{R}} & \to & [-1,1] \\ t & \mapsto & \frac{t}{\sqrt{1+t^2}} \end{array} \right. \quad \textit{be a diffeomorphism.}$$

Additionally, given arbitrary endpoints $x, y \in M$ we define the set of smooth, compact curves $C^\infty_{x,y}$ as

$$C^\infty_{x,y} = C^\infty_{x,y}(\overline{\mathbb{R}}, M) = \left\{ u \in C^\infty(\overline{\mathbb{R}}, M) \,\middle|\, u(-\infty) = x \,, u(+\infty) = y \right\} \ .$$

Owing to this differentiable structure on $\overline{\mathbb{R}}$ we obtain the following characterization of the asymptotic decrease of $C^\infty(\overline{\mathbb{R}})$-smooth functions:

Lemma 2.2 *For each function $f \in C^1(\overline{\mathbb{R}}, \mathbb{R})$ there is a constant $c(f) > 0$, such that the following estimate holds:*

$$|f'(t)| \leqslant \frac{c(f)}{(1+t^2)^{\frac{3}{2}}} \quad \textit{for all } t \in \mathbb{R} \ .$$

Proof. Due to the definition we have $f \circ h^{-1} \in C^1\big([-1,1], \mathbb{R}\big)$. Hence, we are able to choose $c(f)$ to be

$$c(f) = \sup_{[-1,1]} \left|(f \circ h^{-1})'\right| < \infty$$

and conclude the proof with

$$\big((h^{-1})' \circ h\big)(t) = \frac{1}{h'(t)} = (1+t^2)^{\frac{3}{2}} \ .$$

□

Actually, the boundedness of f and f' already yields the following:

Corollary 2.3 *Given $A \in C^1\big(\overline{\mathbb{R}}, \mathrm{GL}(n,\mathbb{R})\big)$, there is a constant $c(A) > 0$ such that the estimate*

$$\|As\|_{1,2} \leqslant c(A)\ \|s\|_{1,2}$$

holds for all $s \in H^{1,2}(\mathbb{R}, \mathbb{R}^n)$, where $(As)(t) = A(t) \cdot s(t)$, $t \in \mathbb{R}$.

Corollary 2.4 *Let $f \in C^1(\overline{\mathbb{R}}, \mathbb{R})$ satisfy the asymptotic condition $f(\pm\infty) = 0$. Then it holds $f \in H^{1,2}(\mathbb{R}, \mathbb{R})$.* [1]

Proof. It is clear from Lemma 2.2 that $f' \in L^2(\mathbb{R}, \mathbb{R})$ holds. The computation for arbitrary $t_0, t \in \mathbb{R}$ relying on Lemma 2.2

$$\begin{aligned} \big|\, f(t) - f(t_0) \,\big| &= \left| \int_{t_0}^{t} f'(\tau)\, d\tau \right| \leqslant \left| \int_{t_0}^{t} \frac{c(f)}{(1+\tau^2)^{\frac{3}{2}}} d\tau \right| \\ &\leqslant c(f) \left| \int_{t_0}^{t} \frac{d\tau}{\tau^3} \right| = c(f)\, \big| t^{-2} - t_0^{-2} \big| \end{aligned}$$

implies the estimate

$$\big|f(t)\big| \leqslant c(f)\, t^{-2},\ t \in \mathbb{R}$$

due to the assumption that $f(\pm\infty) = 0$ as $t_0 \to \pm\infty$. This proves $f \in L^2(\mathbb{R}, \mathbb{R})$. □

1We may even derive $f \in H^{1,1}$ as becomes clear in the proof.

Definition 2.5 *Now let* $\xi \in \mathrm{Vec}(\overline{\mathbb{R}})$ *be a* $C^\infty(\overline{\mathbb{R}})$*-smooth, finite-dimensional vector bundle on* $\overline{\mathbb{R}}$ *and let* $\phi : \xi \xrightarrow{\cong} \overline{\mathbb{R}} \times \mathbb{R}^n$ *be a smooth trivialization.*[2] *Then using the induced one-to-one mapping* ϕ_* *between the associated vector spaces of sections we are able to define*

$$H^{1,2}_{\mathbb{R}}(\xi) = \phi_*^{-1}\big(H^{1,2}(\mathbb{R},\mathbb{R}^n)\big) = \big\{\, \phi_*^{-1}(s) \,\big|\, s \in H^{1,2}(\mathbb{R},\mathbb{R}^n) \,\big\} \ .$$

It is immediate from Corollary 2.3 that ϕ_* *induces a Banach space topology on the vector space* $H^{1,2}_{\mathbb{R}}(\xi)$ *in a way which is independent of the particular choice of the trivialization* ϕ*, because the change of trivialization from* ϕ *to* $\tilde{\phi}$ *under the assumption of smoothness is represented by some* $A = \tilde{\phi} \circ \phi^{-1} \in C^\infty\big(\overline{\mathbb{R}}, \mathrm{GL}(n,\mathbb{R})\big)$.

Once again, it seems worthwhile to underline that we have not deduced this space $H^{1,2}_{\mathbb{R}}(\xi) \subset H^{1,2}_{\mathrm{loc}}(\xi)$ from a finite measure on $\overline{\mathbb{R}}$ but from toplinear isomorphisms with $H^{1,2}(\mathbb{R},\mathbb{R}^n)$. The discussion of the properties of this so-called section functor $H^{1,2}_{\mathbb{R}}$, which are fundamental with respect to the development of the Banach manifold we desire, shall be deferred to the section on manifolds of maps in the appendix. At this stage, we shall be satisfied with a brief definition of this Banach manifold $\mathcal{P}^{1,2}_{x,y}$ as we claimed it to be. Starting off with the exponential map on the complete Riemannian manifold M,

$$\exp : TM \supset \mathcal{D} \to M \ ,$$

where $\mathcal{D}$ stands for an open and convex neighbourhood of the zero section, we denote by $h^*\mathcal{D}$ the induced open and convex neighbourhood of the zero section in the pull-back bundle h^*TM, for any smooth, compact curve $h \in C^\infty(\overline{\mathbb{R}}, M)$.

Definition 2.6 *Regarding the Sobolev embedding* $H^{1,2}_{\mathrm{loc}} \hookrightarrow C^0$ *we are led to the map*

$$\begin{aligned} \exp_h :\ & H^{1,2}_{\mathbb{R}}(h^*\mathcal{D}) \to C^0(\overline{\mathbb{R}}, M) \\ & s \mapsto \exp \circ s, \quad (\exp \circ s)(t) = \exp_{h(t)} \cdot s(t) \end{aligned} \tag{2.2}$$

which is well-defined for all $h \in C^\infty(\overline{\mathbb{R}}, M)$*. Thus, we are able to define*

$$\begin{aligned} \mathcal{P}^{1,2}_{x,y} &= \mathcal{P}^{1,2}_{x,y}(\mathbb{R}, M) \\ &= \big\{\, \exp \circ s \in C^0(\overline{\mathbb{R}}, M) \,\big|\, s \in H^{1,2}_{\mathbb{R}}(h^*\mathcal{D}) \,,\, h \in C^\infty_{x,y}(\overline{\mathbb{R}}, M) \,\big\} \ . \end{aligned}$$

The following proposition will be the first result of the discussion in Appendix A concerning the structure of the set $\mathcal{P}^{1,2}_{x,y}$:

---

2 $\overline{\mathbb{R}}$ is contractible.

Proposition 2.7 *The set of curves* $\mathcal{P}^{1,2}_{x,y} \subset C^0_{x,y}(\overline{\mathbb{R}}, M)$ *with given endpoints* $x, y \in M$ *is equipped with a Banach manifold structure via the atlas of charts*

$$\{ H^{1,2}_{\mathbb{R}}(h^*\mathcal{D}) \, , \, \exp_h \}_{h \in C^\infty_{x,y}(\overline{\mathbb{R}}, M)} \ .$$

Additionally, the following inclusion relations hold:

$$C^\infty_{x,y}(\overline{\mathbb{R}}, M) \overset{\text{dense}}{\subset} \mathcal{P}^{1,2}_{x,y}(\mathbb{R}, M) \overset{\text{dense}}{\subset} C^0_{x,y}(\overline{\mathbb{R}}, M) \ . \tag{2.3}$$

Moreover, there is a countable sub-atlas.

We shall obtain the following representation of the tangent space,[3]

$$T\mathcal{P}^{1,2}_{x,y} = H^{1,2}_{\mathbb{R}}(\mathcal{P}^{1,2\,*}_{x,y} TM) = \bigcup_{s \in \mathcal{P}^{1,2}_{x,y}} H^{1,2}_{\mathbb{R}}(s^*TM) \ . \tag{2.4}$$

This is a Banach bundle on $\mathcal{P}^{1,2}_{x,y}$ with $H^{1,2}(\mathbb{R}, \mathbb{R}^n)$ as characteristic fiber. By strict analogy to $H^{1,2}_{\mathbb{R}}$ we are able to define a section functor (as defined in the appendix)

$$L^2_{\mathbb{R}} : \mathrm{Vec}_{C^\infty}(\overline{\mathbb{R}}) \to \mathrm{Ban}$$

which is endowed with a Banach space topology given by $L^2(\mathbb{R}, \mathbb{R}^n)$. This lends itself to the Banach bundle

$$L^2_{\mathbb{R}}(\mathcal{P}^{1,2\,*}_{x,y} TM) = \bigcup_{s \in \mathcal{P}^{1,2}_{x,y}} L^2_{\mathbb{R}}(s^*TM) \ . \tag{2.5}$$

Then, the second fundamental proposition developed in the appendix is

Proposition 2.8 *Let* $f \in C^\infty(M, \mathbb{R})$ *be an arbitrary, smooth real function on* M. *Then, given critical points* $x, y \in \operatorname{Crit} f$ *as endpoints, the gradient field* ∇f *induces a smooth section in the* L^2*-Banach bundle,*

$$\begin{aligned} F : \mathcal{P}^{1,2}_{x,y} &\to L^2_{\mathbb{R}}(\mathcal{P}^{1,2\,*}_{x,y} TM) \\ s &\mapsto \dot{s} + \nabla f \circ s \ . \end{aligned}$$

Actually, the study of the trajectory manifold $\mathcal{M}^f_{x,y}$ is founded upon exactly this section F. If we substitute the time-independent vector field ∇f by the time-dependent, suitably scaled vector field

$$X_t = \frac{\nabla h_t}{\sqrt{1 + |\dot{h}_t|^2 |\nabla h_t|^2}} \ ,$$

3The extension of $H^{1,2}_{\mathbb{R}}$ and $L^2_{\mathbb{R}}$ to the set of curves $\mathcal{P}^{1,2}_{x,y}$ will also be studied in the appendix.

the associated, more general section F_{X_t} yields the manifold of time-dependent trajectories in an analogous manner. It should be emphasized that by the definition of h_t (see Morse homotopy) X_t is again a gradient field for $|t| > R$. In this most general situation throughout the whole treatment we obtain the local representation of the section F

$$\begin{aligned} &F_{\text{loc}} : H^{1,2}_{\mathbb{R}}(\gamma^*\mathcal{D}) \to L^2_{\mathbb{R}}(\gamma^*TM), \ \ \gamma \in C^\infty_{x_\alpha,x_\beta}(\overline{\mathbb{R}}, M) \\ &F_{\text{loc}}(\xi)(t) = \nabla_t\xi(t) + g\big(t, \xi(t)\big) \ , \end{aligned} \tag{2.6}$$

from the local trivializations of the bundles $H^{1,2}_{\mathbb{R}}(\mathcal{P}^{1,2*}_{x,y}TM)$ and $L^2_{\mathbb{R}}(\mathcal{P}^{1,2*}_{x,y} TM)$. Here, $g : \overline{\mathbb{R}} \times \gamma^*\mathcal{D} \to \gamma^*TM$ is smooth and fibre respecting at each t together with $g(\pm\infty, 0) = 0$. Moreover, the asymptotical fibre derivatives $F_2 g(\pm\infty, 0)$ are conjugated to the Hessians of f^α and f^β at x_α and x_β, respectively. In the following section we will concentrate on this type of map F_{loc}, between Banach spaces. There, it will turn out that it is merely the asymptotic behaviour at the ends of the trajectories which proves significant.

The fundamental link with the trajectory spaces of the (negative) gradient flow is now provided by the following simple proposition. Given critical points x and y of $f \in C^\infty(M, \mathbb{R})$, it holds that

Proposition 2.9 *The zeroes of the section $F : \mathcal{P}^{1,2}_{x,y} \to L^2(\mathcal{P}^{1,2*}_{x,y}TM)$ are exactly the smooth curves which solve the ordinary differential equation*

$$\dot{s} = -\nabla f \circ s$$

and which submit to the convergence condition

$$\lim_{t\to-\infty} s(t) = x, \ \ \lim_{t\to\infty} s(t) = y \ .$$

Proof. Due to the assumption, each zero of this section represents a weakly differentiable and continuous curve with

$$\dot{\gamma} = -\nabla f \circ \gamma$$

as its weak derivative. The asserted property follows by means of iterative insertion and differentiation from the smoothness of f. In principle, this is nothing more than elliptic regularity. In fact, compared to the analogous but less trivial result in Floer homology, this principle should be mentioned.

For the converse, due to the definition of $\mathcal{P}^{1,2}_{x,y}$ and Lemma 2.2, it is sufficient to show that for any solution s of the differential equation obeying the convergence condition demanded above, the composition

$$\tilde{s} = s \circ h^{-1} : [-1, 1] \to M$$

is continuously differentiable. $\tilde{s}$ satisfies

$$(2.7) \qquad (\tilde{s}' \circ h)(t) = \dot{s}(t) \cdot (1+t^2)^{\frac{3}{2}} \ .$$

Thus, we have accomplished the proof if we have shown $\lim_{t \to \pm\infty} \dot{s}(t) \cdot (1+t^2)^{\frac{3}{2}} = 0$. This property of the asymptotic behaviour of the gradient trajectories is immediate from

Lemma 2.10 *Let $X : U(0) \to \mathbb{R}^n$ be a continuously differentiable vector field defined on a neighbourhood of $0 \in \mathbb{R}$ and let 0 be a critical point of X such that the linearization $DX(0)$ is non-degenerate and symmetric. Then, there is an $\epsilon > 0$ such that every solution*

$$(2.8) \qquad \dot{s} = X(s) \quad \textit{with} \quad \lim_{t \to \infty} s(t) = 0$$

satisfies the following estimate: There are constants $c > 0$ and $t_0 \in \mathbb{R}$ depending on s and satisfying

$$(2.9) \qquad |s(t)| \leqslant c\, e^{-\epsilon t} \ \textit{for all} \ t \geqslant t_0 \ .$$

Proof of the lemma. We start by defining the non-negative function

$$\alpha(t) = \frac{1}{2} \langle\, s(t),\, s(t) \,\rangle \ ,$$

that is by means of (2.8): $\alpha'(t) = \Big\langle X\big(s(t)\big), s(t) \Big\rangle$. One more differentiation yields

$$(2.10) \qquad \alpha''(t) = \Big\langle DX\big(s(t)\big) \cdot \dot{s}(t), s(t) \Big\rangle + \Big\langle X\big(s(t)\big), X\big(s(t)\big) \Big\rangle \ .$$

The continuous differentiability of X at 0 gives rise to the identity

$$(2.11) \quad \big|X\big(s(t)\big) - DX(0) \cdot s(t)\big| \ = R\big(s(t)\big) \cdot \ |s(t)| \quad \text{with} \quad \lim_{x \to 0} R(x) = 0$$

and thus to the following estimate:

$$\begin{aligned}
\alpha''(t) \quad &\overset{(2.8),(2.10)}{\geqslant} \quad \Big\langle DX\big(s(t)\big) \cdot X\big(s(t)\big), s(t) \Big\rangle \\
&= \quad \Big\langle DX(0) \cdot \big[X\big(s(t)\big) - DX(0)s(t) + DX(0)s(t)\big]\ , s(t) \Big\rangle \\
&\qquad + \Big\langle\ \big[DX\big(s(t)\big) - DX(0)\big] \cdot X\big(s(t)\big), s(t) \Big\rangle \\
&\overset{(2.11)}{\geqslant} \quad \Big\langle DX(0)s(t), DX(0)s(t) \Big\rangle - \ \|DX(0)\|\ R\big(s(t)\big)\langle s(t), s(t)\rangle \\
&\qquad - \delta\big|X\big(s(t)\big)\big|\,\big|s(t)\big|
\end{aligned}$$

for $t \geqslant t_0(\delta)$, so that $s(t) \leqslant r_0$ is small enough. That is

$$\alpha''(t) \ \geqslant \ |DX(0)s(t)|^2 \ - \epsilon\, |s(t)|^2 \ \ \text{for arbitrary} \quad t \geqslant t_0(\epsilon),\ \epsilon > 0 \ .$$

Given any arbitrary $\delta > 0$, the step before the last stems from the continuity of $DX\big(s(t)\big)$ and the symmetry of $DX(0)$. Applying (2.11) once again and choosing a suitable $\delta = \delta(\epsilon)$ leads to the last estimate.

Due to the fact that $DX(0) \in \mathrm{GL}(n, \mathbb{R})$, we have

$$c = \inf_{x \in S^{n-1}} |DX(0) \cdot x| > 0 \ .$$

If we now choose $\epsilon = \frac{c}{2}$, we obtain the crucial inequality

$$\alpha''(t) \geqslant \frac{c}{2} \, |s(t)|^2 = c\,\alpha(t) \qquad \text{for all} \quad t \geqslant t_0(c, s) \ . \tag{2.12}$$

Hence, the assertion is concluded from the following maximum principle: Setting

$$\alpha_0(t) = \alpha(t_0)\, e^{-\sqrt{c}(t-t_0)} \qquad \text{and} \qquad \Delta(t) = \alpha(t) - \alpha_0(t)$$

we reproduce the estimate (2.12) in the form

$$\Delta''(t) \geqslant c \cdot \Delta(t) \ \text{ for all } t \geqslant t_0$$

with $\Delta(t_0) = 0$ and $\lim\limits_{t \to \infty} \Delta(t) = 0$.

According to these properties the function Δ cannot have a positive local maximum at any $\tau_0 > t_0$. This leads to the estimate

$$\alpha(t) \leqslant \alpha(t_0)\, e^{-\sqrt{c}(t-t_0)} \quad \text{for all } t \geqslant t_0(c, s)$$ [4]

concluding the proof. □

Actually, because of the recursive relation given by $\dot{s} = X(s)$ the gradient trajectories turn out to have an exponential asymptotic decrease in each derivative. Therefore, they are $\overline{\mathbb{R}}$-differentiable infinitely many times and Proposition 2.9 has been proven. □

This proposition also holds immediately for the time-dependent trajectories, because for large times we have to deal with ordinary gradient fields. After these preparations, we have managed to represent the trajectory spaces as zero sets of smooth sections in Banach bundles on Banach manifolds. In order to extract the fundamental result about $\mathcal{M}^f_{x,y}$, namely the manifold property, we need a further transversality condition:

$$F : \mathcal{P}^{1,2}_{x,y} \to L^2(\mathcal{P}^{1,2*}_{x,y} TM)$$

has to intersect transversally the zero section in the L^2-bundle. This so-called Morse-Smale condition is not fullfilled in general. But if we admit minor variations of the Riemannian metric on M, this regularity condition appears to be generic with respect to the parameter set of Riemannian metrics on M. This result shall be investigated in the section on transversality.

4This lower bound t_0 depends on s solely via the condition $\big|s(t)\big| < r_0$, for $t \geqslant t_0(s)$ and given a small $r_0(X)$.

2.2 Fredholm Theory

The aim of this section is to develop the Fredholm properties of the operators which are needed in order to define the trajectory manifolds $\mathcal{M}^f_{x,y}$, $\mathcal{M}^{h^{\alpha\beta}}_{x^\alpha,x^\beta}$ and $\mathcal{M}^{H^{\alpha\beta}}_{x^\alpha,y^\beta}$. At first, we shall be content with considering linear operators on the trivial vector bundle $\overline{\mathbb{R}} \times \mathbb{R}^n$, of the type

$$(F_A s)(t) = \dot{s}(t) + A(t) \cdot s(t) \ ,$$

with $s \in H^{1,2}(\mathbb{R}, \mathbb{R}^n)$ and $A \in C^0_b(\mathbb{R}, \mathrm{End}(\mathbb{R}^n))$.

2.2.1 The Fredholm Operator on the Trivial Bundle

Definition 2.11 *An operator $A \in \mathrm{End}(\mathbb{R}^n)$ is called conjugated self-adjoint if it is self-adjoint with respect to some scalar product. Additionally, we shall henceforth use the following notations:*

$$\begin{aligned} X &= H^{1,2}(\mathbb{R}, \mathbb{R}^n), \qquad Y = L^2(\mathbb{R}, \mathbb{R}^n), \\ \mathcal{S} &= \{A \in GL(n, \mathbb{R}) \mid A \text{ is conjugated self-adjoint}\}, \\ \mathcal{A} &= \left\{A \in C^0\left(\overline{\mathbb{R}}, \mathrm{End}(\mathbb{R}^n)\right) \mid A^{\pm} = A(\pm\infty) \in \mathcal{S}\right\}. \end{aligned}$$

Here, $\mathcal{S}$ describes the set of the Hessians appearing in Morse theory. Thus, the following definition makes sense: Let

$$\mu(a) = \#\left(\sigma(A) \cap \mathbb{R}^-\right) \quad \text{for } A \in \mathcal{S}$$

be the so-called Morse index of A, where $\sigma(A)$ denotes the finite spectrum of A. Moreover, we regard $\mathcal{A}$ as normed vector space with respect to $\|\cdot\|_\infty$.

We now wish to study the following map:

$$\begin{aligned} &C^0_b\left(\mathbb{R}, \mathrm{End}(\mathbb{R}^n)\right) \ni A \mapsto (F_A : X \to Y) \ , \\ &(F_A s)(t) = \dot{s}(t) + A(t) \cdot s(t) \ . \end{aligned}$$

Here,

$$F : C^0_b\left(\mathbb{R}, \mathrm{End}(\mathbb{R}^n)\right) \to \mathcal{L}(X; Y)$$

describes an affine map, which is continuous due to the estimate

$$\begin{aligned} \|(F_A - F_B)(s)\|_0 &= \left(\int_{\mathbb{R}} |(A-B)s|^2 \, dt\right)^{\frac{1}{2}} \\ \leqslant \|A - B\|_\infty \|s\|_0 &\leqslant \|A - B\|_\infty \|s\|_1 \ , \end{aligned}$$

hence,

$$\|F_A - F_B\|_{\mathcal{L}(X;Y)} \leqslant \|A - B\|_\infty \ .$$

Actually, the central statement of this section is that F associates a Fredholm operator[5] $F_A \in \mathcal{F}(X;Y)$ to each $A \in \mathcal{A}$, so that the following formula holds:

$$\operatorname{ind} F_A = \mu(A^-) - \mu(A^+) \ .$$

This is the so-called relative Morse index, which is determined by the spectral flow of A.

Proposition 2.12 *Given $A \in \mathcal{A}$, the linear operator $F_A : X \to Y$ is a Fredholm operator.*

The essential ingredient of the proof of this Fredholm property is provided by the following fundamental lemma about semi-Fredholm operators, bounded linear operators with closed range and kernel both of finite dimensions.

Lemma 2.13 *Let X, Y and Z be Banach spaces and $F \in \mathcal{L}(X;Y)$, $K \in \mathcal{K}(X;Z)$ and $c > 0$ with*

$$\|x\|_X \leqslant c(\,\|Fx\|_Y + \|Kx\|_Z\,)\,, \text{ for all } x \in X \ . \tag{2.13}$$

Then F is a semi-Fredholm operator.

Although this lemma is a well-known, standard technique for investigating Fredholm operators, we nevertheless give a proof for the reader who is not quite familiar with this type of analysis.

Proof. First, let us regard an arbitrary sequence $(x_k)_{k\in\mathbb{N}}$ within the set

$$S = \{x \in \ker F | \ \|x\|_X = 1\} \ .$$

Then assumption (2.13) and the identity

$$Fx_k = 0 \text{ for all } k \in \mathbb{N}$$

yield the estimate

$$\|x_k - x_l\|_X \leqslant c\ \|K(x_k - x_l)\|_Z\,, \text{ for all } k, l \in \mathbb{N} \ .$$

Owing to the compactness of K, the sequence $(Kx_k)_{k\in\mathbb{N}}$ and therefore also the sequence $(x_k)_{k\in\mathbb{N}}$ have convergent subsequences. Thus, S is compact and the kernel of F has finite dimension.

Second, due to this finite dimension the theorem of Hahn-Banach implies the existence of a closed sub-space $X_0 \subset X$ satisfying the relation

$$\ker F \oplus_{\text{top}} X_0 = X \ .$$

5This is a bounded linear operator with both kernel and cokernel having finite dimensions.

Let us now start with a sequence $(Fx_k)_{k\in\mathbb{N}} \subset R(F)$ converging in Y; that is to say, without loss of generality, we can assume

$$(x_k)_{k\in\mathbb{N}} \subset X_0, \quad Fx_k \to y \in Y \ .$$

If $(x_k)_{k\in\mathbb{N}}$ is unbounded, we switch to the normed sequence $\frac{x_k}{\|x_k\|}$. Thus, it holds without loss of generality that

$$(x_k)_{k\in\mathbb{N}}, \quad \|x_k\|_X = 1, \quad \|Fx_k\|_Y \leqslant \frac{1}{k} \text{ for all } k \in \mathbb{N} .$$

According to assumption (2.13) there is a convergent subsequence of $(x_k)_{k\in\mathbb{N}}$ because of the compactness of K and the convergence of $(Fx_k)_{k\in\mathbb{N}}$. Upon consideration of a further subsequence, this leads to

$$x_k \to x, \quad \|x\|_X = 1 \text{ and } Fx = 0$$

in contradiction to the construction of X_0. Therefore, the sequence $(x_k)_{k\in\mathbb{N}}$ must be bounded, so that the same argument, based on (2.13), again yields a convergent subsequence. This implies the identity

$$y = Fx \quad \text{for some } x \in X_0 \ .$$

Hence, $R(F)$ is closed in Y. □

Proof of proposition 2.12: The proof consists of four steps. We shall obtain the semi-Fredholm property from the asymptotic behaviour via a compact operator as in Lemma 2.13 by expressing the action of the operator in question during a compact interval of time. After this main part of the proof, the finiteness of the dimension of the cokernel follows from a formal adjunction of F_A.

Step 1: Let $A \in \mathcal{A}$ be a constant map from $\overline{\mathbb{R}}$ to $\mathcal{S}$. Then, there is a constant $c > 0$, such that the estimate

$$\|s\|_1 \leqslant c \ \|F_A s\|_0 \quad \textit{holds for all } s \in X \ .$$

Proof. By $\mathcal{F} : Y \xrightarrow{\cong} Y$ we denote the Fourier isometry, where we use the notations $X = H^{1,2}(\mathbb{R}, \mathbb{C}^n)$ and $Y = L^2(\mathbb{R}, \mathbb{C}^n)$ throughout this calculation. In particular, it holds that

$$\mathcal{F}(\dot{s})(t) = i\,t\,\mathcal{F}(s)(t), \ t \in \mathbb{R} \quad \text{for all } s \in X \ .$$

Additionally, let $\omega : \mathcal{F}(X) \to Y$ be the operator $\omega(s)(t) = t \cdot s(t)$, $t \in \mathbb{R}$ so that we obtain the identity

$$\mathcal{F}\big(F_A(s)\big) = (i\,\omega + A) \circ \mathcal{F}(s)$$

and therefore

$$F_A = \mathcal{F}^{-1} \circ (i\,\omega + A) \circ \mathcal{F} : X \to Y \ . \tag{2.14}$$

Now let us assume $A \in \mathcal{S}$, so that $\lambda_0 = \min|\sigma(A)| > 0$ holds. We further consider

$$\begin{aligned} &B_1(\omega),\, B_2(\omega) : \mathbb{C}^n \to \mathbb{C}^n\,,\\ &B_1(\omega)\cdot x = (1+\omega^2)^{\frac{1}{2}} x\,,\ \omega \in \mathbb{R}\,,\\ &B_2(\omega) = i\,\omega + A,\ \omega \in \mathbb{R}\ , \end{aligned}$$

and hence

$$0 \notin \sigma\big(B_2(\omega)\big) = i\,\omega + \sigma(A)\ .$$

As a consequence, the inverse $B_2(\omega)^{-1}$ exists and satisfies

$$\big\|B_2^{-1}(\omega)\big\| = \sup_{\lambda\in\sigma\left(B_2(\omega)\right)} \big|\lambda^{-1}\big| = \sup_{\lambda\in\sigma(A)+i\,\omega} \frac{1}{|\lambda|} = \frac{1}{\sqrt{\lambda_0^2+\omega^2}}\ .$$

Summing up we obtain

$$\big\|B_1(\omega)\cdot B_2^{-1}(\omega)\big\| \leqslant \sqrt{\frac{1+\omega^2}{\lambda_0^2+\omega^2}} \leqslant \max\left(\frac{1}{\lambda_0},\, 1\right) = c(A)\ .$$

We henceforth conceive $B_1, B_2^{-1} \in C^0\left(\mathbb{R}, \mathrm{End}(\mathbb{C}^n)\right)$ as multiplication operators in $\mathcal{L}(X;Y)$, so that we obtain the inequality

$$\big\|B_1\, B_2^{-1}\big\|_{L^\infty} \leqslant c(A)\ . \tag{2.15}$$

Putting all this together we are led to the required estimate by the combination of

$$\|s\|_1^2 = \big\|\sqrt{1+\omega^2}\,\mathcal{F}\,s\big\|_0^2 = \big\|\mathcal{F}^{-1} B_1\, B_2^{-1} \mathcal{F}\,\mathcal{F}^{-1} B_2 \mathcal{F}\,s\big\|_0^2$$

with

$$\big\langle \mathcal{F}^{-1} B_1\, B_2^{-1} \mathcal{F}\,\xi,\, \eta\big\rangle_0 = \big\langle B_1\, B_2^{-1}\mathcal{F}\,\xi,\, \mathcal{F}\,\eta\big\rangle_0 \leqslant \big\|B_1\, B_2^{-1}\big\|_\infty\ \|\xi\|_0\ \|\eta\|_0\ .$$

Due to (2.15) this yields

$$\big\|\mathcal{F}^{-1} B_1\, B_2^{-1} \mathcal{F}\big\|_{\mathcal{L}(X;Y)} \leqslant c(A)\ ,$$

such that we conclude the estimate

$$\|s\|_1 \leqslant c(A)\ \big\|\mathcal{F}^{-1} B_2 \mathcal{F}\,s\big\|_0$$

and owing to (2.14) the inequality which we asserted above.

Step 2: Given an arbitrary $A \in \mathcal{A}$, we find constants $T > 0$, $c(T) > 0$, such that it holds that

$$\|s\|_1 \leqslant c(T)\ \|F_A s\|_0 \quad \textit{for all } s \in X,\ s_{|\,[-T,T]} = 0\ .$$

Proof. Step 1 provides us with the estimates

$$\|s\|_1 \leqslant c(A^{\pm})\, \|F_{A^{\pm}} s\|_0 \quad \text{for all } s \in X\ .$$

We therefore define $c = \max\left(c(A^+), c(A^-)\right)$. Given an arbitrary $\epsilon > 0$ and $A \in \mathcal{A}$, there is a $T_\epsilon > 0$ large enough such that

$$\begin{aligned} \|A^- A(t)\| &\leqslant \epsilon \text{ for all } t \leqslant -T \text{ and} \\ \|A^+ - A(t)\| &\leqslant \epsilon \text{ for all } t \geqslant T\ . \end{aligned}$$

Restricting ourselves to $s^{\pm} \in X$ with $s^-|_{[-T,\infty)} = 0$ and $s^+|_{(-\infty,T]} = 0$, respectively, we obtain the inequalities

$$\|F_{A^{\pm}} s^{\pm}\|_0 \leqslant \|F_A s^{\pm}\|_0 + \|(F_{A^{\pm}} - F_A)\, s^{\pm}\|_0 \leqslant \|F_A s^{\pm}\|_0 + \epsilon\, \|s^{\pm}\|_0\ .$$

Let us now consider $s \in X$ equipped with the property $s|_{[-T,T]} = 0$. Then we find $s^{\pm}$ as above fulfilling $s = s^+ + s^-$ such that

$$\begin{aligned} \|s\|_1 &= \|s^-\|_1 + \|s^+\|_1 \leqslant c\,(\, \|F_{A^-} s^-\|_0 + \|F_{A^+} s^+\|_0\,) \\ &\leqslant c\,(\, \|F_A s^-\|_0 + \|F_A s^+\|_0\,) + c\epsilon\,(\, \|s^-\|_0 + \|s^+\|_0\,) \\ &= c\, \|F_A s\|_0 + c\epsilon\, \|s\|_0\ . \end{aligned}$$

This at last implies

$$\|s\|_1 \leqslant c\, \|F_A s\|_0 + c\epsilon\, \|s\|_1\ ,$$

and for an $\epsilon < \frac{1}{c}$ together with an appropriate $T(\epsilon)$

$$\|s\|_1 \leqslant \frac{c}{1 - c\epsilon}\, \|F_A s\|_0, \quad \text{for all } s \in X \text{ with } s|_{[-T(\epsilon),T(\epsilon)]} = 0\ .$$

Step 3: Given any $A \in \mathcal{A}$, there is a Banach space Z and a $K \in \mathcal{K}(X; Z), c > 0$ satisfying

$$\|x\|_X \leqslant c\,(\, \|F_A\|_Y + \|K\, x\|_Z\,) \text{ for all } x \in X\ .$$

In fact, F_A is a semi-Fredholm operator.
Proof. Let $T(A)$ be as provided in Step 2. Then it holds that

$$\begin{aligned} \int_{-T}^{T} |\dot{s} + A\, s|^2\, dt &= \int_{-T}^{T} \left(|\dot{s}|^2 + 2\langle \dot{s}, A\, s\rangle + |A\, s|^2 \right) dt \\ &\geqslant \int_{-T}^{T} \left(\tfrac{1}{2}\, |\dot{s}|^2 - |A\, s|^2 \right) dt \end{aligned}$$

due to the computation of

$$|\dot{s}|^2 + 2\,\langle \dot{s}, A\, s\rangle + |A\, s|^2 \geqslant \frac{1}{2}\, |\dot{s}|^2 - |A\, s|^2$$

from

$$|\dot{s} + 2\,A\,s|^2 \geqslant 0 \ .$$

Therefore by means of $|A(t) \cdot s(t)| \leqslant \|A(t)\| \cdot |s(t)|$ and setting $\tilde{c} = \max_{[-T,T]} \|A(t)\|$, we conclude

$$\int_{-T}^{T} |\dot{s} + A\,s|^2 \, dt \geqslant \frac{1}{2} \int_{-T}^{T} |\dot{s}|^2 \, dt \; - \; \tilde{c} \int_{-T}^{T} |s|^2 \, dt \ .$$

Hence, there is a $c > 0$ satisfying

$$\int_{-T}^{T} \Big(|s|^2 + |\dot{s}|^2 \Big) \, dt \leqslant c \int_{-T}^{T} \Big(|s|^2 + |\dot{s} + A\,s|^2 \Big) \, dt \ . \tag{2.16}$$

Defining a cut-off function $\beta \in C^\infty(\mathbb{R}, [0,1])$ with the proprties

$$\beta(t) = \begin{cases} 0, & |t| \geqslant T+1 \\ 1, & |t| \leqslant T \end{cases} \quad \text{and} \quad \dot{\beta}(t) \neq 0 \ \text{ for } |t| \in (T, T+1)$$

we are able to combine the estimate (2.16) with Step 2:

$$\begin{aligned} \|s\|_1 &= \|\beta\,s + (1-\beta)\,s\|_1 \leqslant \|\beta\,s\|_1 + \|(1-\beta)\,s\|_1 \\ &\leqslant c\,(\,\|\beta\,s\|_0 + \|F_A(\beta\,s)\|_0 + \|F_A\,((1-\beta)\,s)\|_0\,) \end{aligned}$$

for a $c > 0$ large enough. That is

$$\begin{aligned} \|s\|_1 &\leqslant c\,\Big(\|\beta\,s\|_0 + 2\,\big\|\dot{\beta}\,s\big\|_0 + \|\beta\,F_A\,s\|_0 + \|(1-\beta)\,F_A\,s\|_0 \Big) \\ &\leqslant \tilde{c}\,\Big(\|s\|_{L^2\left([-T-1,T+1]\right)} + \|F_A\,s\|_0 \Big) \ . \end{aligned}$$

Considering the following composition of a continuous restriction map and the Rellich compact embedding,

$$\begin{aligned} K : H^{1,2}(\mathbb{R}, \mathbb{R}^n) &\overset{\text{rest}}{\longrightarrow} H^{1,2}\left([-T-1, T+1], \mathbb{R}^n\right) \\ &\overset{\text{cpt.}}{\hookrightarrow} L^2\left([-T-1, T+1], \mathbb{R}^n\right) = Z \ , \end{aligned}$$

we obtain

$$\begin{aligned} K : X &\to Z \\ s &\mapsto s_{|\,[-T-1,T+1]} \in L^2\left([-T-1, T+1]\right) \end{aligned}$$

as a compact operator. Thus, Lemma 2.13 provides the asserted semi-Fredholm property.

Step 4: F_A is a Fredholm operator.
Proof. Up to this point, we know that F_A has a finite-dimensional kernel and a range that is closed in Y. Hence, the cokernel of F_A is a Banach space satisfying the relation

$$\operatorname{coker} F_A \cong R(F_A)^\perp ,$$

because Y is a Hilbert space. Thus, let $r \in R(F_A)^\perp$, that is

$$\langle r\, ,\, \dot{s} + A\, s\rangle_0 = 0, \ \text{ for all } s \in H^{1,2} \ .$$

In particular, we can deduce

$$\langle r\, ,\, \dot{\phi}\rangle_0 = -\langle A^t\, r\, ,\, \phi\rangle_0, \ \text{ for all } \phi \in C_0^\infty(\mathbb{R}, \mathbb{R}^n) \ .$$

But by definition, this means the weak differentiability of r. Hence, $r \in L^2$ is once weakly differentiable with $\dot{r} = A^t \cdot r \in L^2$. Summing up, we obtain

$$r \in W^{1,2} \quad \text{with} \quad \dot{r} = A^t \cdot r \ .$$

Consequently, we know that $r \in X$ and $r \in \ker F_{-A^t}$, because $-A^t \in \mathcal{A}$. Therefore, there is a toplinear isomorphism

$$\operatorname{coker} F_A \cong \ker F_{-A^t} \ .$$

By strict analogy with Step 3 we conclude $\dim \ker F_{-A^t} < \infty$ thus proving the proposition. □

Having seen that the operator F_A is a Fredholm operator, the following analysis will result in a proposition expressing the associated Fredholm index in terms of the relative Morse index. At this point it should be metioned that only the spectral flow of $A \in \mathcal{A}$ appears to be significant. This means the 'number of eigenvalues which change the sign'. In the proof we shall transform A successively without altering the Fredholm index into an operator of a shape that can be analysed easily. Concerning the focal class of operators we shall henceforth use the following notation:

Definition 2.14 *From Proposition 2.12 we consider the subset*

$$\Sigma = F(\mathcal{A}) = \{F_A \in \mathcal{L}(X;Y)\,|\,A \in \mathcal{A}\} \subset \mathcal{F}(X;Y)$$

and denote the equivalence class of operators from Σ with respect to the relation $B^\pm = A^\pm$ by

$$\Theta_{F_A} = \left\{F_B \in \Sigma \,|\, B^\pm = A^\pm\right\},\ A \in \mathcal{A} \ .$$

The following lemma will play a decisive role for the index theorem.

Lemma 2.15 *Given any $F \in \Sigma$, the class Θ_F is contractible within Σ as a subspace of $\mathcal{F}(X;Y)$.*

Proof. Let $F = F_{A_0} \in \Sigma$ be arbitrary and define $\Theta = \Theta_F$. Then we study the map

$$\begin{aligned}
&\kappa : [0,1] \times \Theta \to \Theta \ \text{ with } \ \kappa(\tau, F_A) = F_{A(\tau)}\,,\\
&A(\tau) = (1-\tau)\cdot A + \tau \cdot A_0, \ \text{ that is}\\
&F_{A(\tau)} \in \Theta \ \text{ for all } \tau \in [0,1], \text{ as } A(\tau)^{\pm} = A^{\pm} = A_0^{\pm}\,.
\end{aligned}$$

We obviously have

$$\kappa(0,\cdot) = \mathrm{Id}_\Theta \quad \text{and} \quad \kappa(1,\cdot) = F_{A_0} \ .$$

Thus, we must show the continuity of κ in both variables, so, let us start with

$$\text{(2.17)} \qquad \lim_{n\to\infty} \tau_n = \tau \quad \text{and} \quad \lim_{n\to\infty} F_{A_n} = F_A \quad \text{within } \Theta \ ,$$

that is

$$\lim_{n\to\infty} \|F_{A_n} - F_A\|_{\mathcal{L}(X;Y)} = 0 \ .$$

We must verify

$$\lim_{n\to\infty} \left\|F_{A(\tau)} - F_{A_n(\tau_n)}\right\|_{\mathcal{L}(X;Y)} = 0 \ .$$

Assuming an $\epsilon > 0$ and a subsequence $(n_k)_{k\in\mathbb{N}}$ satisfying

$$\left\|F_{A(\tau)} - F_{A_{n_k}(\tau_{n_k})}\right\|_{\mathcal{L}(X;Y)} \geqslant \epsilon, \ \text{ for all } \ k \in \mathbb{N} \ ,$$

we find a sequence $(u_{n_k})_{k\in\mathbb{N}} \subset X$, such that

$$\|u_{n_k}\|_1 = 1 \quad \text{and} \quad \left\|\big(A(\tau) - A_{n_k}(\tau_{n_k})\big)\cdot u_{n_k}\right\|_0 \geqslant \epsilon \ \text{for all } \ k \in \mathbb{N} \ .$$

Owing to assumption 2.17 we are led to convergence by

$$\begin{aligned}
& \left\|\big(A(\tau) - A_{n_k}(\tau_{n_k})\big)\cdot u_{n_k}\right\|_0 \\
= \ & \|[(1-\tau)\cdot A + \tau A_0 - (1-\tau_{n_k})\cdot A_{n_k} - \tau_{n_k} A_0]\cdot u_{n_k}\|_0 \\
= \ & \|\,(A - A_{n_k})\cdot u_{n_k} + (\tau - \tau_{n_k})A_0 \cdot u_{n_k} \\
& + (\tau_{n_k} A - \tau A + \tau_{n_k} A_{n_k} - \tau_{n_k} A)\cdot u_{n_k}\,\|_0 \\
\leqslant \ & \left\|F_A - F_{A_{n_k}}\right\|_{\mathcal{L}(X;Y)} + |\tau - \tau_{n_k}| \ \|A_0\|_{\mathcal{L}(X;Y)} \\
& + |\tau - \tau_{n_k}| \ \|A\|_{\mathcal{L}(X;Y)} + |\tau_{n_k}| \ \left\|F_A - F_{A_{n_k}}\right\|_{\mathcal{L}(X;Y)} \\
\to \ & 0, \ \text{for } k \to \infty
\end{aligned}$$

in contradiction to the assumption of the above ϵ. □

As a conclusion from this lemma, the index map $\operatorname{ind} : \Sigma \to \mathbb{Z}$ turns out to be constant when restricted to the class Θ_{F_A}. In other words, $\operatorname{ind}(F_A)$ is determined uniquely by the endpoints $A^\pm \in \mathcal{S}$. With this in mind we now come to the final step of the calculation of the Fredholm index. Actually, we carry out an equivalence transformation, namely an appropriate conjugation of the operator F_A :

Let $A \in \mathcal{A}$, that is, $A^\pm \in \mathcal{S}$. Then there exists a $C^\pm \in \mathrm{GL}(n, \mathbb{R})$ satisfying

$$C^\pm A^\pm (C^\pm)^{-1} = \operatorname{diag}(\lambda_1^\pm, \dots, \lambda_n^\pm) \ ,$$

so that the eigenvalues $\lambda_i^\pm \in \sigma(A^\pm)$ are ordered by sign, i.e.

$$\operatorname{sgn} \lambda_i^\pm \geqslant \operatorname{sgn} \lambda_{i+1}^\pm \ .$$

Additionally, the ends $C^\pm$ can be chosen in order to satisfy $\det C^\pm > 0$, so that both C^+ and C^- lie in the same pathwise connected component of $\mathrm{GL}(n, \mathbb{R})$. Thus, there is a curve

$$C \in C^\infty\left(\overline{\mathbb{R}}, \mathrm{GL}(n, \mathbb{R})\right) \text{ with ends } C(\pm\infty) = C^\pm \ ,$$

which additionally is asymptotically constant. This means

$$C(t) = \begin{cases} C^+, & t \geqslant T \\ C^-, & t \leqslant -T \end{cases} \quad \text{for some } T > 0 \ .$$

We shall henceforth denote by C both multiplication operators

$$C_X : X \to X \text{ and } C_Y : Y \to Y \,, \qquad s \mapsto C \cdot s \ .$$

It is obvious that C represents a toplinear isomorphism and that the following identities hold:

$$\begin{aligned} & (C_Y F_A C_X^{-1})(s)(t) = C(t) \cdot \left(\tfrac{\partial}{\partial t} + A(t)\right) \cdot \left(C^{-1}(t) \cdot s(t)\right) \\ = \ & \dot{s}(t) + \left(C(t) \cdot \tfrac{\partial}{\partial t}(C^{-1})(t) + C(t)A(t)C^{-1}(t)\right) \cdot s(t) \\ = \ & \left(F_{C\frac{\partial}{\partial t}(C^{-1}) + CAC^{-1}} s\right)(t) \ . \end{aligned}$$

Here, $\frac{\partial}{\partial t}(C^{-1})(t) = 0$ holds for $|t| \geqslant T$. Thus we compute the ends as

$$(C\tfrac{\partial}{\partial t}(C^{-1}) + CAC^{-1})^\pm = \operatorname{diag}(\lambda_1^\pm, \dots, \lambda_n^\pm) = D^\pm$$

and we obtain the relation

$$C_Y F_A C_X^{-1} \in \Theta_{D^\pm} \ .$$

Owing to the fact that $C_X : X \overset{\cong}{\to} X$ and $C_Y : Y \overset{\cong}{\to} Y$ are isomorphisms, the identity

$$\operatorname{ind}(C_Y F_A C_X^{-1}) = \operatorname{ind} F_A$$

follows from the composition rule of Fredholm indices. So, to sum up, it is sufficient to compute $\operatorname{ind} F_D$ for an element $D \in \mathcal{A}$ having ends of $D^{\pm} = \operatorname{diag}(\lambda_1^{\pm}, \ldots, \lambda_n^{\pm})$. This enables us to reduce the proof of the following proposition to a nearly explicit computation.

Proposition 2.16 *Given any $A \in \mathcal{A}$, the Fredholm index of F_A equals the relative Morse index,*

$$\operatorname{ind} F_A = \mu(A^-) - \mu(A^+) \ .$$

Proof. It is sufficient to compute $\ker F_A$ and $\operatorname{coker} F_A$ when A is of the shape

$$\begin{aligned} &A(t) = \operatorname{diag}\big(\lambda_1(t), \ldots, \lambda_n(t)\big) \text{ where} \\ &\operatorname{sgn} \lambda_i(\pm\infty) \geqslant \operatorname{sgn} \lambda_{i+1}(\pm\infty),\ i = 1, \ldots, n-1\,, \\ &\text{and} \quad \lambda_i(t) = \text{const for } |t| \geqslant 1 \ . \end{aligned}$$

Under these circumstances $s \in H^{1,2}(\mathbb{R}, \mathbb{R}^n)$, $s(t) = \big(s_1(t), \ldots, s_n(t)\big)$ lies within the kernel of F_A, $s \in \ker F_A$, exactly if it represents a global solution of the system of differential equations

$$\dot{s}_i(t) = -\lambda_i(t) \cdot s_i(t), \ \ i = 1, \ldots, n \ ,$$

with the bounded functions $\lambda_i : \mathbb{R} \to \mathbb{R}$. Now, explicit calculation for large times shows that

$$\dot{s}_i(t) = -\lambda_i(t) \cdot s_i(t),\ t \in \mathbb{R} \quad \text{and } 0 \neq s_i \in H^{1,2}(\mathbb{R}, \mathbb{R})$$

holds if and only if

$$\lambda_i^- < 0 \quad \text{and} \quad \lambda_i^+ > 0, \text{ as} \quad s_i(t) = \begin{cases} e^{-\lambda_i^- t}, & t < 1 \\ e^{-\lambda_i^+ t}, & t > 1 \end{cases} \ .$$

Having in mind the ordering of the eigenvalues by sign, we compute

$$\begin{aligned} \dim \ker F_A &= \#\big\{k \in \{1, \ldots, n\} \,|\, \lambda_k^- < 0 \text{ und } \lambda_k^+ > 0\big\} \\ &= \max\big(\mu(A^-) - \mu(A^+), 0\big) \ . \end{aligned}$$

By the same method of computing we obtain the formula

$$\dim \operatorname{coker} F_A = \max\big(\mu(-A^-) - \mu(-A^+), 0\big) = \max\big(\mu(A^+) - \mu(A^-), 0\big)$$

from the isomorphism $\operatorname{coker} F_A \cong \ker F_{-A^t} = \ker F_{-A}$. Computing the difference between the dimensions of the kernel and the cokernel amounts to proving the asserted formula $\operatorname{ind} F_A = \mu(A^-) - \mu(A^+)$. □

It seems worthwhile to point out that by this method of calculating the index, we have executed transformations in order to simplify the operator

F_A. These transformations preserve the Fredholm index but do not leave the dimensions of the kernel and the cokernel invariant in general. Regarding the question of such invariance transformations we refer to Section B in the appendix, where we shall make use of another method to obtain a diagonal shape comparable to Proposition 2.16. In fact, that technique will preserve $\dim\ker F_A$ and $\dim\operatorname{coker} F_A$, too, and we obtain the formulas

$$\begin{aligned}\dim\ker F_A &= \#\left\{k \in \{1,\dots,n\} \,|\lambda_k^- < 0 \text{ and } \lambda_k^+ > 0\right\}\\ \dim\operatorname{coker} F_A &= \#\left\{k \in \{1,\dots,n\} \,|\lambda_k^- > 0 \text{ and } \lambda_k^+ < 0\right\} \ .\end{aligned}$$

2.2.2 The Fredholm Operator on Non-Trivial Bundles

We now discuss the general case of smooth vector bundles ξ on $\overline{\mathbb{R}}$ endowed with a Riemannian metric, so that $H^{1,2}(\xi)$ and $L^2(\xi)$ are well-defined. The problem is that we have to generalize the time-derivation by a covariant derivation which is a priori noncanonical and which might disturb the Fredholm property.

Definition 2.17 *Suppose that we are given the Riemannian bundle $\overline{\mathbb{R}}$ with a covariant derivation ∇ induced by a connection. Then we denote by $\Sigma_{\xi,\nabla}$ the class of operators*

$$\begin{aligned}&F_A : H^{1,2}(\xi) \to L^2(\xi) \ ,\\ &F_A s = \nabla_t s + A\, s \ ,\end{aligned}$$

with $A \in C^0\big(\operatorname{End}(\xi)\big)$, such that $A(\pm\infty)$ is non-degenerate and conjugated self-adjoint on $\xi|_{\{\pm\infty\}}$.

The first question is, whether the Fredholm property is well-defined for all $F_A \in \Sigma_{\xi,\nabla}$. Choosing any trivialization of ξ on $\overline{\mathbb{R}}$, we obtain the following representation for the covariant derivative:

$$\nabla_t^{\text{triv}} s = \tfrac{\partial}{\partial t} s + \Gamma s, \quad s \in H^{1,2}(\mathbb{R}, \mathbb{R}^n) \ , \tag{2.18}$$

where $\Gamma \in C^\infty\big(\overline{\mathbb{R}}, \operatorname{End}(\mathbb{R}^n)\big)$ stems from the Christoffel symbol associated to ∇ and the respective trivialization. It is now decisive how ∇_t^{triv} transforms under a change of the trivialization $\Phi \in C^\infty\big(\overline{\mathbb{R}}, \operatorname{Gl}(n)\big)$. We compute

$$\phi^{-1}\nabla_t^{\text{triv}}(\phi s) = \left(\tfrac{\partial}{\partial t} + \phi^{-1}\dot{\phi} + \phi^{-1}\Gamma\phi\right) s \ . \tag{2.19}$$

Therefore, the trivialized operator $F_{A,\nabla}^{\text{triv}}$ transforms as

$$\phi^{-1} F_{A,\nabla}^{\text{triv}} \phi = \tfrac{\partial}{\partial t} + \phi^{-1}\dot{\phi} + \phi^{-1}\Gamma\phi + \phi^{-1}A\phi \ . \tag{2.20}$$

Since due to the $\overline{\mathbb{R}}$-differentiability $\dot{\phi}$ vanishes asymptotically, i.e. $\dot{\phi}(\pm\infty) = 0$ according to emma 2.2, the only problematic issue which might afflict the Fredholm property is the asymptotic behaviour of Γ, i.e. $\Gamma(\pm\infty)$.

Definition 2.18 *We call a covariant derivation ∇ on a $\overline{\mathbb{R}}$-bundle ξ Fredholm admissible if any trivialization gives rise to asymptotically vanishing Γ-terms, i.e. $\Gamma(\pm\infty) = 0$. As we observed above, this property holds for every trivializations if it holds for one.*

Lemma 2.19 *Let $\xi = u^*E$ be a pull-back bundle with respect to a $\overline{\mathbb{R}}$-curve $u \in C^\infty(\overline{\mathbb{R}}, M)$, where $\pi : E \to M$ is a smooth Riemannian bundle on the manifold M with a fixed covariant derivation ∇. The induced covariant derivation $u^*\nabla$ on ξ is Fredholm admissible.*

Proof. For the sake of simplicity we also denote the curve in local coordinates by u. With respect to these local coordinates on M together with a local trivialization of E the covariant derivation ∇ on E is represented by means of the Christoffel symbol Γ, i.e.

$$\nabla_v w(p) = Dw(p) \cdot v + \Gamma(p)(v, w) \ .$$

This yields the following local representation for the induced covariant derivation $u^*\nabla$, where we denote $u^*\nabla_{\frac{d}{dt}}$ by ∇_t :

$$\nabla_t \xi(t) = \dot{\xi}(t) + \Gamma\big(u(t)\big)\big(\dot{u}(t), \xi(t)\big) \ . \tag{2.21}$$

For these identities concerning the covariant derivation the reader is also referred to [Kli]. Since the above Γ-term is now given by

$$\Gamma(t) = \Gamma\big(u(t)\big)\big(\dot{u}(t), \cdot\big) \ ,$$

we immediately deduce the identity $\Gamma(\pm\infty) = 0$ from $\dot{u}(\pm\infty) = 0$. Hence, $u^*\nabla$ is Fredholm admissible. □

Thus, provided an admissible ∇, the Fredholm property of the above defined operator $F_A \in \Sigma_{\xi,\nabla}$ depends solely on the asymptotic behaviour of A, i.e. $A(\pm\infty)$. We obtain this property for F_A via an arbitrary trivialization $\phi : \overline{\mathbb{R}} \times \mathbb{R}^n \xrightarrow{\cong} \xi$, that is

$$\begin{aligned} \phi^{-1} F_A \phi &\in \Theta_{F_B} \subset \Sigma_{\text{triv}} \ , \\ B^{\pm} &= \phi^{-1}(\pm\infty) A(\pm\infty) \phi(\pm\infty) \ . \end{aligned}$$

Here, ϕ induces the toplinear isomorphisms

$$\begin{aligned} \phi_X &: X \xrightarrow{\cong} H^{1,2}(\xi), \quad (\phi_X s)(t) = \phi(t) \cdot s(t) \ , \\ \phi_Y &: Y \xrightarrow{\cong} L^2(\xi) \qquad \text{analogously} \ , \end{aligned}$$

which are again denoted by ϕ. Finally, we are able to transfer all results from the trivial case in a manner which is independent of the particular trivialization:

Proposition 2.20 *Given a Fredholm admissible covariant derivation ∇ on ξ, every $F_A \in \Sigma_{\xi,\nabla}$ is a Fredholm operator with the index*

$$\operatorname{ind} F_A = \mu(A^-) - \mu(A^+) \ .$$

2.2.3 Generalization to Fredholm maps

Finally we are able to state the result which we need for the class of Banach space maps upon which the entire analysis is founded.

Definition 2.21 *Let ξ be a Riemannian $C^\infty(\overline{\mathbb{R}})$-vector bundle with a Fredholm admissible covariant derivation as above and let $\mathcal{O} \subset \xi$ be open and satisfying the following:*

> *There is a $s \in C_0^\infty(\xi)$ with $s(\overline{\mathbb{R}}) \subset \mathcal{O}$, and $\mathcal{O}_t = \mathcal{O} \cap \xi_t$ is open and convex for all $t \in \overline{\mathbb{R}}$.*

Further, let us assume a map $f : \mathcal{O} \to \xi$ that is smooth and fibre respecting, such that it holds $f(\pm\infty, 0) = 0$ and the fibre derivatives are non-degenerate and conjugated self-adjoint, i.e. $\mu\big(D_2 f(\pm\infty, 0)\big)$ is well-defined. Considering such maps f we are able to define

$$\mathrm{Fred}(\xi,\nabla) = \left\{ F : H^{1,2}(\mathcal{O}) \to L^2(\xi) \;\middle|\; \begin{array}{c} (Fs)(t) = \nabla_t s(t) + f\big(t, s(t)\big)\,, \\ f : \mathcal{O} \to \xi \text{ as above} \end{array} \right\} .$$

Additionally, let us denote by $H^{1,2}(\mathcal{O}) = \big\{ s \in H^{1,2}(\xi) \,|\, \xi(\overline{\mathbb{R}}) \subset \mathcal{O} \big\}$ the open subset of $H^{1,2}(\xi)$.

Proposition 2.22 *Each $F \in \mathrm{Fred}(\xi, \nabla)$ is a Fredholm map with its differential*

$$DF(s_0) \in \Sigma_{\xi,\nabla} \quad \text{for all } s_0 \in H^{1,2}(\mathcal{O})\,,$$

so that the local Fredholm index is determined by the formula

$$\mathrm{ind}\, DF(s_0) = \mu\big(D_2 f(-\infty, 0)\big) - \mu\big(D_2 f(+\infty, 0)\big) \;.$$

Proof. It is clear from the properties assumed for $f : \mathcal{O} \to \xi$ that the following identity holds for all $s_0 \in H^{1,2}(\mathcal{O})$:

$$DF(s_0) = F_{D_2 f \circ s_0} \in \Sigma_{\xi,\nabla} \;.$$

Hence the proof is immediate from Proposition 2.20. □

2.3 Transversality

In this section we shall obtain the result that the trajectory spaces come equipped with the structure of finite-dimensional manifolds. As it has been already mentioned above, we obtain this property only if we restrict to a generic set of metrics or homotopies of Morse homotopies, which in this situation operate as parameters. As regards this genericity result the fundamental technique for the proof is based on a parametrized version of the theorem of Sard-Smale, which is adapted to the given situation of sections in Banach bundles.

2.3.1 The Regularity Conditions

Definition 2.23 *We call a subset $\Sigma \subset X$ of a Baire space $\mathcal{G}$-set, if it is a countable intersection of open and dense subsets and therefore dense, also. We then call a set $G \subset X$ generic with respect to a condition of the points of X if this condition is fulfilled for a $\mathcal{G}$-set $\Sigma \subset G$.*

Proposition 2.24 *Let G and M be Banach manifolds and $\tau : E \to M$ be a Banach bundle on M with fiber $\mathbb{E}$. Additionally, let $\Phi : G \times M \to E$ be a smooth G-parameter section, i.e. Φ is smooth and $\Phi_g = \Phi(g, \cdot) : M \to E$ is a section in E for each $g \in G$. Let Φ satisfy the condition:*
There is a countable trivialization $\{(U, \psi)\}$, $\psi : E|_U \xrightarrow{\cong} U \times \mathbb{E}_U$, such that for each (U, ψ)

- *0 is a regular value of $pr_2 \circ \psi \circ \Phi : G \times U \to \mathbb{E}_U$ and*
- *$pr_2 \circ \psi \circ \Phi_g : U \to \mathbb{E}_U$ is a Fredholm map with index r for all $g \in G$.*

Then there is a $\mathcal{G}$-set $\Sigma \subset G$, such that the set $Z_g = \Phi_g^{-1}(0) = \{ m \in M \mid \Phi_g(m) = 0 \}$ is a closed submanifold of M for all $g \in \Sigma$.

Another way of putting this would be to say:

> Given a generic parameter set $\Sigma \subset G$, the section Φ_g parametrized by $g \in \Sigma$ intersects the zero section in E transversally.

We call the main result of this section the transversality result. The reader may notice that transversality in this sense and concerning the trajectory spaces is equivalent to the Morse-Smale condition, which is also defined in a way that describes a kind of transversality. The essential step of the proof is based on the following technical lemma, which can be proved easily by straightforward methods on Banach spaces.

Lemma 2.25 *Let Φ be a bounded, linear map of Banach spaces, which is onto and of the form*

$$\Phi : E \times F \to G, \quad \Phi(e, f) = \Phi_1(e) + \Phi_2(f) \ .$$

Let Φ_1 and Φ_2 be continuously linear and let $\Phi_2 : F \to G$ be a Fredholm operator. Then there is a decomposition

$$E \times F = \ker \Phi \oplus_{\text{top}} H \ ,$$

such that H is a closed subspace of $E \times F$.

It should be emphasized that the proof relies crucially upon the Fredholm property of Φ_2. If we could assume E and F to be Hilbert spaces, this lemma would appear to be trivial.

Proof of proposition 2.24: Owing to the given countable trivialization of E we are able to assume without loss of generality that E is a trivial bundle. This is to say that we have a smooth map

$$\Phi : G \times M \to \mathbb{E}$$

with respect to a Banach space $\mathbb{E}$, such that $0 \in \mathbb{E}$ is a regular value and the maps $\Phi_g : M \to \mathbb{E}$ are Fredholm maps with index r for all $g \in G$. This assumption can be made because $\Sigma = \bigcap_U \Sigma_U$ is again a $\mathcal{G}$-set if we consider the countable trivialization $\{U, \psi_U\}$. Thus,

$$Z = \Phi^{-1}(0) \subset G \times M \tag{2.22}$$

is a smooth Banach manifold with associated tangent spaces given by

$$T_z Z = \ker D\Phi(z) \qquad \text{for all } z = (g, m) \in Z \ , \tag{2.23}$$

as follows via local coordinate charts from Lemma 2.25 (!) and from the implicit function theorem.

Now the main step of the proof is to conclude that the restriction of the projection map to this Banach manifold Z,

$$\pi : Z \to G \ ,$$

is a Fredholm map endowed with the same index as Φ_g, if we consider $z = (g, m) \in Z$. First, let us focus on the kernels of the maps $D\pi(x) : T_z Z \to \mathbb{E}$ and $D_2\Phi(z) : T_m M \to \mathbb{E}$. We obtain the identity

$$\ker D\pi(z) = T_z Z \cap T_m M = \ker D_2\Phi(z) \ , \tag{2.24}$$

of finite-dimensional vector spaces with the assumption that $D_2\Phi(z)$ is Fredholm. Second, we consider the following homomorphism between the at first merely algebraic quotient vector spaces:

$$\begin{array}{rccc} \widetilde{D_1\Phi} : & T_g G / R(D\pi) & \longrightarrow & E / R(D_2\Phi) \\ & [v]_{R(D\pi)} & \mapsto & [D_1\Phi\, v]_{R(D_2\Phi)} \ . \end{array} \tag{2.25}$$

Having in mind the identities

$$v = D\pi \cdot (v, w) \quad \text{and} \quad D_1\Phi \cdot v = -D_2\Phi \cdot w$$

for each pair $(v, w) \in T_z Z \subset T_g G \times T_m M$, we immediately deduce the isomorphism property of $\widetilde{D_1\Phi}$ from the fact that $D\Phi(z)$ is onto. Knowing that the cokernel of $D_2\Phi(z)$ has a finite dimension, $\widetilde{D_1\Phi}$ appears to be an isomorphism between finite-dimensional cokernels,

$$\widetilde{D_1\Phi} : \operatorname{coker}\big(D\pi(z)\big) \xrightarrow{\cong} \operatorname{coker} D_2\Phi(z) \ . \tag{2.26}$$

Hence, the equations (2.24) and (2.26) yield the Fredholm property of the projection map π. Finally, we are able to apply the well-known theorem of Sard-Smale to this smooth Fredholm map between Banach manifolds. It leads to a $\mathcal{G}$-set $\Sigma \subset G$ of regular values of π.

Now let $b \in \Sigma$. If $\Phi(b,m)$ does not vanish for any m from M, 0 is trivially a regular value of Φ_b. Therefore, let us consider an $m \in M$ satisfying $\Phi(b,m) = 0$. We then have to show that the operator

$$D\Phi_b(m) = D_2\Phi(b,m) : T_mM \to \mathbb{E} \tag{2.27}$$

is onto. Let $\gamma \in \mathbb{E}$ be given arbitrarily. Due to the regularity of b with respect to Φ, there is a pair $(\alpha,\beta) \in T_bG \times T_mM$ satisfying

$$\gamma = D\Phi(b,m)\cdot(\alpha,\beta) = D_1\Phi\cdot\alpha + D_2\Phi\cdot\beta \ . \tag{2.28}$$

On the other hand, b is also regular with respect to $\pi : Z \to G$, that is, given any $\alpha \in T_bG$ we find an $(\alpha',\beta') \in T_{(b,m)}Z$, such that it holds

$$\alpha' = D\pi(b,m)\cdot(\alpha',\beta') = \alpha \ . \tag{2.29}$$

Hence, (α,β') belongs to the kernel of $D\Phi(b,m)$, i.e.

$$0 = D\Phi(b,m)\cdot(\alpha,\beta') = D_1\Phi\cdot\alpha + D_2\Phi\cdot\beta' \ . \tag{2.30}$$

Subtracting equation (2.30) from (2.28), we find a preimage of γ under the map $D_2\Phi(b,m)$ of the form

$$\gamma = D_2\Phi(b,m)\cdot(\beta-\beta') \ .$$

This yields the proof. □

It is the aim of this section to obtain a generic set of smooth Riemannian metrics, each of which satisfies the Morse-Smale condition. Within the given framework it is sufficient to fix an arbitrary metric g_0 and to find a generic set of variations with respect to g_0. By this we mean a set of smooth sections $A \in \operatorname{End}(TM)$ which are fibrewise self-adjoint with respect to g_0, positively definite and which differ from the section of identities by a sufficiently small amount. We then vary ∇f in terms of $A\cdot\nabla f$. In order to be able to apply Proposition 2.24 to our transversality problem, we need a Banach manifold structure upon this parameter set of smooth sections in $\operatorname{End}(TM)$.

Definition 2.26 *Let ξ be a smooth vector bundle on M endowed with a bundle norm $|\cdot|$ and a covariant derivation. Given a sequence $(\epsilon_n)_{n\in\mathbb{N}}$ of positive real numbers, we define*

$$\begin{aligned}&\|s\|_\epsilon = \textstyle\sum_{k\geqslant 0} \epsilon_k \sup_M \left|\nabla^k s\right| \quad \textit{for all } s \in C^\infty(\xi)\ , \\ &\left|\nabla^k s\right|(p) = \max\left\{ |\nabla_{x_1} \dots \nabla_{x_k} s(p)| \,\middle|\, \|x_1\| = \dots = \|x_k\| = 1\,,\ x_i \in T_pM \right\}\ .\end{aligned}$$

This norm $\|\cdot\|_\epsilon$ gives rise to the vector space

$$C^\epsilon(\xi) = \left\{ s \in C^\infty(\xi) \,\middle|\, \|s\|_\epsilon < \infty \right\}\ .$$

It is easy to verify that $\left(C^\epsilon(\xi)\,,\ \|\cdot\|_\epsilon\right)$ is a Banach space if $C^\epsilon(\xi)$ is non-void. An explicit construction of a sufficiently large set of smooth sections with suitably chosen compact support, which the reader can find in [F3], gives rise to the following lemma.

Lemma 2.27 *There is a sequence $(\epsilon_n)_{n\in\mathbb{N}}$ such that the Banach space $C^\epsilon(\xi)$ is dense in $L^2(\xi)$.*

Remark Obviously, the following inequality holds:

$$\|\cdot\|_\epsilon \leqslant \|\cdot\|_{\epsilon'}\,, \ \text{ if } 0 < \epsilon_n \leqslant \epsilon_n' \text{ for all } n \in \mathbb{N}\ .$$

In other words, we have a continuous embedding of $C^{\epsilon'}$ in C^ϵ. Summing up, we get the sequence of embeddings

$$C^{\epsilon'} \overset{\text{cont.}}{\hookrightarrow} C^\epsilon \overset{\text{cont.}}{\hookrightarrow} C^k \overset{\text{cont.}}{\hookrightarrow} C^0 \ \text{ for all } k \in \mathbb{N}\ . \tag{2.31}$$

Definition 2.28 *Setting*

$$\begin{aligned} E_{g_0} &= \mathrm{End}_{\mathrm{sym},g_0}(TM) \\ \mathcal{T}_{g_0} &= \{ (p, T_p) \in E_{g_0} \mid T_p \textit{ positively definite w.r.t. } \langle\cdot,\cdot\rangle_p = g_0(p) \} \end{aligned}$$

we now define the vector bundle E_{g_0} on M. It consists of the endomorphisms which are self-adjoint with respect to g_0, so that it forms a subbundle of $\mathrm{End}(TM) \to M$. Additionally, $\mathcal{T}_{g_0}$ represents an open neighbourhood of the submanifold $\mathbb{1}_M$ in $E_{g_0} \subset \mathrm{End}(TM)$. Due to the above remark $C^\epsilon(\xi) \hookrightarrow C^0(\xi)$ is a continuous embedding and therefore we obtain

$$\mathcal{G}_{g_0} = C^\epsilon(\mathcal{T}_{g_0}) = \left\{ X \in C^\epsilon(E_{g_0}) \,\middle|\, X_p \in (\mathcal{T}_{g_0})_p \textit{ f.a. } p \in M \right\}$$

as an open neighbourhood of $\mathbb{1} \in C^\epsilon(E_{g_0})$.

Thus, $\mathcal{G}_{g_0}$ is a suitable Banach manifold of parameters with regard to our regularity analysis. It should be mentioned that each tangent space $T_X\mathcal{G}_{g_0}$ at $X \in \mathcal{G}_{g_0}$ has the form

$$T_X\mathcal{G}_{g_0} = C^{\epsilon}(E_{g_0}) \ .$$

In order to prove the central regularity result already for the time-dependent case which includes the time-independent one as a special case[6] we still need a further restriction on the homotopy $h^{\alpha\beta}$:

Definition 2.29 *A finite homotopy* $h^{\alpha\beta} : \mathbb{R} \times M \to \mathbb{R}$, *i.e. satisfying*

$$h(t,\cdot) = \begin{cases} f^{\alpha}, & t \leqslant -R \\ f^{\beta}, & t \geqslant R \end{cases} \quad \textit{for some } R \geqslant 0 \ ,$$

is called regular, if it fulfills the condition:

Given any $\mathbb{R}$*-critical point* x_0 *of* h, *i.e.*

$$\nabla h_t(x_0) = 0 \ \textit{for all } t \in \mathbb{R} \ ,$$

the consequently well-defined operator from $\Sigma_{x_0^*TM}$

$$\frac{\partial}{\partial t} + H^2 h_t(x_0) : H^{1,2}(x_0^*TM) \to L^2(x_0^*TM)$$

is onto.

In this framework, $H^2 h_t(x_0) \in \mathrm{End}(T_{x_0}M)$ *denotes the well-defined (!) Hessian of* h_t *at* x_0.

Generally, this regularity condition is not fulfilled. But if f^{α} and f^{β} are functions with isolated critical points, we can always find a regular, finite homotopy between them. This is deduced by the following argument. Let us start from a non-regular $\mathbb{R}$-critical point x_0 of $h^{\alpha\beta}$ by assumption. We then choose a function $k \in C_0^{\infty}(\mathbb{R} \times M, \mathbb{R})$ with suitably small support and satisfying $dk_t(x_0) \neq 0$ for some $t \in \mathbb{R}$. Let us now replace the homotopy $h^{\alpha\beta}$ by $h'^{\alpha\beta} = h^{\alpha\beta} + k$. This replacement can be iterated for all at most countably many non-regular $\mathbb{R}$-critical points within a compact interval of time. We thus obtain a regular, finite homotopy. In particular, $h^{\alpha\beta}$ is a regular homotopy if it is trivial, that is to say if $h_t = f^{\alpha}$ for all $t \in \mathbb{R}$.

Let us now study the following map:

$$\begin{aligned} \Phi \ : \ & \mathcal{G}_{g_0} \times \mathcal{P}^{1,2}_{x,y} \to L^2(\mathcal{P}^{1,2}_{x,y}) \\ & (A,\gamma) \mapsto \dot{\gamma} + \frac{A \cdot \nabla_{g_0} h_t}{\sqrt{1 + \left|\dot{h}_t\right|^2 \cdot \left|A \cdot \nabla_{g_0} h_t\right|^2}} \ . \end{aligned}$$

---

[6] $\dot{h}_t = 0$ for all $t \in \mathbb{R}$

Referring to the local coordinates and the trivialization of $L^2(\mathcal{P}^{1,2}_{x,y})$ we compute the representation of Φ as

$$\begin{aligned}\Phi_{\text{loc}} \quad &: \quad \mathcal{G}_{g_0} \times H^{1,2}(\gamma^*\mathcal{O}) \to L^2(\gamma^*TM) \\ &(A,\xi) \mapsto \nabla_t\xi + \Theta(\xi)\cdot\dot\gamma + \nabla_2 \exp(\xi)^{-1} \cdot \left(\frac{A\cdot\nabla h_t}{\sqrt{1+\ldots}} \circ \exp_\gamma \xi\right) \quad .\end{aligned}$$

Since $C^\epsilon(\mathcal{T}_{g_0}) \hookrightarrow C^0(\mathcal{T}_{g_0})$ is a continuous inclusion, Φ_{loc} is also smooth with respect to the parameter variable A. We may now state the decisive proposition, where we denote Φ_{loc} again by Φ.

Proposition 2.30 *Given any parameter* $A \in \mathcal{G}_{g_0}$*,*

$$\Phi_A = \Phi_{\text{loc}}(A,\cdot) : H^{1,2}(\gamma^*\mathcal{O}) \to L^2(\gamma^*TM)$$

is a Fredholm map with the constant index $\mu(x) - \mu(y)$*. Moreover, if* h_t *is a regular homotopy,* $0 \in L^2$ *is a regular value of* Φ*.*

Proof. a) The Fredholm result is an immediate conclusion from Lemma 2.19 and the investigations in Section 2.2.3, because

$$\Phi_A \in \text{Fred}(\gamma^*TM, \gamma^*\nabla) \quad \text{for all } A \in \mathcal{G}_{g_0} \quad .$$

b) The focal regularity result is now obtained as follows:
The differential of Φ has the form

$$D\Phi(a,\xi)\cdot(B,\eta) = D_1\Phi(A,\xi)\cdot B + D_2\Phi(A,\xi)\cdot\eta \quad .$$

Here, $D_2\Phi(A,\xi) \in \Sigma_{\gamma^*TM}$ is a Fredholm operator of the already well-known type. For sake of computational plainness we choose the short notations

$$\begin{aligned}&d = D_1\Phi(A,\xi)\cdot B \in L^2(\gamma^*TM) \ , \\ &d(t) = \nabla_2 \exp(\xi)^{-1}(t) \cdot \\ &\left(\frac{B\nabla h_t + \left|\dot h_t\right|^2 \left(B\nabla h_t \left|A\nabla h_t\right|^2 - A\nabla h_t \langle A\nabla h_t, B\nabla h_t\rangle\right)}{\left(1+\left|\dot h_t\right|^2\cdot\left|A\nabla h_t\right|^2\right)^{\frac{3}{2}}} \circ \exp_\gamma \xi\right)(t)\end{aligned}$$

together with

$$\begin{aligned}\alpha(t) &= \left(\left|\dot h_t\right|^2 \circ \exp_\gamma \xi\right)(t) \\ x(t) &= \left(\nabla h_t \circ \exp_\gamma \xi\right)(t) \\ y(t) &= \left((A\nabla h_t)\circ \exp_\gamma \xi\right)(t) \\ B(t) &= \left(B \circ \exp_\gamma \xi\right)(t) \quad .\end{aligned}$$

Now let $(A,\xi) \in \mathcal{G}_{g_0} \times H^{1,2}$ satisfy $\Phi(A,\xi) = 0$. Then ξ is smooth, too. Since due to a) $R\big(D_2\Phi(A,\xi)\big)$ is a closed subspace of L^2 with finite codimension, the same is true for $R(D\Phi)$. Thus, let us start from any vector $c \in L^2$ within the finite-dimensional subspace of L^2 satisfying

$$\begin{gathered} \big\langle D\Phi \cdot (B,\eta)\,,\, c\big\rangle_{L^2} = 0 \\ \text{for all} \quad (B,\eta) \in C^\epsilon(E_{g_0}) \times H^{1,2}(\gamma^*TM) \ . \end{gathered} \tag{2.32}$$

It remains to show that $c(t)$ vanishes for all $t \in \mathbb{R}$. The relation (2.32) leads in particular to

$$\big\langle D_2\Phi \cdot \eta\,,\, c\big\rangle_{L^2} = 0 \quad \text{for all } \eta \in H^{1,2} \ . \tag{2.33}$$

As we already analysed in the Fredholm section, due to the smoothness $\xi \in C^\infty$, this implies that c can be expressed as the solution of some non-autonomous linear differential equation

$$\dot c(t) = X(t) \cdot c(t) \tag{2.34}$$

with respect to a suitable trivialization of γ^*TM. If c is not the zero section, the uniqueness result for ordinary differential equations implies that c vanishes nowhere, i.e. $c(t) \neq 0$, for all $t \in \mathbb{R}$. Briefly, given $c \neq 0$, we obtain

$$\begin{gathered} \big\langle D_1\Phi(A,\xi) \cdot B\,,\, c\big\rangle_{L^2} = 0 \quad \text{for all } B \in C^\epsilon(E_{g_0}), \\ c \in C^\infty(\gamma^*TM) \text{ with } c(t) \neq 0 \text{ for all } t \in \mathbb{R} \ . \end{gathered} \tag{2.35}$$

Due to the above discussion about C^ϵ this is generally true for all $B \in L^2(E_{g_0})$, that is especially for all B with compact support. Choosing B with a suitably small support we may also give a pointwise formulation of (2.35) for a single $t \in \mathbb{R}$: Let us fix any arbitrary $t \in \mathbb{R}$. Then there exists a

$$\tilde c \in T_pM \ , \ \ \tilde c \neq 0 \ \text{with } p = \exp_\gamma \xi(t) \ ,$$

satisfying the equation

$$\Big\langle Bx + \alpha\big(Bx\langle y,y\rangle - y\langle y,Bx\rangle\big),\, \tilde c\Big\rangle = 0 \tag{2.36}$$

for all $B \in \operatorname{End}(T_pM)$, which are symmetric with respect to $\langle\cdot,\cdot\rangle = g_0(\cdot,\cdot)$ on T_pM. Let P be the symmetric operator

$$Pz = z\,\langle y,\, y\rangle - y\,\langle y,\, z\rangle \ .$$

Then (2.36) may be written as

$$\big\langle (\mathbb{1} + \alpha P)Bx,\, \tilde c\big\rangle = 0 \quad \text{for all } B \in \operatorname{End}_{\mathrm{sym}}(T_pM) \ . \tag{2.37}$$

Since α is positive and P has a non-negative spectrum, it holds that

$$(\mathbb{1} + \alpha P)\,\tilde{c} \neq 0 \ .$$

Therefore, there is a $z \neq 0$ satisfying

$$\langle Bx, z\rangle = 0 \quad \text{for all } B \in \mathrm{End}_{\mathrm{sym}}(T_pM) \ . \tag{2.38}$$

A simple computation shows that the space of symmetric endomorphisms complies with the property:

For any pair $v,\ w \neq 0$ there exists a symmetric B with $\langle Bv, w\rangle \neq 0$.

Thus (2.38) gives rise to the conclusion

$$x(t) = (\nabla h_t \circ \exp_\gamma \xi)(t) = 0 \quad \text{for all } t \in \mathbb{R} \ . \tag{2.39}$$

However, due to the identity $\Phi(A, \xi) = 0$, it follows that $s = \exp_\gamma \xi$ must be a constant trajectory. This leads to a contradiction, as we assumed h_t to be a regular homotopy thus implying the surjectivity of $D_2\Phi$. □

2.3.2 The Regularity Results

Combining Propositions 2.24 and 2.30 we now are led to the result about transversality.

Theorem 1 *Given a generic set of smooth metrics, the trajectory spaces* $\mathcal{M}^f_{x,y}$ *and* $\mathcal{M}^{h^{\alpha\beta}}_{x,y}$ *are closed submanifolds of* $\mathcal{P}^{1,2}_{x,y}$ *of the finite dimension* $\mu(x) - \mu(y)$.

Remarks.

- A priori, we can only speak of local dimensions, because they are induced by the Fredholm indices of the local linearizations of the Fredholm section. These indices are only local analytic invariants which depend continuously on the underlying point of the zero set. The fact that they are equal on each component of this trajectory space follows from the mere dependence of the index formula on the fixed endpoints x and y in this special framework. In fact, if we look at the analogous Fredholm problem in Floer homology, we observe an additional parameter in the index formula, which involves a certain Chern number as a strictly local invariant, in general. In contrast to our situation, the trajectory spaces in Floer homology may have different dimensions for different components.

- There is a corresponding result for the time-dependent case of homotopy trajectories, when the Riemannian metric is homotoped over a finite interval of time along with $h^{\alpha\beta}$. Let arbitrary generic metrics associated to the functions f^α and f^β be given. Then there is a generic set of associated homotopies $g_t^{\alpha\beta}$, such that $\mathcal{M}^{h^{\alpha\beta}, g^{\alpha\beta}}_{x,y}$ has the property of a manifold.

Finally, we have to study the case of λ-parametrized trajectories: Let $h_0^{\alpha\beta}$ and $h_1^{\alpha\beta}$ be finite homotopies of the form

$$h_i^{\alpha\beta}(t,\cdot) = \begin{cases} f^\alpha, & t \leqslant -R \\ f^\beta, & t \geqslant R \end{cases}, \quad i = 0, 1 ,$$

and let $H^{\alpha\beta}$ be a λ-homotopy satisfying

$$\begin{aligned} &H^{\alpha\beta} : [0,1] \times \mathbb{R} \times M \to \mathbb{R} \\ &H^{\alpha\beta}(\lambda, t, \cdot) = \begin{cases} f^\alpha, & t \leqslant -R \\ f^\beta, & t \geqslant R \end{cases}, \\ &H^{\alpha\beta}(0,\cdot,\cdot) = h_0^{\alpha\beta}, \quad H^{\alpha\beta}(1,\cdot,\cdot) = h_1^{\alpha\beta} . \end{aligned}$$

Then we consider the 1-parameter family of sections,

$$\begin{aligned} G^{\alpha\beta} \quad : \quad & [0,1] \times \mathcal{P}^{1,2}_{x_\alpha, y_\beta} \to L^2(\mathcal{P}^{1,2\,*}_{x_\alpha, y_\beta} TM) \\ & (\lambda, \gamma) \mapsto \left(\dot\gamma(t) + \left(\frac{\nabla H^{\alpha\beta}(\lambda, t, \cdot)}{\sqrt{1 + \ldots}} \circ \gamma \right)(t) \right)_{t \in \mathbb{R}} , \end{aligned}$$

with $x_\alpha \in \mathrm{Crit}(f^\alpha)$ and $y_\beta \in \mathrm{Crit}(f^\beta)$. Corresponding to the first remark above, the Riemannian metric determining ∇ may likewise be (λ, t)-dependent. The following transversality result holds:

Theorem 2 *Let $h_0^{\alpha\beta}$ and $h_1^{\alpha\beta}$ be regular, finite homotopies together with its associated Morse-Smale metrics. Then there is a generic set of λ-homotopies $H^{\alpha\beta}$ of the above form and a generic set of suitable homotopies of the Riemannian metric, such that $(G^{\alpha\beta})^{-1}(0)$ is a $\big(\mu(x_\alpha) - \mu(y_\beta) + 1\big)$-dimensional submanifold of $[0,1] \times \mathcal{P}^{1,2}_{x_\alpha, y_\beta}$.*

Proof. Apart from the Fredholm property which is needed according to Proposition 2.24 with respect to the fixed local trivialization of the Banach bundle $L^2(\mathcal{P}^{1,2\,*}_{x_\alpha, y_\beta} TM)$, the proof of this regularity follows in a way that is analogous to Proposition 2.30. And this Fredholm property is concluded as follows: Given the local trivialization at γ, we consider the linearized map

$$\begin{aligned} G \quad : \quad & \mathbb{R} \times H^{1,2}(\gamma^* TM) \to L^2(\gamma^* TM) \\ & (\tau, \xi) \mapsto \big(G_1(\lambda) \cdot \tau + G_2(\lambda) \cdot \xi \big) , \\ & G_2(\lambda) \cdot \xi = \nabla_t \xi + A(\lambda) \cdot \xi , \\ \text{with} \quad & \begin{matrix} G_1 : & [0,1] \to & L^2(\gamma^* TM), \\ G_2 : & [0,1] \to & \Sigma_{\gamma^* TM} \end{matrix} \qquad \text{both smooth} . \end{aligned}$$

First, let us observe that the regularity assumption for the homotopies $h_0^{\alpha\beta}$ and $h_1^{\alpha\beta}$ implies that the operators $G_2(0)$ and $G_2(1)$ are onto. Second, $G_2(\lambda)$

being a Fredholm operator for each λ yields the finiteness of the codimension of $R(G)$. Additionally, it is immediate from the continuity argument that the index of $G(\lambda)$ is constant on the entire interval $[0,1]$. Hence, we compute the index of G by means of the index formula

$$\operatorname{ind} G_2(0) = \dim\ker G_2(0) = \mu(x_\alpha) - \mu(y_\beta)$$

and due to the fact that $G_2(0)$ is onto, as

$$\operatorname{ind} G(\lambda) = \operatorname{ind} G(0) = 1 + \mu(x_\alpha) - \mu(y_\beta)\ .$$

□

It should be mentioned at this stage that, generally, $\mathcal{M}^{H^{\alpha\beta}}_{x_\alpha,y_\beta}$ is a bounded manifold and that it is due to the regularity assumption at $\lambda = 0, 1$, that the boundary satisfies the relation

$$\partial\mathcal{M}^{H^{\alpha\beta}}_{x_\alpha,y_\beta} = \mathcal{M}^{H^{\alpha\beta}}_{x_\alpha,y_\beta} \cap \{0,1\} \times \mathcal{P}^{1,2}_{x_\alpha,y_\beta}\ .$$

Thus, there are no interior points $u_\lambda \in \mathcal{M}^{H^{\alpha\beta}}_{x_\alpha,y_\beta}$ which belong to the parameters $\lambda = 0$ or $\lambda = 1$. For a schematic picture of this situation the reader is referred to Figure 4.4 on page 146.

2.4 Compactness

In this section we wish to study the compactness properties of the trajectory spaces in question. The fundamental feature which gives rise to the special results in this context is formed by the discrepancy between the $H^{1,2}$-topology prescribed by the analytic methods and the C^∞_{loc}-topology on the set of the solutions of the ordinary differential equation

$$\dot\gamma = -\nabla f \circ \gamma\ .$$

Actually, this C^∞_{loc}-topology appears to convey geometrical significance rather than the $H^{1,2}$-topology. Due to the deduction from the implicit function theorem the finite-dimensional manifolds $\mathcal{M}^f_{x,y}$ and $\mathcal{M}^{h^{\alpha\beta}}_{x^\alpha,x^\beta}$ are obviously manifolds without boundary. Thus, in relation to the dimension one is naturally led to the question of compactness. In the case of relative Morse index $\mu(x) - \mu(y) = 1$ this would immediately imply the finiteness of $\widehat{\mathcal{M}}^f_{x,y}$, and the case of dimensions 1 and 2 would allow an easy classification of the diffeomorphism type.

As it will turn out throughout this investigation, the essential obstruction to the convergence of subsequences in the $H^{1,2}$-topology is formed by the geometrical splitting-up into the so-called broken trajectories. Actually, it is in relation with the $\mathbb{R}$-action on $\mathcal{M}^f$ that the C^∞_{loc}-topology appears to be appropriate for the description of this phenomenon. Due to the converse analysis concerning the gluing operation in the next section, the broken trajectories play the role of boundary points with regard to the trajectory manifolds. Briefly, we shall conclude the central results aimed at the construction of the canonical ∂-operator and of homomorphisms within the Morse homology from a kind of oriented cobordism equivalence for isolated, simply broken trajectories.

2.4.1 The Space of Unparametrized Trajectories

The aim of this subsection is to analyse the essential property of the trajectory manifold $\mathcal{M}^f_{x,y}$ in the time-independent case, namely the symmetry with respect to additive reparametrization of the trajectories. This action is called 'time-shifting'. We shall henceforth use the designation 'time-independent' or 'unparametrized' trajectories for the orbits of this time-shifting operation.

Let $\gamma : \mathbb{R} \to M$ be a solution of the differential equation $\dot{\gamma} = -\nabla f \circ \gamma$. Then, obviously, the same property holds for the shifted curve $\gamma \bullet \tau = \gamma_\tau = \gamma(\cdot + \tau)$, as we compute

$$\frac{\partial}{\partial t}(\gamma \bullet \tau) = \dot{\gamma}(\cdot + \tau) \ .$$

Moreover, it holds that $\gamma(\overline{\mathbb{R}}) = (\gamma \bullet \tau)(\overline{\mathbb{R}})$ for all $\gamma \in \mathcal{M}^f_{x,y}$, that is the unparametrized set of points $\gamma(\overline{\mathbb{R}})$ remains unchanged. It is in this sense of point sets that we shall use the expression 'geometrical' behaviour. We make this observation of symmetry precise by the following proposition.

Proposition 2.31 *The additive group $\mathbb{R}$ acts smoothly, freely and properly on the manifold $\mathcal{M}^f_{x,y}$ by*

$$\begin{aligned} \mathbb{R} \times \mathcal{M}^f_{x,y} &\to \mathcal{M}^f_{x,y} \\ (\tau, \gamma) &\mapsto \gamma \bullet \tau = \gamma(\cdot + \tau) \ , \end{aligned}$$

provided that the endpoints are distinct, $x \neq y$, so that $\mathcal{M}^f_{x,y}$ consists of non-constant trajectories.

Proof. Notwithstanding the purpose of developing the Morse homology in strict analogy to Floer homology, that is to say by means of the Fredholm analysis which solely takes into account relative Morse theory, we shall build a bridge to classical Morse theory at this stage. Let us deviate here from the purely relative concept in order to elucidate the geometrical relations with the dynamical system which is provided by the (negative) gradient flow. This proof shall make use of the identification $\mathcal{M}^f_{x,y} \approx W^u(x) \cap W^s(y)$. We consider the smooth evaluation map

$$\begin{aligned} E_0 : \quad \mathcal{P}^{1,2}_{x,y}(\mathbb{R}, M) &\to M \\ \gamma &\mapsto \gamma(0) \ . \end{aligned}$$

The smoothness of this map follows immediately from the representation by suitable local coordinates,

$$\begin{aligned} E_{0,\mathrm{loc}} : \quad H^{1,2}_{\mathbb{R}}(h^*\mathcal{D}) &\to U\big(h(0)\big) \subset T_{h(0)}M \\ \xi &\mapsto \Big(\exp^{-1}_{h(0)} \circ E_0 \circ \exp_h\Big)(\xi) = \xi(0) \ , \end{aligned}$$

where $E_{0,\mathrm{loc}}$ is continuously linear. If we restrict the evaluation map E_0 to the trajectory manifold $\mathcal{M}^f_{x,y}$, we obtain an embedding

$$E_0 : \mathcal{M}^f_{x,y} \hookrightarrow M \ , \tag{2.40}$$

because the differential

$$\begin{aligned} &DE_0(\gamma) : T_\gamma \mathcal{M}^f_{x,y} \to T_{\gamma(0)}M \\ &DE_0(\gamma) \cdot \xi = \xi(0) \end{aligned}$$

proves injective at each $\gamma \in \mathcal{M}^f_{x,y}$. Specifically, let us consider a tangent vector $\xi \in T_\gamma \mathcal{M}^f_{x,y} = \ker DF(\gamma)$ satisfying $\xi(0) = 0$. This argument analogous to (2.34) in Proposition 2.30, that ξ may be treated as solution of a linear, ordinary differential equation. Thus the identity $\xi \equiv 0$ is concluded from the uniqueness with respect to the initial value $\xi(0) = 0$. As a consequence the evaluation map leads to the diffeomorphism between the manifolds

$$\begin{aligned} E_0 : \quad \mathcal{M}^f_{x,y} &\xrightarrow{\approx} W^u(x) \cap W^s(y) \\ \gamma &\mapsto \gamma(0) \end{aligned} \tag{2.41}$$

as it was already mentioned in the introduction. It is obvious that E_0 identifies the group action $(\tau, \gamma) \mapsto \gamma \bullet \tau$ with the negative gradient flow,

$$\begin{aligned} \mathbb{R} \times \big(W^u(x) \cap W^s(y)\big) &\to W^u(x) \cap W^s(y) \\ (t\,, p) &\mapsto \Psi_t(p) \ . \end{aligned}$$

This gradient flow restricted to $W^u(x) \cap W^s(y)$ represents a smooth, free and proper $\mathbb{R}$-action. Thus the equivalence

$$\gamma \bullet \tau = \big(E_0^{-1} \circ \Psi_\tau \circ E_0\big)(\gamma)$$

yields the proof. □

In order to describe the orbit space $\widehat{\mathcal{M}}^f_{x,y} = \mathcal{M}^f_{x,y}/\mathbb{R}$ of the so-called unparametrized[7] or time-independent trajectories, we define the smooth function

$$\varphi = f \circ E_0 : \mathcal{M}^f_{x,y} \to \mathbb{R} \quad \text{with} \tag{2.42}$$
$$d\varphi(\gamma)\cdot\xi = df\big(\gamma(0)\big)\cdot\xi(0) = \big\langle \nabla f\big(\gamma(0)\big), \xi(0)\big\rangle \ .$$

Definition 2.32 *Let $f(y) < a < f(x)$, such that $\gamma(0)$ is not critical for any $\gamma \in \varphi^{-1}(a)$. Then a is a regular value of φ and*

$$\mathcal{M}^{f,a}_{x,y} = \varphi^{-1}(a)$$

is a $\big(\mu(x)-\mu(y)-1\big)$-dimensional submanifold of $\mathcal{M}^f_{x,y}$ and thus of $\mathcal{P}^{1,2}_{x,y}$, too.

From a geometrical point of view the orbit space $\widehat{\mathcal{M}}^f_{x,y}$ represents exactly this transversal intersection of a surface associated to a regular level of f with the manifold $W^u(x)\cap W^s(y)$, which gives an equivalent description of the trajectory space $\mathcal{M}^f_{x,y}$ as we have already seen. The reader is also referred to [Bo] and [S]. The following proposition yields an appropriate description for the orbit space.

Proposition 2.33 *The map*

$$\begin{array}{rccc} \Psi^a : \mathbb{R}\times\mathcal{M}^{f,a}_{x,y} & \to & \mathcal{M}^f_{x,y} \\ (\tau,\gamma) & \mapsto & \gamma\bullet\tau \end{array}$$

represents a $\mathbb{R}$-equivariant diffeomorphism with respect to the trivial $\mathbb{R}$-action on the left and to the action studied in Proposition 2.31 on the right side.

Proof. Due to Proposition 2.31 the $\mathbb{R}$-equivariance

$$\Psi^a(\tau+\sigma,\gamma) = \Psi^a(\tau,\gamma)\bullet\sigma$$

is obvious. It is similarly straightforward to verify the bijectivity of Ψ^a by means of Proposition 2.31 and the uniqueness for solutions of ordinary differential equations. Thus it is sufficient to prove that

$$D\Psi(\tau,\gamma) : \mathbb{R}\times T_\gamma\mathcal{M}^{f,a}_{x,y} \to T_{\gamma\bullet\tau}\mathcal{M}^f_{x,y}$$

is an isomorphism. Let $(s,\xi)\in\mathbb{R}\times T_\gamma\mathcal{M}^{f,a}_{x,y}$. Then (2.42) yields the identity

$$df\big(\gamma(0)\big)\cdot\xi(0) = 0 \ . \tag{2.43}$$

7This parametrization by the time variable t must not be confused with the dependence on an additional parameter λ as will be considered below.

Given $(s,\xi) \in \ker D\Psi^a(0,\gamma)$, the equation

$$0 = D\Psi^a(0,\gamma)\cdot(s,\xi) = \dot\gamma\cdot s + \xi$$

follows. From $\dot\gamma = -\nabla f\circ\gamma$ we obtain $\xi(0) = s\cdot\nabla f\big(\gamma(0)\big)$, such that (2.43) implies the identity $\xi(0) = 0$. Hence we conclude the injectivity of $D\Psi^a(0,\gamma)$ together with the surjectivity due to the dimension argument. Finally, we obtain the proof of the assertion from the identity

$$\Psi^a(\tau,\gamma) = \phi_\tau\left(\Psi^a\left(0,\gamma\right)\right) \ ,$$

because ϕ_τ is a diffeomorphism according to Proposition 2.31. □

Summing up this geometrical investigation we obtain Ψ^a as a diffeomorphism

$$\widehat{\mathcal{M}}^f_{x,y} \equiv \mathcal{M}^f_{x,y}/\mathbb{R} \approx \mathcal{M}^{f,a}_{x,y} \ ,$$

which appears to be quite suitable in order to describe the differentiable structure on the space of unparametrized trajectories provided by the negative gradient flow of f.

2.4.2 The Compactness Result for Unparametrized Trajectories

First of all, let us state once again the fundamental assumptions for the investigation in the following text. Let M be a complete finite-dimensional Riemannian manifold and let f be a smooth function on M satisfying suitable conditions as studied above such that

$$\widehat{\mathcal{M}}^f_{x,y} = \mathcal{M}^f_{x,y}/\mathbb{R}$$

represents a $\big(\mu(x)-\mu(y)-1\big)$-dimensional manifold. The first observation with respect to this statement is that the function f does not necessarily need to be a Morse function. It is sufficient for f to give rise to a transversal intersection of *df* and M as a zero section within the cotangent bundle T^*M, and we need a metric which is generic with respect to the Morse-Smale condition.

If we now wish to study the compactness properties of the trajectory manifold associated to f, we have to introduce further conditions on this function. In fact, we do not really need f to be a Morse function in the sense of coercivity, but to satisfy a more general condition as is fixed in the following definition.

Definition 2.34 *We say that the function f fulfills the Palais-Smale condition, which generalizes coercivity, if the following holds:*

(P-S) *Every sequence $(x_n)_{n\in\mathbb{N}} \subset M$, such that $\big(|f(x_n)|\big)_{n\in\mathbb{N}}$ is bounded and $|\nabla f(x_n)| \to 0$ converges, has a convergent subsequence.*

A subset $K \subset \widehat{\mathcal{M}}^f_{x,y}$ is called compact up to broken trajectories of order ν or up to $(\nu-1)$-fold broken trajectories exactly if for all $(\hat{u}_n)_{n\in\mathbb{N}} \subset K$:
Either $(\hat{u}_n)_{n\in\mathbb{N}}$ possesses a convergent subsequence, or there are critical points

$$x = y_0, \dots, y_i = y \in \operatorname{Crit} f,\ 2 \leqslant i \leqslant \nu$$

and connecting trajectories together with associated reparametrization times

$$v_j \in \mathcal{M}^f_{y_j, y_{j+1}}, \quad (\tau_{n,j})_{n\in\mathbb{N}} \subset \mathbb{R}\,,\ j = 0, \dots, i-1\ ,$$

such that we obtain the convergence

$$u_{n_k} \cdot \tau_{n_k,j} \xrightarrow{C^\infty_{\text{loc}}} v_j$$

for some subsequence $(n_k)_{k\in\mathbb{N}}$.

This convergence of trajectories $w_j \xrightarrow{C^\infty_{\text{loc}}} w$ in $C^\infty_{x,y}(\overline{\mathbb{R}}, M)$ shall henceforth be called weak convergence, and the weak convergence of unparametrized trajectories with respect to suitably provided reparametrization times is denoted by geometrical convergence.

The geometrical convergence is sketched in Figure 2.1. Referring to these terms

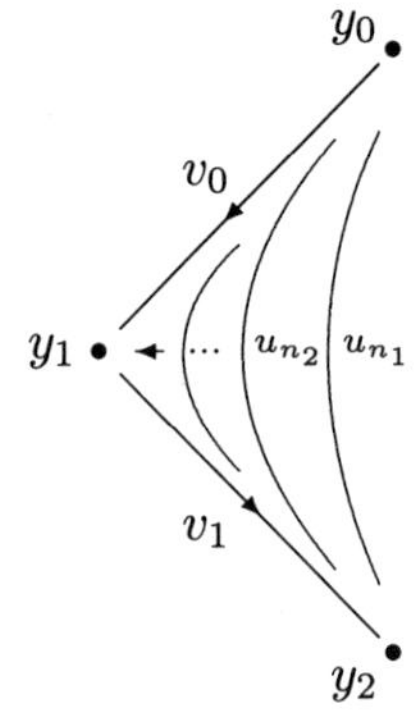

Figure 2.1: Weak or geometrical convergence toward a simply broken trajectory

the central statement of this section is given by

Proposition 2.35 *If f satisfies the Palais-Smale condition, the manifold $\widehat{\mathcal{M}}^f_{x,y}$ is compact up to broken trajectories of order $\mu(x) - \mu(y)$.*

We are immediately led to the conclusion:

Corollary 2.36 *If, moreover, it holds that* $\mu(x) = \mu(y) + 1$, *then* $\widehat{\mathcal{M}}^f_{x,y}$ *is a finite set of unparametrized trajectories.*

In order to give a proof for the central compactness result we first state an immediate consequence of the Palais-Smale condition.

Auxiliary Proposition 2.37 *If* f *satisfies (P-S), there is a constant* $\epsilon > 0$ *depending merely on* f, x *and* y, *such that the set*

$$K^{x,y}_\epsilon = \{ z \in M \mid f(y) \leqslant f(z) \leqslant f(x),\ \|\nabla f(z)\| \leqslant \epsilon \} \quad \textit{is relatively compact}.$$

The following lemma shall express the 'geometrical compactness' of the unparametrized trajectory spaces in terms of compactness of the parametrized spaces with respect to the C^∞_{loc}-topology.

Lemma 2.38 *Every sequence* $(u_n)_{n\in\mathbb{N}} \subset \mathcal{M}^f_{x,y}$ *possesses a weakly convergent subsequence*

$$u_{n_k} \overset{C^\infty_{\text{loc}}}{\to} v \in C^\infty(\mathbb{R}, M) \ .$$

Proof. Essentially, the strategy for proving this lemma is to execute a transition from the $H^{1,2}$-topology of the Hilbert submanifold $\mathcal{M}^f_{x,y}$ to the C^0_{loc}-topology. It appears to be of crucial importance that we have a uniform bound on the $H^{1,2}$-norm for the trajectories with fixed endpoints $x, y \in \operatorname{Crit} f$. The C^∞_{loc}-convergence will, in principle, follow iteratively by elliptic regularity.

The fact that the trajectories $(u_n)_{n\in\mathbb{N}}$ arise from the negative gradient flow with respect to the fixed endpoints x and y yields the uniform estimate

$$\int_s^t |\dot{u}_n(\tau)|^2 d\tau = \int_s^t \langle \dot{u}_n, -\nabla f \circ u_n \rangle \, d\tau \leqslant f(x) - f(y) \quad \text{for all } s \leqslant t \ . \tag{2.44}$$

Let d be the Riemannian distance on M. Then the estimate

$$\begin{aligned} d\big(u_n(t), u_n(s)\big) \quad &\leqslant \quad \int_s^t |\dot{u}_n(\tau)| \, d\tau \\ &\overset{\text{Hölder}}{\leqslant} \quad \sqrt{|t-s|} \sqrt{\int_s^t |\dot{u}_n(\tau)|^2 \, d\tau} \\ &\overset{(2.44)}{\leqslant} \quad \sqrt{|t-s|} \sqrt{f(x) - f(y)} \end{aligned}$$

gives rise to the equicontinuity of $(u_n)_{n\in\mathbb{N}}$ in $C^0(\mathbb{R}, M)$. In order to be able to apply the theorem of Arzela-Ascoli we still have to analyse the pointwise convergence. At this stage the uniform L^2-bound on the derivatives $\dot{u}_n$ as well as the (P-S)-condition appear to be the decisive arguments. Thus let us choose a fixed $t_0 \in \mathbb{R}$. According to Auxiliary Proposition 2.37 we can assume without loss of generality the inequality

$$|(\nabla f \circ u_n)(t_0)| \geqslant \epsilon \quad \text{for all } n \in \mathbb{N} .$$

Due to the asymptotic decrease $\lim\limits_{|t|\to\infty} \left|\nabla f\big(u_n(t)\big)\right| = 0$ we can find a sequence $(t_n)_{n\in\mathbb{N}} \subset \mathbb{R}$, such that for all $n \in \mathbb{N}$:

$$\text{(2.45)} \quad \left|\nabla f\big(u_n(t_n)\big)\right| = \epsilon \quad \text{and} \quad \left|\nabla f\big(u_n(s)\big)\right| \geqslant \epsilon \quad \text{for all } s \in [t_n, t_0] .$$

Thus the auxiliary proposition helps us in fixing the points $\big(u_n(t_n)\big)_{n\in\mathbb{N}}$ within the bounded set $K_\epsilon^{x,y}$. In other words, there is an $r_0 > 0$ such that the estimate $d\big(u_n(t_n), x\big) < r_0$ holds for all $n \in \mathbb{N}$. Referring to the inequality

$$d\big(x, u_n(t_0)\big) < r_0 + l_n(t_0) \quad \text{with} \quad l_n(s) = \int_{t_n}^{s} |\dot{u}_n(\tau)|\, d\tau$$

the only step remaining in order to achieve pointwise convergence is to find an upper bound for $\big(l_n(t_0)\big)_{n\in\mathbb{N}}$. Calculating

$$\frac{dl_n}{ds}(\tau) = |\nabla f \circ u_n(\tau)| \geqslant \epsilon ,$$
$$\frac{d(f \circ u_n)}{ds}(\tau) = -\,|\nabla f \circ u_n(\tau)|^2$$

by means of (2.45) for $t_n \leqslant \tau \leqslant t_0$ gives rise to the estimate

$$\frac{dl_n}{ds}(\tau) \leqslant \left(-\frac{1}{\epsilon}\right) \frac{d}{ds}(f \circ u_n)(\tau) .$$

Hence, we obtain the upper bound from

$$\begin{aligned} l_n(t_0) &= \int_{t_n}^{t_0} \frac{dl_n}{ds}(\tau) d\tau \leqslant \frac{1}{\epsilon} \int_{t_0}^{t_n} \frac{d(f \circ u_n)}{ds}(\tau) d\tau = \frac{f\big(u_n(t_n)\big) - f\big(u_n(t_0)\big)}{\epsilon} \\ &\leqslant \frac{f(x) - f(y)}{\epsilon} . \end{aligned}$$

Since $\big(u_n(t)\big)_{n\in\mathbb{N}}$ is a relatively compact set for each fixed instant $t \in \mathbb{R}$, the theorem of Arzela-Ascoli provides a subsequence $(u_{n_k})_{k\in\mathbb{N}}$ converging on compact intervals. This amounts to stating that there is a

$$v \in C^0(\mathbb{R}, M) \quad \text{with} \quad v(t) = \lim_{k\to\infty} u_{n_k}(t),\ t \in \mathbb{R}, \text{ s.t.}$$
$$u_{n_k\,|\,[-R,R]} \overset{C^0([-R,R])}{\longrightarrow} v_{|\,[-R,R]} \quad \text{for all } R \geqslant 0 \ .$$

The fact that the trajectories $(u_{n_k})_{k\in\mathbb{N}}$ are solutions of the ordinary differential equation $\dot{u} = -\nabla f(u)$ associated to the smooth gradient field implies the C^k-convergence

$$u_{n_k\,|\,[-R,R]} \overset{C^k([-R,R])}{\longrightarrow} v_{[-R,R]} \quad \text{with} \quad \dot{v} = -\nabla f \circ v$$

iteratively for all $k \in \mathbb{N}$, thus concluding the proof. □

Lemma 2.39 *Let* $(u_n)_{n\in\mathbb{N}} \subset \mathcal{M}^f_{x,y}$ *be a weakly convergent sequence*

$$u_n \overset{C^\infty_{\text{loc}}}{\longrightarrow} v \in C^\infty(\mathbb{R}, M) \ ,$$

such that $v \in \mathcal{M}^f_{x,y}$ *is a trajectory in the same trajectory space. Then the sequence* $(u_n)_{n\in\mathbb{N}}$ *is also* $H^{1,2}$*-convergent,*

$$u_n \overset{\mathcal{P}^{1,2}_{x,y}}{\longrightarrow} v \in \mathcal{M}^f_{x,y} \ .$$

Proof. The first step of the proof is to observe that it is sufficient to show that the elements u_n of the sequence converge uniformly toward x and y, respectively, for $|t| \to \infty$. If this is true, Lemma 2.10 and the associated marginal remark on the dependence $t_0 = t_0(s)$ imply the existence of a fixed $T \geqslant 0$, such that we observe uniform exponential asymptotic convergence with respect to local coordinates at x and y:

$$\left|u_{n,\text{loc}_{x,y}}(t)\right| \leqslant c_{x,y}\mathrm{e}^{-\lambda_{x,y}|t|} \quad \text{for } |t| \geqslant T \text{ and for all } n \in \mathbb{N} \ . \tag{2.46}$$

Without loss of generality we can assume that the entire sequence lies within a coordinate neighbourhood of v in $\mathcal{P}^{1,2}_{x,y}$, $(u_n)_{n\in\mathbb{N}} \subset \mathcal{O}^{1,2}(v)$. Then referring to the local coordinates of the Hilbert manifold

$$u_n = \exp_v \xi_n, \quad \xi_n \in H^{1,2}_{\mathbb{R}}(v^*TM) \ ,$$

the weak convergence

$$\xi_n \overset{C^\infty_{\text{loc}}}{\longrightarrow} 0$$

together with the uniform estimate (2.46) finally imply the asserted $H^{1,2}$-convergence.

Thus we have to show the following uniform asymptotic convergence at y (and analogously at x):

Given any neighbourhood $U(y)$, there is an associated $t_0 \in \mathbb{R}$ such that the relation $u_n(t) \in U(y)$ holds for all $t \geqslant t_0$ and $n \in \mathbb{N}$.

Let $B(y)$ be an isolating, open neighbourhood of y, that is $B(y) \cap \operatorname{Crit} f = \{y\}$. Then the (P-S) condition on f guarantees that

$$N_\epsilon = \{\, p \in B(y) \mid |f(p) - f(y)| < \epsilon,\, |\nabla f(p)| < \epsilon \,\}, \quad \epsilon > 0$$

forms a fundamental system of neighbourhoods of y. Moreover, we notice that

$$U = \bigcup_{n \in \mathbb{N}} u_n(\mathbb{R})$$

is a bounded set. This is clear from the decomposition

$$U = \bigcup_{n \in \mathbb{N}} \{\, u_n(t) \mid |\nabla f \circ u_n(t)| \geqslant \epsilon \,\} \cup \bigcup_{n \in \mathbb{N}} \{ u_n(t) \mid |\nabla f \circ u_n(t)| < \epsilon \,\} \; .$$

In analogy to the proof of Lemma 2.38 the sets $\{\, u_n(t) \mid |\nabla f \circ u_n(t)| \geqslant \epsilon \,\}$ are uniformly bounded and

$$\bigcup_{n \in \mathbb{N}} \{\, u_n(t) \mid |\nabla f \circ u_n(t)| < \epsilon \,\} \subset K_\epsilon^{x,y} \quad \text{is relatively compact} \; .$$

In consequence of the boundedness of U, the vector field ∇f is globally Lipschitz continuous on $\overline{U}$. This fact together with the equicontinuity of $(u_n)_{n \in \mathbb{N}}$ from Lemma 2.38 gives rise to the existence of a constant c such that the estimate

$$\Big| \big|\nabla f \circ u_n(t)\big| - \big|\nabla f \circ u_n(s)\big| \Big| < c \cdot \sqrt{|t - s|}$$

holds uniformly for all $n \in \mathbb{N}$. Hence we conclude the inequality

$$\text{(2.47)} \quad \big|\nabla f \circ u_n(s)\big| > \big|\nabla f \circ u_n(t)\big| - c\sqrt{|t - s|} \quad \text{for all } s, t \in \mathbb{R},\ n \in \mathbb{N} \; .$$

Let us now prove the uniform convergence asserted above by contradiction: If we assume that there is an $\epsilon > 0$ together with sequences $n_k \to \infty$ and $t_k \to \infty$ satisfying $\big|\nabla f \circ u_{n_k}(t_k)\big| > \epsilon$ for all $k \in \mathbb{N}$, then we first deduce, due to (2.47),

$$\big|\nabla f \circ u_{n_k}(s)\big| > \frac{\epsilon}{2} \quad \text{for} \quad |t_k - s| < \frac{\epsilon^2}{4c^2} =: \delta \; ,$$

and thus, second,

$$f\big(u_{n_k}(t_k)\big) - f\big(u_{n_k}(t_k + \delta)\big) = \int_{t_k}^{t_k + \delta} \big|\nabla f \circ u_{n_k}(s)\big|\, ds \; > \; \delta \cdot \frac{\epsilon}{2} \; = \; \frac{\epsilon^3}{8c^2} \; .$$

From this we extract the estimate

$$f\big(u_{n_k}(t_k)\big) - f(y) > \frac{\epsilon^3}{8c^2} \quad \text{for all } k \in \mathbb{N} \; ,$$

in contradiction to the assumption of $t_k \to \infty$ and $u_{n_k} \overset{C^\infty_{\text{loc}}}{\longrightarrow} v \in \mathcal{M}^f_{x,y}$. □

Proof of proposition 2.35: At this stage we finally explore the obstruction to the strong $H^{1,2}$-convergence of the given sequence $(u_n)_{n\in\mathbb{N}}$ of unparametrized trajectories. As we remarked above, Lemma 2.38 leads us to the weak convergence

$$u_{n_k} \overset{C^\infty_{\text{loc}}}{\longrightarrow} v \in C^\infty(\mathbb{R}, M) \tag{2.48}$$

and thus to a trajectory associated to the same negative gradient flow,

$$\left(\frac{\partial}{\partial t} + \nabla f\right) \circ v = 0 \ ,$$

but we cannot necessarily conclude that this eventually new trajectory lies in the same trajectory space $\widehat{\mathcal{M}}^f_{x,y}$. In fact, it is the $\mathbb{R}$-invariance with respect to the parametrization by time which gives rise to the possibility of the so-called geometrical splitting-up of the sequence of trajectories. Given any convergent subsequence of parametrized representatives according to Lemma 2.38, the relation (2.48) implies the estimate

$$f\big(v(t)\big) \in \big[f(y), f(x)\big] \quad \text{for all } t \in \mathbb{R} \ .$$

Consequently, the set of possible manifolds of trajectories $\mathcal{M}^f_{\tilde{x},\tilde{y}}$, which might contain the element v, is already confined by the condition

$$f(y) \leqslant f(\tilde{y}) \leqslant f(\tilde{x}) \leqslant f(x) \ . \tag{2.49}$$

Of course, we have to verify that v is an element of such a trajectory space $\mathcal{M}^f_{x',y'}$ anyway. We deduce this from (2.48) and the fundamental Palais-Smale condition as follows:
First, relation (2.48) implies the confinement by

$$f\big(v(T)\big) \in \big[f(y), f(x)\big] \quad \text{for all } T \in \mathbb{R} \ , \tag{2.50}$$

such that the solution property $\dot{v} = -\nabla f \circ v$ gives rise to the estimate

$$\int_{-T}^{T} \big|\dot{v}(s)\big|^2 ds = f\big(v(T)\big) - f\big(v(-T)\big) \leqslant f(x) - f(y) \quad \text{for all } T \in \mathbb{R} \ . \tag{2.51}$$

Thus we obtain the finite L^2-norm $\int_{-\infty}^{\infty} |\dot{v}|^2 ds < \infty$ and in consequence of this the asymptotic behaviour

$$\lim_{t\to\pm\infty} \nabla f\big(v(t)\big) = \lim_{t\to\pm\infty} \dot{v}(t) = 0 \ . \tag{2.52}$$

Finally, by combining the Palais-Smale condition on f with (2.50) and (2.52) we deduce the relation $v \in \mathcal{M}^f_{\tilde{x},\tilde{y}}$ with respect to critical points which satisfy the estimate (2.49).

Now, we have to study separately the different cases for $\tilde{x}$ and $\tilde{y}$. Considering the first possibility, $v \in \mathcal{M}^f_{x,y}$, we immediately obtain strong $\mathcal{P}^{1,2}_{x,y}$-convergence from Lemma 2.39. All the other cases which lead to the splitting-up into broken trajectories, that is, without loss of generality,

$$v \in \mathcal{M}^f_{x,y'} \quad \text{with } f(y) < f(y') \ ,$$

can now be controlled by means of suitable reparametrizations. Let us choose any regular level a of f satisfying $f(y) < a < f(y')$ and a reparametrization of (u_{n_k}),

$$\tilde{u}_k = u_{n_k} \cdot \tau_k = u_{n_k}(\cdot + \tau_k), \ (\tau_k)_{k \in \mathbb{N}} \subset \mathbb{R} \ ,$$

such that the identity $f\big((u_{n_k} \cdot \tau_k)(0)\big) = a$ holds for all $k \in \mathbb{N}$. If we apply Lemma 2.38 once again to this sequence $\tilde{u}_k$, a further weakly convergent subsequence

$$\tilde{u}_{k_l} \xrightarrow{C^\infty_{\text{loc}}} \tilde{v}$$

is extracted such that it complies with the estimate

$$f(y) \leqslant f\big(\tilde{v}(+\infty)\big) \leqslant f\big(\tilde{v}(-\infty)\big) \leqslant f(y') \ .$$

Iterating these procedures of sorting by the values of f at the ends of the trajectories and executing suitable reparametrizations gives rise to a process which provides us step by step with either $H^{1,2}$-convergent subsequences or merely C^∞_{loc}-convergent subsequences tending toward constant, i.e. trivial, trajectories, or it 'recovers' new critical points of f. It is due to the Morse-Smale estimate for the Morse indices at the ends of the non-constant trajectories,

$$\mu\big(v^{(j)}(-\infty)\big) - \mu\big(v^{(j)}(+\infty)\big) > 0 \ ,$$

that this iterative process of reparametrization stops at broken trajectories of order $\mu(x) - \mu(y)$ at the latest. This can be seen immediately from the inequality

$$\mu(y) < \ldots < \mu\big(v^{(j)}(-\infty)\big) = \mu\big(v^{(j+1)}(+\infty)\big) < \ldots < \mu(x) \ ,$$

which necessarily holds after restriction to non-constant trajectories and suitable renumbering. Thus the assertion of the proposition follows. □

Remark It seems worth emphasizing, once again, that instead of the strong compactness condition for the sublevel-sets $M^a(f) = f^{-1}\big((-\infty, a]\big)$, which is meant by the coercivity for classical Morse functions, the Palais-Smale property defined above is thoroughly sufficient for the compactness result of the trajectory manifolds $\widehat{\mathcal{M}}^{(1)}_{x,y}$, which plays an essential role in the definition of the ∂-operator.

2.4.3 The Compactness Result for Homotopy Trajectories

Essentially, the statements on the compactness of the trajectory manifolds for the ∂-operator may also be transferred to the trajectory spaces of the homotopy morphisms. However, since we restrict our considerations to Morse homology, we shall work out the following proofs for the compactness result under more specialized assumptions. This will simplify, for instance, the step of the uniform L^2-bound.

Definition 2.40 *Let us remember that the coercivity property of the Morse functions appeared as an essential part within the proof of the compactness result for unparametrized trajectories. Turning now to the time-dependent situation of a homotopy between Morse functions we have to guarantee an analogous kind of coercivity. We call a smooth homotopy* $h = h^{\alpha\beta} : \mathbb{R} \times M \to \mathbb{R}$ *between the two Morse functions* f^α *and* f^β *a Morse homotopy, if* $h_t(\cdot) = h(t, \cdot)$ *satisfies the following condition:*

$$(2.53) \quad \lim_{n\to\infty} h_{t_n}(x_n) = \infty, \quad \textit{if } (t_n, x_n)_{n\in\mathbb{N}} \subset \mathbb{R} \times M,\ \mathrm{d}(x_0, x_n) \to \infty\ .^8$$

Now let us consider a smooth, regular Morse homotopy $f^\alpha \overset{h_t}{\simeq} f^\beta$. Apart from the demanded coercivity property it particularly holds that

$$\dot{h}_s(\cdot) = \frac{\partial h}{\partial t}(s, \cdot) = 0 \text{ for all } |s| \geqslant R\ .$$

Thus, the $\big(\mu(x_\alpha) - \mu(x_\beta)\big)$-dimensional manifold of the so-called homotopy trajectories (h-trajectory) comprises the solution curves $u_{\alpha\beta}$ of the ordinary non-autonomous differential equation

$$(2.54) \qquad \dot{\gamma}(t) = -\frac{\nabla h_t}{\sqrt{1 + |\dot{h}_t|^2 |\nabla h_t|^2}} \circ \gamma(t)$$

with the ends fixed at the critical points x_α and x_β of f^α and f^β, respectively.

As to this manifold we are again provided with a compactness result similar to that in the time-independent case. Once again the only obstruction against the strong $H^{1,2}$-compactness consists of the C^∞_{loc}-convergence toward broken trajectories. This time we additionally must distinguish between the $\mathbb{R}$-invariant trajectories with respect to f^α and f^β and the $h^{\alpha\beta}$-trajectories. The convergence behaviour of homotopy trajectories is expressed by

8Note that this definition relies crucially on the completeness of the Riemannian manifold.

Proposition 2.41 *Let* $(u_n)_{n\in\mathbb{N}} \subset \mathcal{M}^{h^{\alpha\beta}}_{x^k_\alpha, x^0_\beta}$ *be a sequence of* $h^{\alpha\beta}$*-trajectories. Provided that there is no* $H^{1,2}$*-convergent subsequence, there exist critical points*

$$x_\alpha = x^0_\alpha, \dots, x^k_\alpha \in \operatorname{Crit}(f^\alpha),\ x^0_\beta, \dots, x^l_\beta = x_\beta \in \operatorname{Crit}(f^\beta)\ ,$$
$$1 \leqslant k + l \leqslant \mu(x_\alpha) - \mu(x_\beta)$$

and ($h^{\alpha\beta}$*-)trajectories*

$$v^i_\alpha \in \mathcal{M}^{f^\alpha}_{x^i_\alpha, x^{i+1}_\alpha},\ v^j_\beta \in \mathcal{M}^{f^\beta}_{x^j_\beta, x^{j+1}_\beta},\ v_{\alpha\beta} \in \mathcal{M}^{h^{\alpha\beta}}_{x^\alpha, x^\beta}$$

together with reparametrization sequences

$$(\tau^i_{\alpha,n})_{n\in\mathbb{N}},\ (\tau^j_{\beta,n})_{n\in\mathbb{N}} \subset \mathbb{R}\ ,$$

such that up to the choice of a respectively suitable subsequence weak convergence of the form

$$u_n \cdot \tau^i_{\alpha,n} \xrightarrow{C^\infty_{\mathrm{loc}}} v^i_\alpha\,,\quad u_n \cdot \tau^j_{\beta,n} \xrightarrow{C^\infty_{\mathrm{loc}}} v^j_\beta \quad \textit{and} \quad u_n \xrightarrow{C^\infty_{\mathrm{loc}}} v_{\alpha\beta}$$

with $0 \leqslant i \leqslant k-1$ *and* $0 \leqslant j \leqslant l-1$ *holds. Moreover, the inequality*

$$\mu(x^0_\alpha) < \dots < \mu(x^k_\alpha) \leqslant \mu(x^0_\beta) < \dots < \mu(x^l_\beta)$$

is satisfied.

Provided that we are able to prove a statement on the C^∞_{loc}-convergence in the case of homotopy trajectories, which is analogous to Lemma 2.38, the proof of this compactness result is accomplished according to Proposition 2.35. The weak convergence which proves crucial for this compactness result is obtained as follows:

Auxiliary Proposition 2.42 *Each solution* $\gamma : \big(t^-(t_0, x_0), t^+(t_0, x_0)\big) \to M$ *of the non-autonomous differential equation (2.54) satisfies* $t^+(t_0, x_0) = \infty$, *that is the* 2*-parameter flow is global in forward time direction.*

Proof. Let a maximal solution of (2.54) $\gamma : (t^-, t^+) \to M$ be given. Due to the completeness of the metric space M and the requirement of coercivity (2.53) to the Morse homotopy $h^{\alpha\beta}$ it holds that

$$\lim_{t\to t^+} h\big(t, \gamma(t)\big) = \infty\ ,$$

if t^+ is finite. Thus, provided that t^+ is finite, it follows that

$$\lim_{t\to t^+} \frac{d}{ds} h\big(s, \gamma(s)\big)(t) = \infty\ . \tag{2.55}$$

The computation

$$\begin{aligned} \frac{d}{ds} h(s,\gamma(s))(t) &= \dot{h}(t,\gamma(t)) + \nabla h_t \cdot \dot{\gamma}(t) \\ &\overset{(2.54)}{=} \left(\dot{h}_t - \frac{|\nabla h_t|^2}{\sqrt{1+|\dot{h}_t|2|\nabla h_t|2}} \right) \circ \gamma(t) \end{aligned} \tag{2.56}$$

together with (2.55) yields

$$\lim_{t \to t+} \dot{h}(t,\gamma(t)) = \infty \ .$$

Applying the identity (2.54) once again we deduce

$$\lim_{t \to t+} \dot{\gamma}(t) = 0$$

and hence a contradiction to the supposed maximality of γ. □

Auxiliary Proposition 2.43 *There is a uniform L^2-estimate for the homotopy trajectories with fixed endpoints. It is given as*

$$\int_{-\infty}^{\infty} |\dot{\gamma}|^2 \, dt \; < \; c(x_\alpha, x_\beta), \quad \textit{for all } \gamma \in \mathcal{M}^{h^{\alpha\beta}}_{x_\alpha,x_\beta} \ .$$

Proof. First, we may estimate the integral by means of the identity (2.56). This yields

$$\begin{aligned} \int_{-\infty}^{\infty} |\dot{\gamma}|^2 \, dt &\overset{(2.54)}{=} \int_{-\infty}^{\infty} \frac{|\nabla h_t|^2}{1+|\dot{h}_t|^2|\nabla h_t|^2} \, dt \leqslant \int_{-\infty}^{\infty} \frac{|\nabla h_t|^2}{\sqrt{1+|\dot{h}_t|^2|\nabla h_t|^2}} \, dt \\ &\overset{(2.56)}{=} -\int_{-\infty}^{\infty} \left(\frac{d}{d\tau} h(\tau,\gamma(\tau))(t) - \dot{h}(t,\gamma(t)) \right) dt \\ &= f^\alpha(x_\alpha) - f^\beta(x_\beta) + \int_{-R}^{R} \dot{h}(t,\gamma(t)) dt \ . \end{aligned} \tag{2.57}$$

Since we deal with coercive Morse functions, the compactness of the set $\{ x \in M \mid f^\alpha(x) \leqslant f^\alpha(x_\alpha) \}$ implies the relative compactness of the set $\{ \gamma(-R) \in M \mid \gamma \in \mathcal{M}_{x_\alpha,x_\beta} \}$. As a consequence the continuous 2-parameter flow from Auxiliary Proposition 2.42 yields the fact that

$$\bigcup_{\gamma \in \mathcal{M}_{x_\alpha,x_\beta}} \gamma([-R,R]) \quad \text{is relatively compact} \ .$$

Therefore the uniform bound

$$
\begin{aligned}
K &= K(x_\alpha, x_\beta) \\
&= \sup\left\{ \left|\dot{h}(t,\gamma(t))\right| \mid t \in [-R,R],\, \gamma \in \mathcal{M}^{h^{\alpha\beta}}_{x_\alpha,x_\beta} \right\} < \infty
\end{aligned}
\tag{2.58}
$$

exists. Finally, the uniform L^2-bound is obtained from (2.57) as

$$c(x_\alpha, x_\beta) = f^\alpha(x_\alpha) - f^\beta(x_\beta) + 2KR \ .$$

□

Corollary 2.44 *In particular, we obtain a compactness result for the h-trajectories as point sets from the auxiliary proposition together with the time-constant Morse functions at the ends of the Morse homotopy:*

$$\bigcup_{\gamma \in \mathcal{M}_{x_\alpha,x_\beta}} \gamma(\overline{\mathbb{R}}) \quad \textit{is relatively compact} \ . \tag{2.59}$$

Relying on these preparations we are now able to prove the crucial weak convergence also for the homotopy trajectories.

Lemma 2.45 *Let* $(u_n)_{n\in\mathbb{N}} \subset \mathcal{M}^{h^{\alpha\beta}}_{x_\alpha,x_\beta}$ *be a sequence of* $h^{\alpha\beta}$*-trajectories with fixed endpoints. Then there is a weakly convergent subsequence*

$$u_{n_k} \xrightarrow{C^\infty_{\mathrm{loc}}} v \in C^\infty(\mathbb{R}, M) \ .$$

Proof. The proof is accomplished by strict analogy to Lemma 2.38. Auxiliary Proposition 2.43 gives rise to the equicontinuity of the sequence. The subsequent Corollary 2.44 yields the pointwise convergence for approriate subsequences. It is worthwhile to mention that at this point we crucially relied on the coercivity of the asymptotically constant homotopy. The assertion now follows analogous to Lemma 2.38 together with the property

$$v \in \mathcal{M}^{h^{\alpha\beta}}_{x'_\alpha,x'_\beta}, \quad \text{with} \quad f^\alpha(x'_\alpha) \leqslant f^\alpha(x_\alpha),\ f^\beta(x_\beta) \leqslant f^\beta(x'_\beta) \ . \tag{2.60}$$

□

The proof of Proposition 2.41 is carried out now in accordance with principally the same scheme as for the time-independent trajectories. The only difference consists of the distinction between the $\mathbb{R}$-invariant trajectories and the h-trajectories:

Proof of proposition 2.41: It is sufficient to consider the case $f(x'_\alpha) < f(x_\alpha)$. Once again, let a be any fixed regular value from $\big(f(x'_\alpha), f(x_\alpha)\big)$ and choose a reparametrization sequence in accordance with the condition

$$f\big(u_{n_k}(\tau_k)\big) = a \ . \tag{2.61}$$

Obviously the divergence $\tau_k \to -\infty$ follows, that is, without loss of generality, $\tau_k < -R$. Since reparametrization leaves the uniform estimate from Auxiliary Proposition 2.43 invariant and since we can further assume pointwise convergence of subsequences, we are able to deduce the weak convergence

$$u_{n_k} \cdot \tau_k \overset{C^\infty_{\mathrm{loc}}}{\longrightarrow} v' \in \mathcal{M}^{f^\alpha}_{y,y'} \tag{2.62}$$

toward an f^α-trajectory up to the choice of a suitable subsequence. An analogous reparametrization algorithm thus leads us to the assertion. □

Corollary 2.46 *Let us state an important conclusion from Auxiliary Proposition 2.42. Since the set $\{x \in M | f^\alpha(x) \leqslant f^\alpha(x_\alpha)\}$ is compact, the same is true for the set*

$$\left\{ \gamma(+\infty) \mid f^\alpha\big(\gamma(-R)\big) \leqslant f^\alpha(x_\alpha)\,,\ \dot{\gamma} = -\frac{\nabla h_t}{\sqrt{1+\ldots}} \circ \gamma \right\}$$

due to the auxiliary proposition. Thus, each critical point $x_\alpha \in \operatorname{Crit} f^\alpha$ can be connected to only finitely many critical points $x_\beta \in \operatorname{Crit} f^\beta$ via $h^{\alpha\beta}$-trajectories. In fact, this will be the reason that the homotopy morphism $\Phi^{\beta\alpha}$, which will be dealt with in Chapter 4, is indeed well-defined.

2.4.4 The Compactness Result for λ-Parametrized Trajectories

The compactness result for the parameter trajectory manifold $\mathcal{M}^{H^{\alpha\beta}}_{x_\alpha,y_\beta}$ associated to a generic homotopy of Morse homotopies is obtained according to the same scheme as for homotopy trajectories. We recall that this manifold of λ-parametrized trajectories (provided that we deal with a compact discrete set) forms the basis for the definition of the chain homotopy operator $\Psi^{\beta\alpha}$. Since the parameter dependence

$$H_\lambda = H(\lambda,\cdot,\cdot),\ H : [0,1] \times \mathbb{R} \times M \to \mathbb{R}$$

is smooth and the parameter set compact, there are analogous uniform estimates. Once again, the only obstruction to the $H^{1,2}$-compactness occurs as the C^∞_{loc}-convergence toward broken trjaectories. Here, these broken trajectories consist of one $H^{\alpha\beta}_\lambda$-trajectory and several f^α- and f^β-trajectories depending on the dimension and the connected component of the trajectory manifold.

In particular, we deduce the conclusion that $\mathcal{M}^{H^{\alpha\beta}}_{x_\alpha,y_\beta}$ represents a finite set $\big\{(\lambda_1,u_1),\ldots,(\lambda_\nu,u_\nu)\big\}$ with $\lambda_i \neq 0,1$, provided that we deal with the relative Morse index $\mu(x_\alpha) = \mu(y_\beta) - 1$.

2.5 Gluing

In the last section we saw how the broken trajectories associated to the time-independent and time-dependent gradient flow form an obstruction to the compactness of the trajectory spaces. By introducing the gluing operation we now intend to complete the complementary concept of 'gluing' and 'compactness' as was described in the introduction. The gluing operation $\#_\rho$ will map simply broken trajectories, that is, pairs of the form

$$(u, v) \in \mathcal{M}_{x,y} \times \mathcal{M}_{y,z} \ ,$$

into the higher dimensional trajectory space $\mathcal{M}_{x,z}$ in a way depending on an additional positive gluing parameter ρ. At first we restrict ourselves to the trajectory spaces of the time-independent gradient flow associated to a fixed Morse function. Principally, the construction of the gluing map for trajectory spaces in the time-dependent and λ-parametrized case is accomplished similarly with regard to the technical details.

As to notation and the main ideas for the proof we follow the concept of gluing that was developed in [F6] for the symplectic case. For a compact presentation of this principle which in general is related to C. H. Taubes, the reader is referred to [F4].

2.5.1 Gluing for the Time-Independent Trajectory Spaces

The crucial problem concerning the gluing operation we are looking for is its dependence on noncanonical assumptions. For example, it appears that the compactness of the set of broken trajectories $(u,\ v) \in \mathcal{M}_{x,y} \times \mathcal{M}_{y,z}$, which are to be 'glued together' and mapped into $\mathcal{M}_{x,z}$, is an indispensable prerequisite. The crucial problem, however, is the lack of uniqueness. Notwithstanding the fact that given a compact set of broken trajectories we can always construct a gluing map, which embeds the glued trajectories in a unique way, this is performed in a necessarily noncanonical way. Clearly, any suitable composition with certain diffeomorphisms on the space of trajectories provides us with another, different gluing map endowed with the same convergence behaviour with regard to the compactness-gluing cobordism. A consequence of this uncertainty is the lack of a natural, immediate law of associativity,

$$(u\#_{\rho_1}v)\#_{\rho_2}w \ \stackrel{?}{=} \ u\#_{\rho_1}(v\#_{\rho_2}w) \ .$$

The result, which will be sufficient according to our concerns with homology theory, is stated as follows:

Theorem 3 *Given a compact set of simply broken trajectories* $K \subset \mathcal{M}^f_{x,y} \times \mathcal{M}^f_{y,z}$*, there is a lower bound* $\rho_K \geqslant 0$ *and a smooth map*

$$\# : K \times [\rho_K, \infty) \to \mathcal{M}^f_{x,z}$$
$$(u, v, \rho) \mapsto u \#_\rho v$$

satisfying: The map $\#_\rho : K \hookrightarrow \mathcal{M}^f_{x,z}$ *is an embedding for each gluing parameter* $\rho \geqslant \rho_K$*. Moreover, given a compact set* $\widehat{K} \subset \widehat{\mathcal{M}}^f_{x,y} \times \widehat{\mathcal{M}}^f_{y,z}$ *of unparametrized trajectories,* $\#$ *induces a smooth embedding*

$$\widehat{\#} : \widehat{K} \times [\rho_{\widehat{K}}, \infty) \hookrightarrow \widehat{\mathcal{M}}_{x,z} \ ,$$

such that we obtain weak convergence toward the simply broken trajectory

$$\hat{u} \widehat{\#}_\rho \hat{v} \xrightarrow{C^\infty_{\mathrm{loc}}} (\hat{u}, \hat{v})$$

as ρ *tends to infinity,* $\rho \to \infty$*. On the other hand we obtain the converse result that any sequence of unparametrized trajectories converging to a simply broken trajectory finally lies within the range of such a gluing map* $\widehat{\#}$*.*

The gluing procedure may be sketched as in Figure 2.2.

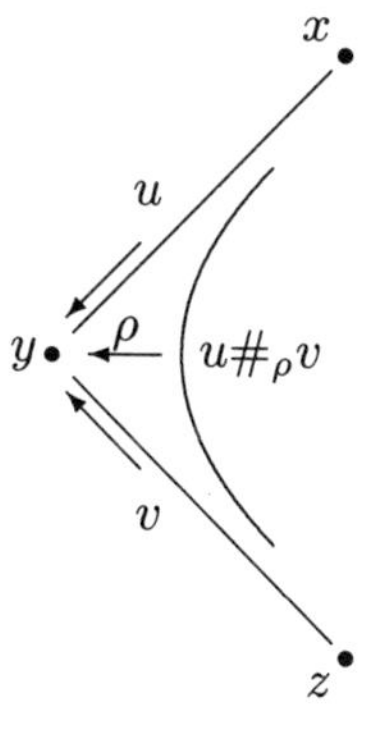

Figure 2.2: Gluing for unparametrized trajectories

In relation to the investigation in the last section this theorem supplies us with a far more detailed characterization of the obstructions to compactness of the unparametrized trajectory spaces $\widehat{\mathcal{M}}_{x,y}$, at least as concerns the relative Morse index 2. Actually, these obstructions are described topologically as cylindrical ends $\widehat{K} \times [\rho_{\widehat{K}}, \infty)$ with the set $\widehat{K}$ of simply broken trajectories as a cross section. It should be mentioned that a treatment of higher dimensional trajectory spaces would require a more involved analysis. But in principle we are led to a kind of cobordism theory as we have already explained in the introduction. This is what we call the 'compactness-gluing' cobordism.

The construction of the gluing map is carried out in three steps: First, we establish a smooth pre-gluing map $\#^o_\rho$, which glues broken trajectories $(u,\, v) \in \mathcal{M}_{x,y} \times \mathcal{M}_{y,z}$ rendering a smooth curve in $\mathcal{P}^{1,2}_{x,y}$, such that the image $(u\#^o_\rho v)(\overline{\mathbb{R}})$ differs from $u(\overline{\mathbb{R}}) \cup v(\overline{\mathbb{R}})$ by a sufficiently small amount and, moreover, such that the convergence toward the unparametrized trajectories

$$u\#^o_\rho v \overset{C^\infty_{\mathrm{loc}}}{\longrightarrow} (\hat{u}, \hat{v})$$

is provided for $\rho \to \infty$. The crucial second step will provide a procedure of associating nearby trajectories from $\mathcal{M}_{x,z}$ to these preparatorily glued curves, which represent approximate solutions of the negative gradient flow, in a unique way for large gluing parameters ρ. In order to guarantee uniqueness we in principle use a normal bundle of $\mathcal{M}_{x,z}$ within $\mathcal{P}^{1,2}_{x,z}$. However, this normal bundle is not supplied canonically. By means of the projection along the noncanonically transversal fibres[9] we find a contracting map. Due to the Banach contraction mapping principle and in consequence the implicit function theorem, we obtain a unique solution of the negative gradient flow if the gluing parameter ρ is large enough, that is to say that we differ little enough from the original trajectories up to time-shifting. Third and last, we have to verify the embedding property. This is largely owing to the uniqueness[10] guaranteed by the Banach fixed point theorem, if ρ is large enough.

The Pre-Gluing Map and the Uncanonical Normal Bundle

Let us assume as in Theorem 3 that $K \subset \mathcal{M}^f_{x,y} \times \mathcal{M}^f_{y,z}$ is a compact subset. Then given any pair $(u,\, v) \in K$, we choose smooth curves $(\tilde{u},\, \tilde{v}) \in C^\infty_{x,y} \times C^\infty_{y,z}$ appropriately such that within the Banach manifolds $\mathcal{P}^{1,2}_{x,y}$ and $\mathcal{P}^{1,2}_{y,z}$ there are coordinate neighbourhoods $\mathcal{U}_{\tilde{u}}$ and $\mathcal{U}_{\tilde{v}}$ containing the previously given trajectories:

$$u \in \mathcal{U}_{\tilde{u}} \quad \text{and} \quad v \in \mathcal{U}_{\tilde{v}} \ .$$

In particular, let us choose $\tilde{u}$ and $\tilde{v}$ asymptotically constant at the common endpoint y:

$$\begin{matrix} \tilde{u}(t) = y, & t \geqslant T \\ \tilde{v}(t) = y, & t \leqslant -T \end{matrix} \ , \text{ for some } T = T(\tilde{u},\, \tilde{v}) \ .$$

9In fact, this can be formulated as a retraction.

10still referring to the noncanonical normal bundle

Due to the compactness of K we find a finite covering by such neighbourhoods $\mathcal{U}_{\tilde{u}} \times \mathcal{U}_{\tilde{v}}$ yielding the constant

$$T(K) = \max_{\{(\tilde{u},\tilde{v})\}} T(\tilde{u},\tilde{v}) \ .$$

In the following chapter, on orientation and gluing with respect to Fredholm operators, especially such asymptotically constant curves will play a significant role. These curves allow us to define the simplest kind of gluing operation,[11] namely

$$(\tilde{u}\#_\rho\tilde{v})(t) = \begin{cases} \tilde{u}(t+\rho), & t \leqslant 0 \\ \tilde{v}(t-\rho), & t \geqslant 0 \end{cases} \ . \tag{2.63}$$

We thus obtain

$$\tilde{u}\#_\rho\tilde{v} \in C^\infty_{x,z} \quad \text{for all } \ \rho \geqslant T+1 \ \text{ with } \ T = T(\tilde{u},\tilde{v}) \ .$$

We shall henceforth use the symbol $\#$ for the gluing operation concerning trajectories as well as for the gluing of asymptotically constant curves.

Definition 2.47 *Throughout the whole section we will use the following cut-off functions:*

$$\beta^\pm : \overline{\mathbb{R}} \to [0,1] \ ,$$

which are defined by

$$\beta^-(t) = \begin{cases} 1, & t \leqslant -1 \\ 0, & t \geqslant 0 \end{cases} \quad \text{and} \quad \beta^+(t) = \begin{cases} 0, & t \leqslant 0 \\ 1, & t \geqslant 1 \end{cases} \ .$$

We are now able to define the preliminary, approximative gluing map, that is the so-called pre-gluing map,

$$\begin{aligned} \#^o : \quad & K \times [\rho_0,\infty) \to \mathcal{P}^{1,2}_{x,y} \\ & (u,v,\rho) \mapsto u\#^o_\rho v \end{aligned}$$

by

$$u\#^o_\rho v = \exp_{\tilde{u}\#_\rho\tilde{v}}\left(\beta^- \cdot \left[\exp^{-1}_{\tilde{u}}(u)\right]_\rho + \beta^+ \cdot \left[\exp^{-1}_{\tilde{v}}(v)\right]_{-\rho}\right) \tag{2.64}$$

for $\rho_0 = \rho_0(K) > T(K)+1$. *Here,* $(\tilde{u},\tilde{v})$ *are chosen suitably with* $(u,v) \in \mathcal{U}_{\tilde{u}} \times \mathcal{U}_{\tilde{v}}$ *referring to the above covering. The notation*

$$u_\rho(t) = u(t+\rho), \ \ t \in \mathbb{R}$$

describes the time-shifting that has been already studied above.

11Note that this does not concern non-trivial trajectories of the gradient flow.

It is easy to verify that the definition of $u\#_\rho^o v$ is independent of the individually chosen coordinate charts from the fixed atlas, as the equivalent representation

$$(2.65)\quad (u\#_\rho^o v)(t) = \begin{cases} u_\rho(t), & t \leqslant -1 \\ \exp_y[\beta^- \exp_y^{-1}(u_\rho) + \beta^+ \exp_y^{-1}(v_{-\rho})](t), & |t| \leqslant 1 \\ v_{-\rho}(t), & t \geqslant 1 \end{cases}$$

follows from the asymptotic behaviour of $\tilde{u}$ and $\tilde{v}$ and the definition of $\tilde{u}\#_\rho\tilde{v}$. Hence it is only with respect to the lower bound $\rho_0(K)$ that the definition of $\#^o$ depends on the choice of the covering of K. Moreover, the two equivalent representations (2.64) and (2.65) yield the smoothness of this pre-gluing map $\#^o$. We conclude the differentiability with respect to u and v from (2.64), and formula (2.65) implies the differentiability with respect to ρ yielding the formula[12]

$$(2.66)\quad \frac{\partial}{\partial\rho}(u\#_\rho^o v)(t) = \begin{cases} \dot{u}_\rho(t), & t \leqslant -1 \\ \nabla_2 \exp_y \cdot \begin{bmatrix} \beta^- \nabla_2 \exp_y^{-1} \cdot \dot{u}_\rho \\ - \quad \beta^+ \nabla_2 \exp_y^{-1} \cdot \dot{v}_{-\rho} \end{bmatrix} (t), & |t| \leqslant 1 \\ -\dot{v}_{-\rho}(t), & t \geqslant 1 \end{cases}$$
$$\in L^2\left((u\#_\rho^o v)^* TM\right) .$$

Remark As we will see in Proposition 2.50 below, already this pre-gluing map $\#_\rho^o$ has the embedding property if ρ is large enough, i.e. $\rho \geqslant \tilde{\rho}(K)$ for $\tilde{\rho}(K)$ large enough. This can be seen from the fact that the differential associated to the map (2.64)

$$\begin{aligned} \ker DF_u \times \ker DF_v \quad &\to \quad H^{1,2}((u\#_\rho^o v)^* TM) \\ (\xi, \zeta) \quad &\mapsto \quad \nabla_2 \exp\left(\beta^- \left(\nabla_2 \exp^{-1} \xi\right)_\rho + \beta^+ \left(\nabla_2 \exp^{-1} \zeta\right)_{-\rho}\right) \end{aligned}$$

will prove injective for large ρ.

In the following considerations we provide the fundamental structures which will serve to find the actual gluing map according to the Banach contraction principle. We shall construct suitable Hilbert bundles, a linear gluing version, which is injective, and a normal bundle which is to guarantee the uniqueness of the gluing operation.

It is worth mentioning that if we could cover K by only one single pair of coordinate neighbourhoods, we would be able to reduce the whole problem to the trivialized framework. Then we could renounce this somewhat extensive bundle constructions. The trouble in the non-trivial case, however,

12For the definition of $\nabla_2 \exp$ the reader is referred to Definition A.15 in Appendix A.2.

consists of determining the correction term between the actual gluing of trajectories and the already defined approximate pre-gluing in a way which is independent of the trivialization. Actually, this amounts to the below construction of a normal bundle $L^{\perp}$.

Throughout the following investigations we will use the notations

$$\begin{aligned} \chi &= (u,v,\rho) \in K \times [\rho_0,\infty), \\ w_\chi &= u\#^o_\rho v \in \mathcal{P}_{x,z} = \mathcal{P}^{1,2}_{x,z}, \\ F_o : \ & \mathcal{P}_{x,z} \to L^2(\mathcal{P}^*_{x,z}TM) \\ & \gamma \mapsto \dot{\gamma} + \nabla f \circ \gamma \ , \end{aligned}$$

where we refer to the local representation with respect to $w \in C^\infty_{x,z}$:

$$\text{(2.67)}\quad F_w = \nabla_2 \exp_w^{-1} \circ F_o \circ \exp_w : H^{1,2}(w^*TM) \subset \mathcal{O}_w \to L^2(w^*TM) \ .$$

Since the broken trajectories $(u,v) \in K$ are composed of smooth curves, the nonlinear operator $F_{w_\chi} \in \operatorname{Fred}(w_\chi^*TM)$ is well-defined as well. The Riemannian metric $\langle \cdot,\cdot \rangle$ on M induces well-defined Riemannian metrics

$$\begin{aligned} \langle \xi, \zeta \rangle^{0,2}_\gamma &= \int_{\mathbb{R}} \langle \xi(t), \zeta(t) \rangle_{\gamma(t)} dt \\ \langle \xi, \zeta \rangle^{1,2}_\gamma &= \langle \nabla_t \xi, \nabla_t \zeta \rangle^{0,2}_\gamma + \langle \xi, \zeta \rangle^{0,2}_\gamma \\ \text{for} \quad \gamma &\in C^\infty_{x,z} \subset \mathcal{P}^{1,2}_{x,z} \end{aligned}$$

together with the associated fibre norms $\|\cdot\|_{1,2}$ and $\|\cdot\|_{0,2}$ on the bundles (see the appendix)

$$H^{1,2}(\mathcal{P}^*_{x,z}TM) \subset L^2(\mathcal{P}^*_{x,z}TM) \ .$$

Thus the smooth map $\#^o : K \times [\rho_0,\infty) \to \mathcal{P}_{x,z}$ supplies us with the Hilbert bundles

$$H = \#^{o*} H^{1,2}(\mathcal{P}^*_{x,z}TM) \text{ and } L = \#^{o*} L^2(\mathcal{P}^*_{x,z}TM)$$

on the base space $K \times [\rho_0,\infty)$. Additionally, the inclusion

$$i : K \hookrightarrow \mathcal{M}_{x,y} \times \mathcal{M}_{y,z}$$

induces the Hilbert bundle

$$\left(H^{1,2}\right)^2 = i^* \left(H^{1,2}_{|\mathcal{M}_{x,y}} \times H^{1,2}_{|\mathcal{M}_{y,z}} \right)$$

on $K \times [\rho_0,\infty)$. According to these bundles H and L we are able to identify the map F_{w_χ} from (2.67) for $w_\chi \in \#^o\big(K \times [\rho_0,\infty)\big)$ via the pre-gluing map $\#^o : \chi \mapsto w_\chi$ with the fibre restriction of a smooth bundle map

$$
\begin{array}{ccc}
H \supset \mathcal{O} & \xrightarrow{F} & L \\
& \searrow \quad \oplus \quad \swarrow & \\
& K \times [\rho_0, \infty) &
\end{array}
$$

In this context $\mathcal{O}$ denotes an open neighbourhood of the zero section, such that the boundary is properly bounded away from the zero section with respect to the norm $\|\cdot\|_{1,2}$. This is concluded by means of the injectivity neighbourhood associated to the exponential map, the compactness of K and the construction of the pre-gluing map, because the image of the curves $u\#_\rho^o v(\overline{\mathbb{R}})$ remains within a compact neighbourhood. This bundle map $F : H \supset \mathcal{O} \to L$ together with the identification $F_\chi = F_{w_\chi}$ thus has the explicit representation

$$
\begin{aligned}
F_{w_\chi}(\xi)(t) &= \begin{cases} \nabla_2 \exp_{u_\rho}(\xi')^{-1} \cdot \left[\left(\frac{\partial}{\partial t} + \nabla f\right) \circ \exp_{u_\rho} \xi'\right](t), & t \leqslant -1 \\ \nabla_2 \exp_{v_{-\rho}}(\xi'')^{-1} \cdot \left[\left(\frac{\partial}{\partial t} + \nabla f\right) \circ \exp_{v_{-\rho}} \xi''\right](t), & t \geqslant 1 \end{cases} \\
&= \begin{cases} F_{u_\rho}(\xi')(t), & t \leqslant -1 \\ F_{v_{-\rho}}(\xi'')(t), & t \geqslant 1 \end{cases}
\end{aligned}
$$

with

$$
\begin{array}{ll}
\xi' \in H^{1,2}_{u_\rho}, & \xi'_{|(-\infty,-1]} \equiv \xi_{|(-\infty,-1]} \ , \\
\xi'' \in H^{1,2}_{v_{-\rho}}, & \xi''_{|[1,\infty)} \equiv \xi_{|[1,\infty)} \ ,
\end{array}
$$

and we obtain the identity

$$F_{w_\chi}(0) = F_o(w_\chi) \ .$$

Moreover, the fibre derivative

$$D_2 F(\chi)(0) = DF_{w_\chi}(0) : H_\chi \to L_\chi$$

at 0 is exactly the linearization of F_o, which has been already studied as a Fredholm operator of the type

$$\frac{\partial}{\partial t} + X(t) \ .$$

We shall henceforth use the shorter notation

$$D_\chi = DF_{w_\chi}(0), \ \ \chi \in K \times [\rho_0, \infty) \ .$$

Definition 2.48 *We are able to define the linear version of the pre-gluing operation as a bundle homomorphism by*

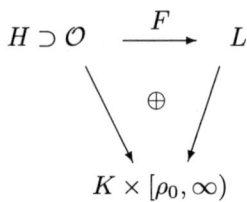

$$\begin{array}{ccc} \left(H^{1,2}\right)^2 & \xrightarrow{\ \#\ } & H \\ & \searrow \ \oplus \ \swarrow & \\ & K \times [\rho_0, \infty) & \end{array}$$

$$(\xi \#_\chi \zeta)(t) \quad = \quad \begin{cases} \xi_\rho(t), & t \leqslant -1 \\ \nabla_2 \exp \cdot \left[\begin{array}{ll} & \beta^- \nabla_2 \exp^{-1} \cdot \xi_\rho \\ + & \beta^+ \nabla_2 \exp^{-1} \cdot \zeta_{-\rho} \end{array} \right](t), & |t| \leqslant 1 \\ \zeta_{-\rho}(t), & t \geqslant 1 \end{cases}$$

The fibre derivative $\nabla_2 \exp$ *is taken at the points*

$$\exp_y^{-1}\left(u \#_\rho^o v\,(t)\right), \quad \exp_y^{-1}\left(u_\rho\,(t)\right) \quad \textit{and} \quad \exp_y^{-1}\left(v_{-\rho}\,(t)\right), \quad \textit{respectively}$$

This corresponds exactly to the linearization of $\#^o$ *as it is given by equation (2.65).*

The crucial point of the construction of the gluing map can be found in the process which deduces the linearization of the trajectory gluing from this linear version of the approximate gluing $\#^o$. By the linearization of the actual trajectory gluing we mean a projection onto the tangent bundle of $\mathcal{M}_{x,z}$ as a subbundle of $H^{1,2}_{|\mathcal{M}_{x,z}}$. But it is exactly this projection map which cannot be canonical within our construction. We need the concept of orthogonality, which is global with respect to K, namely the normal bundle of $\mathcal{M}_{x,z}$ as has been already described above. So, let us consider the following candidate for a normal bundle:

Definition 2.49 *Since* $D_u = DF_u$ *and* $D_v = DF_v$ *associated to* $(u, v) \in \mathcal{M}_{x,y} \times \mathcal{M}_{y,z}$ *are surjective Fredholm operators,*

$$(\ker)^2 = \bigcup_{\chi \in K \times [\rho_0, \infty)} \ker D_u \times \ker D_v \subset \left(H^{1,2}\right)^2$$

is a subbundle with constant and finite dimension. Then the following vector space is well-defined for any $\chi \in K \times [\rho_0, \infty)$*:*

$$L_\chi^\perp = \left\{ v_\chi \in H^{1,2}_{w_\chi} = H_\chi \;\middle|\; \begin{array}{c} \langle v_\chi, \xi \#_\rho \zeta \rangle_\chi^{0,2} = 0 \\ \textit{for all } (\xi, \zeta) \in \ker D_u \times \ker D_v \end{array} \right\}.$$

The following first, fundamental proposition within this discussion of the gluing problem states that these vector spaces represent the fibres of the normal bundle we are looking for.

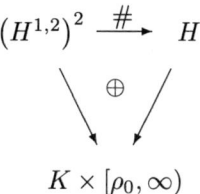

Proposition 2.50 *There is a lower parameter bound $\rho_1 \geqslant \rho_0$, such that for any gluing parameter $\rho \geqslant \rho_1$ and broken trajectory $(u, v) \in K$ the Fredholm operator $D_\chi : H_\chi \to L_\chi$ is onto and the composition of $\#_\chi$ with the L^2-orthogonal projection onto* $\ker D_\chi$ *induces an isomorphism*

$$\phi_\chi = \mathrm{Proj}^{L_\chi}_{\ker D_\chi} \circ \#_\chi : \ker D_u \times \ker D_v \xrightarrow{\cong} \ker D_\chi \ .$$

We deduce from this first result that

$$\ker D = \bigcup_{\chi \in K \times [\rho_1, \infty)} \ker D_\chi$$

is a subbundle of H on $K \times [\rho_1, \infty)$ with the finite dimension $\mu(x) - \mu(z)$ and that

$$\#_{|(\ker)^2} : (\ker)^2 \to H$$

represents a bundle homomorphism. Hence, we obtain

$$L^\perp = \bigcup_{\chi \in K \times [\rho_1, \infty)} L^\perp_\chi$$

as a finite-codimensional vector bundle on $K \times [\rho_1, \infty)$ satisfying the relation

$$L^\perp \oplus_{\mathrm{top}} R\left(\#_{|(\ker)^2}\right) = H \ .$$

In fact, as we shall see below, we are led to the decomposition

$$L^\perp \oplus_{\mathrm{top}} \ker D = H \ . \tag{2.68}$$

Summing up these arguments we may say that $L^\perp$ can be considered to be the normal bundle which is needed for the gluing construction.

Proof of proposition 2.50: Concerning the central feature, the proof of this nonlinear gluing theorem is essentially patterned on the proof for an analogous gluing operation in Floer homology as can be found in [F-H]. As we already noticed above, the operator $D_\chi = DF_{w_\chi}$ referring to

$$w_\chi = u \#^o_\rho v \in \mathcal{P}^{1,2}_{x,z}$$

is a Fredholm operator satisfying the identity

$$\mathrm{ind}\, D_\chi = \mu(x) - \mu(z) \ .$$

According to the assumption that the Morse-Smale condition be fulfilled, the Fredholm operators D_u and D_v are onto. Hence we obtain the expressions

$$\dim \ker D_u = \mathrm{ind}\, D_u = \mu(x) - \mu(y),$$
$$\dim \ker D_v = \mathrm{ind}\, D_v = \mu(y) - \mu(z) \ .$$

Therefore it is sufficient to show that, given any broken trajectory $(u,v) \in K$, we can find a lower parameter bound $\rho(u,v)$, such that ϕ_χ is onto for all $\rho \geqslant \rho(u,v)$. This is due to the assumed compactness of K which implies a uniform constant $\rho(K)$, such that ϕ_χ is onto for all $\chi \in K \times [\rho(K),\infty)$. Hence we can conclude

$$\begin{aligned} & \dim\ker D_\chi \leqslant \dim\ker D_u + \dim\ker D_v \\ \overset{\text{(cf.)}}{=} \quad & \mu(x) - \mu(z) = \operatorname{ind} D_\chi \leqslant \dim\ker D_\chi \ , \end{aligned}$$

which gives rise immediately to the identity

$$\dim\ker D_\chi = \dim\ker D_u + \dim\ker D_v$$

and which consequently implies the asserted isomorphism property of ϕ_χ for all $\chi \in K \times [\rho(K),\infty)$.

Therefore, let us choose an arbitrary, fixed broken trajectory $(u,\ v) \in K$. Then it suffices to find a constant $C < 0$ and a lower parameter bound $\tilde{\rho}$, such that the uniform estimate

$$\|D_\chi \xi\|_{L^2_\chi} \geqslant c\,\|\xi\|_{L^2_\chi} \tag{2.69}$$

holds for any $\rho \geqslant \tilde{\rho}$ and $\xi \in L^\perp_\chi$. This is clear, because there is a

$$\xi \in \ker D_\chi \ \text{ with } \ \langle\, \xi,\ R(\#_\chi)\,\rangle_{L^2_\chi} = 0\ ,$$

if ϕ_χ should fail to be surjective. But owing to (2.69) this $\xi \in L^\perp_\chi$ has to be the zero vector.

Let us now prove the estimate (2.69) indirectly. We assume sequences $(\rho_n)_{n\in\mathbb{N}}$ and $(\xi_n)_{n\in\mathbb{N}}$ satisfying the conditions

$$\rho_n \to \infty,\ \xi_n \in L^\perp_{\chi_n},\ \|\xi_n\|_{L^2_{\chi_n}} = 1 \quad \text{and} \quad \|D_{\chi_n}\xi_n\| \to 0\ . \tag{2.70}$$

First, we wish to deduce the convergence

$$\xi_n \overset{H^{1,2}_{\text{loc}}}{\longrightarrow} 0$$

from these assumptions. As a matter of simplification we shall transfer the situation to the trivialized case of $\mathbb{R}^n$. Because we have chosen the broken trajectory $(u,v) \in K$ to be fixed and due to the fact that, provided any large gluing parameter ρ, the piece of curve $u\#^\circ_\rho v_{|[-1,1]}$ lies within the normal neighbourhood $N(y)$ of y with respect to the exponential map, we obtain

$$\bigcup_{\rho\in[\rho_0,\infty)} (u\#^\circ_\rho v)^* TM$$

as a trivial vector bundle on $\overline{\mathbb{R}} \times [\rho_0, \infty)$. Thus starting from a trivialization $\Phi_y : TM_{|N(y)} \to N(y) \times \mathbb{R}^n$ we find further trivializations Φ_u and Φ_v of u^*TM and v^*TM respectively together with

$$\Phi : \bigcup_{\rho \in [\rho_0, \infty)} (u \#_\rho^o v)^* TM \longrightarrow [\rho_0, \infty) \times \overline{\mathbb{R}} \times \mathbb{R}^n$$

satisfying the relations

$$\Phi_{|\,[1,\infty)} \equiv \Phi_{u|\,[1,\infty)} \text{ and } \Phi_{|\,(-\infty,-1]} \equiv \Phi_{v|\,(-\infty,-1]} \ ,$$

such that we obtain the uniform estimate[13]

$$0 < c_1 \leqslant \sup_{t \in \mathbb{R}} \left| \Phi_{u \#_\rho^o v}(t) \right| \leqslant c_2 \quad \text{for all } \rho \geqslant \rho_0 \ ,$$

and for all $\rho \geqslant \rho_0$ the isomorphisms

$$\Phi_\chi : H^{1,2}_\chi \xrightarrow{\cong} H^{1,2}(\mathbb{R}, \mathbb{R}^n) \quad \text{with} \quad 0 < \tilde{c}_1 \leqslant \|\Phi_\chi\|_{\text{Op}} \leqslant \tilde{c}_2 \ .$$

Concerning the class L^2 we are led to an analogous result.

Next, these uniform estimates enable us to consider assumptions (2.70) above without loss of generality for the trivial case

$$D_{\chi_n} = \frac{\partial}{\partial t} + A_{\chi_n} : H^{1,2}(\mathbb{R}, \mathbb{R}^n) \to L^2(\mathbb{R}, \mathbb{R}^n) \ ,$$

that is

$$(2.71) \quad (\xi_n)_{n \in \mathbb{N}} \subset H^{1,2}(\mathbb{R}, \mathbb{R}^n), \ \|\xi_n\|_{L^2} = 1 \quad \text{with} \quad \|D_{\chi_n} \xi_n\|_{L^2} \to 0 \ .$$

Here, A_{χ_n} is a curve

$$A_{\chi_n} \in C^\infty\big(\overline{\mathbb{R}}, \operatorname{End}(\mathbb{R}^n)\big)$$

determined by the trivialization Φ_{χ_n}. Referring to the trivialization $\Phi_y : T_yM \to \mathbb{R}^n$ we obtain

$$A_y = \Phi_y \cdot D\big(\nabla_2 \exp_y^{-1} \cdot (\nabla f \circ \exp_y)\big) \cdot \Phi_y^{-1}$$

as a constant operator which is conjugated to the Hessian $H^2 f(y)$, so that the Fredholm operator

$$D_y = \frac{\partial}{\partial t} + A_y$$

induces an isomorphism

$$(2.72) \qquad D_y : H^{1,2} \xrightarrow{\cong} L^2 \ .$$

---

13 $|\Phi_p|$ is the norm of $\Phi_p : T_pM \xrightarrow{\cong} \mathbb{R}^n$ with respect to the Riemannian metric $\langle \cdot, \cdot \rangle$ upon TM.

This is immediate from the trivial spectral flow associated to the constant curve A_y. Now let $\beta : \mathbb{R} \to [0,1]$ be a smooth cut-off function of the type

$$\beta(t) = \begin{cases} 1, & |t| \leqslant \frac{1}{2} \\ 0, & |t| \geqslant 1 \end{cases}$$

and let $r_n \to \infty$ be the sequence $r_n = \frac{1}{2}\rho_n$. Additionally, we choose the short notation

$$\beta^{r_n}(t) = \beta(\frac{t}{r_n}), \text{ s.t. } \frac{d}{dt}(\beta^{r_n})(t) = \frac{1}{r_n}\dot{\beta}(\frac{t}{r_n}) \ .$$

These preparations give rise to the following series of estimates:

$$\begin{aligned} & \|D_y(\beta^{r_n} \cdot \xi_n)\|_{L^2} \\ = \ & \left\|(\dot{\beta^{r_n}}) \cdot \xi_n + \beta^{r_n} \cdot D_y \xi_n\right\|_{L^2} \\ \leqslant \ & \frac{1}{r_n}\left\|\dot{\beta}(\tfrac{1}{r_n}\cdot)\xi_n\right\|_{L^2} + \|D_y\xi_n\|_{L^2\left([-r_n,r_n]\right)}, \quad c = \max_{[-1,1]}|\dot{\beta}| \\ \leqslant \ & \frac{1}{r_n}\cdot c + \|D_{\chi_n}\xi_n\|_{L^2} + \|(D_y - D_{\chi_n})\xi_n\|_{L^2\left([-r_n,r_n]\right)} \\ \leqslant \ & \frac{1}{r_n}\cdot c + \|D_{\chi_n}\xi_n\|_{L^2} + \sup_{t\in[-r_n,r_n]} \|A_{\chi_n}(t) - A_y\|_{\mathrm{End}(\mathbb{R}^n)} \ \|\xi_n\|_{L^2} \ . \end{aligned}$$

Due to assumption (2.71) and the estimate

$$\begin{aligned} \sup_{[-r_n,r_n]} \|A_{\chi_n} - A_y\|_{\mathrm{End}(\mathbb{R}^n)} \ \leqslant \ & \sup_{t\in[-1,1]} \|A_{\chi_n}(t) - A_y\| \\ & + \sup_{t\in[-r_n,-1]} \|A_u(t+\rho_n) - A_y\| \\ & + \sup_{t\in[1,r_n]} \|A_v(t-\rho_n) - A_y\| \ , \end{aligned}$$

where all terms on the right side tend to zero as u and v converge exponentially to the critical point y, we conclude

$$\lim_{n\to\infty} \|D_y(\beta^{r_n} \cdot \xi_n)\|_{L^2} = 0 \ .$$

Therefore, the isomorphism D_y from (2.72) yields the $H^{1,2}_{\mathrm{loc}}$-convergence we wish to verify. Finally, this result can be transferred back to the non-trivial case, that is

$$\xi_n \overset{H^{1,2}_{\chi_n}\left([-r_n,r_n]\right)}{\longrightarrow} 0 \ . \tag{2.73}$$

By this conclusion, that the supports of ξ_n become more and more concentrated at the asymptotic ends as n increases, we are now able to transfer the estimate

$$\|D_{\chi_n}\xi_n\|_{L^2_{\chi_n}} \to 0$$

for the varying Fredholm operators to the fixed operators D_u and D_v:

$$\begin{aligned}
\left\|D_u(\beta^-_{1-\rho_n}\xi_{n,-\rho_n})\right\|_{L^2_u} &\leqslant \left\|(\dot{\beta^-})_{1-\rho_n}\xi_{n,-\rho_n}\right\|_{L^2_u} + \left\|\beta^-_{1-\rho_n}D_u\xi_{n,-\rho_n}\right\|_{L^2_u} \\
&= \left\|\dot{\beta}^-_1\xi_n\right\|_{L^2_{u_{\rho_n}}} + \left\|\beta^-_1 D_{u_{\rho_n}}\xi_n\right\|_{L^2_{u_{\rho_n}}} \\
&= \left\|\dot{\beta}^-_1\xi_n\right\|_{L^2_{\chi_n}} + \left\|\beta^-_1 D_{\chi_n}\xi_n\right\|_{L^2_{\chi_n}} \\
&\leqslant \left\|\xi_n\right\|_{L^2_{\chi_n}([-2,-1,])} + \left\|D_{\chi_n}\xi_n\right\|_{L^2_{\chi_n}} .
\end{aligned}$$

According to assumption (2.70) and the above estimate (2.73) we are led to the convergence

$$\left\|D_u(\beta^-_{1-\rho_n}\xi_{n,-\rho_n})\right\|_{L^2_u} \to 0 .$$

Since D_u is a Fredholm operator, this amounts to the existence of an $\alpha \in \ker D_u$ with

$$\beta^-_{1-\rho_n}\cdot\xi_{n,-\rho_n} \xrightarrow{H^{1,2}_u} \alpha .$$

In particular we obtain the convergence

$$\left\|\beta^-_1\cdot\xi_n - \alpha_{\rho_n}\right\|_{L^2_{u_{\rho_n}}} \to 0 .$$

In an analogous manner we are provided with a $\gamma \in \ker D_v$ satisfying the relation

$$\left\|\beta^+_{-1}\cdot\xi_n - \gamma_{-\rho_n}\right\|_{L^2_{v_{-\rho_n}}} \to 0 .$$

Summing up, we observe the following approximate behaviour:

$$\begin{aligned}
&\left\|\beta^-_1\xi_n - \beta^-\alpha_{\rho_n}\right\|_{L^2_{\chi_n}} \to 0, \quad \alpha\in\ker D_u , \\
&\left\|\beta^+_{-1}\xi_n - \beta^+\gamma_{-\rho_n}\right\|_{L^2_{\chi_n}} \to 0, \quad \gamma\in\ker D_v .
\end{aligned} \tag{2.74}$$

The contradiction with the former assumption is now caused by trying to piece together all these different conclusions in the series of identities:

$$\begin{aligned}
1 &= \lim_{n\to\infty}\left\|\xi_n\right\|_{L^2_{\chi_n}} \\
&= \lim_n \langle\,\beta^-_1\xi_n + \beta^+_{-1}\xi_n + (1-\beta^-_1-\beta^+_{-1})\xi_n,\,\xi_n\,\rangle_{L^2_{\chi_n}} \\
&\overset{(2.73)}{=} \lim_n\langle\,\beta^-_1\xi_n,\,\xi_n\,\rangle_{L^2_{\chi_n}} + \lim_n\langle\,\beta^+_{-1}\xi_n,\,\xi_n\,\rangle_{L^2_{\chi_n}} , \\
&\quad\ \text{as } \operatorname{supp}(1-\beta^-_1-\beta^+_{-1})\subset[-2,2] \\
&\overset{(2.74)}{=} \lim_n\langle\,\beta^-\alpha_{\rho_n},\,\xi_n\,\rangle_{L^2_{\chi_n}} + \lim_n\langle\beta^+\gamma_{-\rho_n},\,\xi_n\,\rangle_{L^2_{\chi_n}} \\
&= \lim_n\langle\,\alpha\#_{\chi_n}\gamma,\,\xi_n\,\rangle_{L^2_{\chi_n}} \\
&= 0, \text{ as } \xi_n\in L^\perp_{\chi_n} \text{ for all } n\in\mathbb{N} .
\end{aligned}$$

The proof of the assertion follows. □

The Trajectory Gluing

Up to now we have merely analysed the pre-gluing $\#^o$ as an approximate version of the gluing operation. However, the thus defined curves $w_\chi = u\#^o_\rho v$ are not yet trajectories for the negative gradient flow. Resorting to the normal bundle $L^\perp$ constructed above we now try to find the correction term for $\#^o$ with respect to the actual trajectory gluing. Moreover, this correction term has to be provided globally on the entire base space $K \times [\rho_1, \infty)$. This requires K to be compact. In other words, our aim is to find a 'correction section' γ in the bundle H on $K \times [\rho_1, \infty)$ in a unique way, so that

$$\exp_{w_\chi} \gamma(\chi) \quad \text{for } \chi \in K \times [\rho_1, \infty)$$

finally represents a gradient flow trajectory. That is to say that we are searching for a section

$$\gamma : K \times [\rho_1, \infty) \to H \quad \text{satisfying} \quad F \circ \gamma = 0$$

with respect to the formerly defined bundle map

$$\begin{array}{ccc} H \supset \mathcal{O} & \xrightarrow{\;F\;} & L \\ & \searrow \;\; \oplus \;\; \swarrow & \\ & K \times [\rho_0, \infty) & \end{array}$$

Additionally, we want $\exp_{w_\chi} \gamma(\chi)$ to comply with the geometrical convergence toward the broken trajectory (u, v) as the gluing parameter ρ tends to infinity, $\rho \to \infty$. Since we are provided with the convergence property for $u\#^o_\rho v$ due to the construction, we shall be satisfied with a sufficiently rapid convergence for the correction term,

$$\gamma(u\#^o_\rho v) \to 0 \quad \text{as } \rho \to \infty \ .$$

We shall derive all these results from the existence and uniqueness theorems which are supplied within the Banach space calculus. In this framework the bundle $L^\perp$ guarantees the (noncanonical!) uniqueness. It is essentially the fixed point calculus due to a contraction mapping which will lead us to the necessary correction term. We will obtain this contraction mapping from Proposition 2.50, that is the bundle decomposition (2.68), which gives rise to the vector bundle isomorphism

$$D_{|L^\perp} : L^\perp \xrightarrow{\cong} L^2 \ .$$

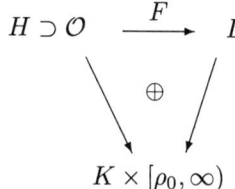

Thus, considering the inverse map[14]

$$G : L^2 \xrightarrow{\cong} L^\perp, \ G = (D_{|L^\perp})^{-1} \ ,$$

we can extract the following estimate from a refinement of the proof of Proposition 2.50:

Lemma 2.51 *There is a lower parameter bound $\rho_2 \geqslant \rho_0$ and a uniform constant $C_{K,1} > 0$, such that the isomorphism G satisfies the estimate*

$$\|G_\chi \xi\|_{1,2,\chi} \leqslant C_{K,1} \, \|\xi\|_{0,2,\chi} \ \textit{for all } \chi \in K \times [\rho_2, \infty), \ \xi \in L^2_\chi \ .$$

Proof. The decisive feature within this proof is the uniformness with respect to the broken trajectories $(u, v) \in K$. An equivalent statement of the assertion via $\zeta = G_\chi \cdot \xi \in L^\perp_\chi$ is given by:

$$\|\zeta\|_{1,2,\chi} \leqslant C_{K,1} \, \|D_\chi \cdot \zeta\|_{0,2,\chi} \quad \text{for all } \zeta \in L^\perp_\chi, \ \chi \in K \times [\rho_2, \infty) \ .$$

In order to carry out an indirect proof similar to Proposition 2.50 we assume a parameter sequence $\rho_n \to \infty$ and associated broken trajectories $\big((u_n, v_n)\big)_{n \in \mathbb{N}} \subset K$ together with

$$\zeta_n \in L^\perp_{\chi_n} \ , \ \chi_n = u_n \#^o_{\rho_n} v_n$$

satisfying the conditions

$$\|\zeta_n\|_{1,2,\chi_n} = 1 \quad \text{and} \quad \|D_{\chi_n} \zeta_n\|_{0,2,\chi_n} \to 0 \ . \tag{2.75}$$

In contrast to Proposition 2.50 the main difficulty lies in the estimate for an inconstant sequence $\big((u_n, v_n,)\big)_{n \in \mathbb{N}}$ instead of a fixed pair (u, v). Therefore we construct an appropriate transformation to the situation of such a fixed pair. Due to the compactness of K, we can assume without loss of generality an $H^{1,2}$-convergent sequence, that is in particular a C^0-convergent sequence

$$(u_n, v_n) \xrightarrow{C^0} (u, v) \in K \ .$$

Considering the contractability of the curves u, v and $u \#^o_\rho v$ we find an open neighbourhood U of $u(\overline{\mathbb{R}}) \cup v(\overline{\mathbb{R}})$, a trivialization $\Phi : TM_{|U} \xrightarrow{\approx} U \times \mathbb{R}^n$ and an $n_0 \in \mathbb{N}$, such that the inclusion

$$\chi_{n,m}(\overline{\mathbb{R}}) = u_n \#^o_{\rho_m} v_n(\overline{\mathbb{R}}) \subset U \quad \text{holds for all } m, n \geqslant n_0 \ .$$

In order to transfer assumption (2.75) to the fixed broken trajectory (u, v), we consider the following homomorphism family between vector bundles on $\overline{\mathbb{R}}$

14The reader is referred to [F6] or [F4] in the case of Floer homology.

$$\begin{array}{ccc} (u_n \#^o_{\rho_m} v_n)^* TM & \xrightarrow{\Phi_{m,n}} & (u \#^o_{\rho_m} v)^* TM \\ & \searrow \quad \oplus \quad \swarrow & \\ & \overline{\mathbb{R}} & \end{array}$$

relying on the definition

$$\Phi_{m,n}\left(\xi_{\chi_{n,m}(t)}\right) = \Phi^{-1}\left(u\#^o_{\rho_m} v(t),\, \mathrm{pr}_2 \circ \Phi\left(\xi_{\chi_{n,m}(t)}\right)\right) \ .$$

Let us henceforth use the notations $\chi_{nn} = u_n \#^o_{\rho_n} v_n$ and $\chi_n = u\#^o_{\rho_n} v$. Then the explicit construction of the pre-gluing map $\#^o_{\rho_n}$ together with the $H^{1,2}$-convergence

$$(u_n, v_n) \overset{H^{1,2}}{\longrightarrow} (u, v)$$

yields the convergence of the induced Banach space isomorphisms

$$\Phi_n : H^{1,2}_{\chi_{nn}} \overset{\cong}{\to} H^{1,2}_{\chi_n} \text{ and } L^2_{\chi_{nn}} \overset{\cong}{\to} L^2_{\chi_n}, \text{ respectively,}$$

with $\Phi_n(s)(t) = \Phi_{nn}\big(s(t)\big)$. This is due to the convergence

$$\begin{array}{rll} & \|\Phi_n\|_{\mathcal{L}(H^{1,2}_{\chi_{nn}}, H^{1,2}_{\chi_n})},\ \|\Phi_n\|_{\mathcal{L}(L^2_{\chi_{nn}}, L^2_{\chi_n})} & \longrightarrow 1 \\ \text{and} & \|\Phi_n \circ D_{\chi_{nn}} - D_{\chi_n} \circ \Phi_n\|_{\mathcal{L}(H^{1,2}_{\chi_{nn}}, L^2_{\chi_n})} & \longrightarrow 0 \end{array}$$

as n goes to infinity, $n \to \infty$.

Moreover, given any $\epsilon \in (0,1)$ the convergence $\Phi_n \to \mathrm{id}_{H^{1,2}_{\chi_n}}$ provides an $n_0(\epsilon) \in \mathbb{N}$, so that we can transform the former assumption

$$\|\zeta_n\| = 1 \quad \text{for all } \zeta_n \in L^\perp_{\chi_{nn}}$$

into the estimate

$$1 - \epsilon \leqslant \left\| \left(\mathrm{pr}_{L^\perp_{\chi_n}} \circ \Phi_n\right)(\zeta_n)\right\|_{1,\chi_n} \leqslant 1 + \epsilon$$

for all $n \geqslant n_0$. Piecing everything together we come to conditions similar to those in the proof of Proposition 2.50. Now confined to the fixed pair of trajectories (u, v) and using the notation

$$\xi_n = \left(\mathrm{pr}_{L^\perp_{\chi_n}} \circ \Phi_n\right)(\zeta_n) \in L^\perp_{\chi_n}$$

we obtain the conditions

$$\left|\, \|\xi_n\|_{1,\chi_n} - 1\right| \leqslant \epsilon,\ \|D_{\chi_n}\xi_n\| \to 0,\ \rho_n \to \infty\ ,$$

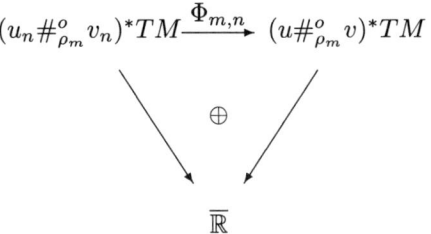

as

$$\left\| \left(\mathrm{pr}_{R(\#_{\chi_n})} \circ \Phi_n \right) (\zeta_n) \right\|_{1,\chi_n} \to 0 \ .$$

Hence we are finally able to conclude as in Proposition 2.50:

$$\lim_{n\to\infty} \|\xi_n\|_{0,\chi_n} = 0 \ .$$

On the other hand, the convergence $\|D_{\chi_n}\xi_n\|_{0,\chi_n} \to 0$ leads to the estimate

$$\lim \left\| \dot{\xi}_n \right\|_{0,\chi_n} \leqslant C \lim \|\xi_n\|_{0,\chi_n}$$

for some suitable constant $C > 0$. But this would imply the inequality

$$\lim \|\xi_n\|_{1,\chi_n} \leqslant (1 + C) \lim \|\xi_n\|_{0,\chi_n} = 0$$

contradicting

$$1 - \epsilon \leqslant \lim \|\xi_n\|_{1,\chi_n}$$

as stated above. Thus we have proven the uniform estimate of the assertion.

□

Let us now introduce the zero point lemma which is central within our Banach space calculus.

Lemma 2.52 *Let $f : E \to F$ be a smooth map between Banach spaces of the form*

$$f(x) = f(0) + Df(0) \cdot x + N(x) \quad \textit{with} \quad \dim\ker Df(0) < \infty \ .$$

Let us further assume $G : F \to E$ to be a right inverse,

$$Df(0) \circ G = \mathrm{id}_F \ ,$$

and let $C > 0$ be a constant such that the contraction property

$$\|GN(x) - GN(y)\| \leqslant C(\|x\| + \|y\|) \|x - y\|$$

is satisfied for all $x, y \in B_\epsilon(0)$ with $\epsilon = \frac{1}{5C}$. Then the initial condition $\|Gf(0)\| \leqslant \frac{\epsilon}{2}$ implies the unique existence of a zero point of the map f

$$x_0 \in B_\epsilon(0) \cap G(F) \ ,$$

which additionally fulfills the estimate

$$\|x_0\| \leqslant 2 \|Gf(0)\| \ .$$

Proof. The proof is accomplished as a direct application of the contraction mapping principle. Considering the mapping

$$\varphi : E \to E, \;\; \varphi(x) = -G\big(f(0) + N(x)\big) \;\; ,$$

we easily calculate the inclusion relation, as $x \in B_\epsilon(0)$ yields the estimate

$$\|\varphi(x)\| \;\leqslant \frac{\epsilon}{2} + C\epsilon^2 = \epsilon\,(\frac{1}{2} + C\epsilon) = \epsilon\,(\frac{1}{2} + \frac{1}{5}) < \epsilon$$

due to the assumption. Moreover, φ contracts on $B_\epsilon(0)$ as we immediately see from

$$\begin{aligned} \|\varphi(x) - \varphi(y)\| \;\; &= \;\; \|GN(x) - GN(y)\| \\ &\leqslant \;\; C\,2\epsilon\;\|x - y\| \;\; = \frac{2}{5}\;\|x - y\| \\ &\leqslant \;\; \frac{1}{2}\;\|x - y\| \;\; . \end{aligned}$$

Hence we find a unique $x_0 \in B_\epsilon(0)$ satisfying the identity $\varphi(x_0) = x_0$. Additionally we obtain the estimate

$$\|x_0\| \;-\; \|\varphi(0)\| \;\leqslant\; \|\varphi(x_0) - \varphi(0)\| \;\leqslant \frac{1}{2}\;\|x_0\| \;\; .$$

Thus we deduce the inequality $\|x_0\| \; < 2\;\|Gf(0)\|$ and complete the proof. □

It is by means of this fundamental lemma that we shall later prove the existence of the necessary correction section $\gamma : K \times [\rho_k, \infty) \to L^\perp$. Let us choose a fixed $\chi \in K \times [\rho_2, \infty)$ and let us consider the smooth mapping between the Banach spaces

$$F_\chi : H^{1,2}_\chi \to L^2_\chi \;\; ,$$

that is, explicitly,[15]

$$F_\chi(x) = \nabla_t x + \Theta(x) \cdot \dot{w}_\chi + \nabla_2 \exp(x)^{-1} \cdot (\nabla f \circ \exp_{w_\chi} x)$$

with $F_\chi(0) = F(w_\chi)$, $DF_\chi(0) = D_\chi$. Considering the expansion

$$F_\chi(x) = F_\chi(0) + D_\chi(0) \cdot x + N_\chi(0, x)$$

we compute the difference of the nonlinear terms as

$$\begin{aligned} N_\chi(x) - N_\chi(y) \;\; &= \;\; F_\chi(x) - F_\chi(y) - D_\chi(0) \cdot (x - y) \\ &= \;\; \big(D_\chi(y) - D_\chi(0)\big) \cdot (x - y) + N_\chi(y, x - y) \;\; . \end{aligned}$$

15For the notations used in this formula the reader is referred to the discussion in Appendix A.2.

Analysis of the nonlinear dependence in the smooth exponential terms together with the compactness of K, that is

$$\bigcup_{\chi\in K\times[\rho_0,\infty)} w_\chi(\overline{\mathbb{R}}) \quad \text{relatively compact} \;, \tag{2.76}$$

yields an estimate which is uniform with respect to χ,

$$\|N_\chi(x) - N_\chi(y)\|_{0,\chi} \leqslant C_{K,2}\left(\|x\|_{1,\chi} + \|y\|_{1,\chi}\right) \|x-y\|_{1,\chi} \;. \tag{2.77}$$

The constant $C_{K,2}$ is a function of the C^2-norm of $\nabla_2 \exp$ on the compact closure of the set in (2.76). The required estimate thus follows from this inequality together with Lemma 2.51 and with $C_K = C_{K,1}\cdot C_{K,2}$,

$$\begin{aligned} &\|G_\chi N_\chi(x) - G_\chi N_\chi(y)\|_{1,\chi} \leqslant C_K\left(\|x\|_{1,\chi} + \|y\|_{1,\chi}\right) \|x-y\|_{1,\chi} \\ &\text{and } \; \|G_\chi F_\chi(0)\|_{1,\chi} = \|G_\chi F(w_\chi)\|_{1,\chi} \leqslant C_{K,1}\,\|F(w_\chi)\|_{0,\chi} \;, \\ &\text{for all } \; \chi\in K\times[\rho_2,\infty) \;. \end{aligned} \tag{2.78}$$

Finally, it remains to show that the deviation of the approximately glued trajectories from the actual gradient flow trajectories tends to zero as the gluing parameter ρ goes to infinity. This follows immediately from the next lemma.

Lemma 2.53 *There are uniform constants* $m>0$ *and* $\alpha>0$*, such that the estimate*

$$\|F(w_\chi)\|_{0,\chi} \leqslant \alpha\,\mathrm{e}^{-m\rho}$$

holds for all $\rho\geqslant\rho_0$ *and* $(u,v)\in K$.

Proof. Due to the construction of the pre-gluing operation $w_\chi = u\#^o_\rho v$ for the broken trajectories we immediately obtain the identity

$$F(w_\chi)(t) = 0 \quad \text{for} \quad |t|\; > 1 \;,$$

that is, without loss of generality,

$$\begin{aligned} u(t) &= \exp_y \xi(t), \qquad & t &\geqslant -\rho-1 \\ v(t) &= \exp_y \zeta(t), \qquad & t &\leqslant \rho+1 \;, \end{aligned}$$

Therefore we may continue this analysis in local coordinates at y, so that we consider the situation on $\mathbb{R}^n$:

$$\begin{aligned} &F = \frac{\partial}{\partial t} + X : H^{1,2}(\mathbb{R},\mathbb{R}^n) \to L(\mathbb{R},\mathbb{R}^n) \;, \\ &X:\mathbb{R}^n\to\mathbb{R}^n,\; DX(0) = \text{Hessian of } F \text{ at } y \;, \\ &w_\chi = \beta^-\cdot\,\xi_\rho + \beta^+\cdot\,\zeta_{-\rho} \;. \end{aligned}$$

Thus it is sufficient to estimate

$$\begin{aligned}\|F(w_\chi)\|_{L^2} &= \|F(w_\chi)\|_{L^2([-1,1])} \leqslant \text{const } \|w_\chi\|_{H^{1,2}([-1,1])} \\ &\leqslant \text{const}\left(\|\xi_\rho\|_{H^{1,2}([-1,0])} + \|\zeta_{-\rho}\|_{H^{1,2}([0,1])} \right) .\end{aligned}$$

According to former discussions of the solutions for the gradient flow equation, we observe an exponential decrease at the critical point 0 of the form

$$|s(t)| \leqslant c(s)\,\mathrm{e}^{-mt},\ t \geqslant 0 ,$$

where the constant m depends merely on $DX(0)$. The constant c depending continuously on s may be estimated uniformly due to the compactness of K, so that the inequality

$$\|\xi_\rho\|_{L^2([-1,0])} = \left(\int_{-1}^{0} |\xi(t+\rho)|^2\, dt \right)^{\frac{1}{2}} \leqslant \text{const}(u,v)\,\mathrm{e}^{-m\rho},\ \text{etc.} ,$$

yields the completion of the proof. □

Summing up, we can state that there is a lower parameter bound ρ_K, such that, given any $\chi \in K \times [\rho_K, \infty)$, the mappings

$$F_\chi : H_\chi \to L_\chi \quad \text{and} \quad G_\chi : L_\chi \to H_\chi,\ R(G_\chi) = L_\chi^\perp$$

satisfy the assumptions for the fundamental Lemma 2.52 and guarantee the unique existence of a correction term

$$\gamma(\chi) \in B_\epsilon(0) \cap L_\chi^\perp$$

complying with

$$F_\chi\big(\gamma(\chi)\big) = 0 \quad \text{and} \quad \|\gamma(\chi)\|_{1,\chi} < 2\,C_{K,1}\,\|F(w_\chi)\|_{0,\chi} ,$$

and thus

$$\|\gamma(\chi)\|_{1,\chi} < c(K)\,\mathrm{e}^{-m\rho} .$$

Therefore it remains only to verify the smoothness of this correction section γ.

Lemma 2.54 *The unique section* $\gamma : K \times [\rho_K, \infty) \to L^\perp \cap B_\epsilon(0)$ *solving* $F \circ \gamma = 0$ *is smooth.*

Proof. Since we have already solved the existence and uniqueness problem, we are now able to regard the same zero point problem in local coordinates, that is with respect to a local trivialization of $L^\perp$ and L^2. Then we immediately obtain the assertion from the implicit function theorem together with the smoothness of F. In particular, we are provided with a uniform estimate for the derivatives

$$\|D\gamma(\chi)\| \leqslant \mathrm{const}(K) \cdot \|\gamma(\chi)\| \quad \text{for all } \chi \in K \times [\rho_K, \infty) \ .$$

□

Piecing together all these results we obtain the trajectory gluing as asserted above in the shape of a smooth mapping

$$\begin{aligned} \#: \quad K \times [\rho_K, \infty) &\to \mathcal{M}_{x,z} \\ \chi &\mapsto \exp_{u\#^o_\rho v} \gamma(\chi) \ . \end{aligned} \tag{2.79}$$

The next step is to derive the embedding property for fixed parameter $\rho \geqslant \rho_K$ from the embedding property of the pre-gluing map $\#^o_\rho$ and the exponential convergence

$$\|D\gamma(\chi)\|, \ \|\gamma(\chi)\| \leqslant \mathrm{const}\, \mathrm{e}^{-m\rho} \ .$$

We shall prove this property in the following section for the gluing operation induced for the unparametrized trajectories

$$(\hat{u}, \hat{v}) \in \widehat{\mathcal{M}}_{x,y} \times \widehat{\mathcal{M}}_{y,z} \ .$$

The Embedding Result and the Convergence

In order to arrive at the statement of Theorem 3 we are finally going to prove the embedding property and compatibility with complementary weak convergence toward the original broken trajectory as regards the spaces of unparametrized trajectories. At first, we define the gluing operation which is induced on the unparametrized trajectory spaces $\widehat{\mathcal{M}}_{x,y}$.

Definition 2.55 *Let us consider the diffeomorphism already analysed above,*

$$\begin{aligned} \phi: \quad \widehat{\mathcal{M}}_{x,y} \times \mathbb{R} &\xrightarrow{\approx} \mathcal{M}_{x,y} \\ (\hat{u}, t) &\mapsto \hat{u}_t \ , \end{aligned}$$

for which we implicitly use the identification

$$\begin{aligned} \widehat{\mathcal{M}}_{x,y} &= \mathcal{M}^a_{x,y}, \\ \hat{u}_0 &\equiv \hat{u} \end{aligned} \tag{2.80}$$

with respect to a fixed regular value a from $\big(f(y), f(x)\big)$. By reverse argumentation we can also consider the projection map

$$[\cdot] : \mathcal{M}_{x,y} \to \widehat{\mathcal{M}}_{x,y} \ ,$$

so that it holds $[\hat{u}_0] = \hat{u}$ *together with* $f\big(\hat{u}(0)\big) = a$. *Given a compact set of unparametrized broken trajectories*

$$\widehat{K} \subset \widehat{\mathcal{M}}_{x,y} \times \widehat{\mathcal{M}}_{y,z} \text{ compact },$$

we define the induced gluing operation on this set by

$$\hat{u}\widehat{\#}_\rho\hat{v} = [\hat{u}\#_\rho\hat{v}] = (\hat{u}\#_\rho\hat{v})_{\tau(\hat{u},\hat{v},\rho)} ,$$

where $\tau(\hat{u},\hat{v},\rho)$ *is determined uniquely via the condition*

$$f\Big(\big(\hat{u}\#_\rho\hat{v}\big)\big(\tau(\hat{u},\hat{v},\rho)\big)\Big) = a .$$

This means that, after the gluing of the parametrized trajectories $(\hat{u}_0,\, \hat{v}_0) \in \mathcal{M}_{x,y} \times \mathcal{M}_{y,z}$, *we have to execute a gauge-shifting or reparametrization by* $\tau(\ldots)$, *in order to get an unparametrized trajectory* $\hat{u}\widehat{\#}_\rho\hat{v} \in \widehat{\mathcal{M}}_{x,z}$ *regarding the above identification.*

Now, we are able to prove the embedding result in version of

Proposition 2.56 *There is a lower parameter bound* $\rho_{\widehat{K}}$, *such that the unparametrized gluing map*

$$\begin{array}{rll} \widehat{\#}: & \widehat{K} \times (\rho_{\widehat{K}}, \infty) & \to \widehat{\mathcal{M}}_{x,z} \\ & (\hat{u},\hat{v},\rho) & \mapsto \hat{u}\widehat{\#}_\rho\hat{v} \end{array}$$

is a smooth embedding.

Proof. The differentiability property follows immediately from the respective result for the gluing map $\#$. Hence we have to show (a) the regularity, that is the property that

$$D\widehat{\#}(\hat{u},\hat{v},\rho) \quad \text{is an isomorphism for each triple} \quad (\hat{u},\hat{v},\rho) ,$$

and (b) the injectivity of the gluing map.

(a): Due to the compactness of $\widehat{K}$ it is sufficient to find a lower parameter bound $\rho(\hat{u},\,\hat{v})$ for each broken trajectory $(\hat{u},\,\hat{v}) \in \widehat{K}$ in such a way that the regularity of $D\widehat{\#}(\hat{u},\hat{v},\rho)$ holds for all $\rho > \rho(\hat{u},\hat{v})$. Since the regularity of the differential of a continuously differentiable map defined upon a finite-dimensional manifold is a so-called open condition,[16] this lower bound can be found as an upper-semicontinuous[17] function on $\widehat{K}$, so that $\rho(\widehat{K}) = \max_{(\hat{u},\hat{v})\in\widehat{K}} \rho(\hat{u},\hat{v})$ exists.

16In other words, each regular point $(\hat{u},\hat{v},\rho) \in \widehat{K} \times [0,\infty)$ with regard to the differential $D\widehat{\#}$ has a neighbourhood in $\widehat{K} \times [0,\infty)$ consisting of regular points.

17ρ upper-semicontinuous means: $x_n \to x \Rightarrow \overline{\lim}\rho(x_n) \leqslant \rho(x)$.

Hence, we shall prove the existence of such a bound for any arbitrarily chosen fixed pair $(\hat{u}, \hat{v})$. Due to the identity

$$\begin{aligned}\dim\left(\widehat{K} \times (\rho_{\widehat{K}}, \infty)\right) &= \mu(x) - \mu(y) - 1 + \mu(y) - \mu(z) - 1 + 1 \\ &= \mu(x) - \mu(z) - 1 \\ &= \dim \widehat{\mathcal{M}}_{x,z}\end{aligned}$$

it is sufficient to verify the injectivity of $D\widehat{\#}(\hat{u}, \hat{v}, \rho)$ for all $\rho > \rho(\hat{u}, \hat{v})$. This will be accomplished indirectly. Let us henceforth use the shorter notation $\hat{u} = u$ referring to the above identification $\widehat{\mathcal{M}} = \mathcal{M}^a$. From the analysis of the unparametrized trajectory spaces we obtain the decomposition of the tangent space

$$T_{\hat{u}\widehat{\#}_\rho\hat{v}}\mathcal{M}_{x,z} = T_{\hat{u}\widehat{\#}_\rho\hat{v}}\widehat{\mathcal{M}}_{x,z} \oplus \mathbb{R}\frac{\partial}{\partial t}(\hat{u}\widehat{\#}_\rho\hat{v}) \ , \tag{2.81}$$

at

$$\hat{u}\widehat{\#}_\rho\hat{v} \in \widehat{\mathcal{M}}_{x,z} \overset{(2.80)}{=} \mathcal{M}^a_{x,z} \subset \mathcal{M}_{x,z} \ .$$

This follows from the construction of the induced unparametrized gluing operation with

$$\frac{\partial}{\partial t}(\hat{u}\widehat{\#}_\rho\hat{v}) = \frac{\partial}{\partial t}(\hat{u}\#_\rho\hat{v})_{\tau(\hat{u},\hat{v},\rho)}, \ \ f\big(\hat{u}\#_\rho\hat{v}(\tau)\big) = a \ .$$

Without loss of generality we can assume $a = f(y)$. Then we compute the differential of $\widehat{\#}$ by means of (2.81) as

$$\begin{aligned}&D\widehat{\#}(\hat{u}, \hat{v}, \rho)(\xi, \zeta, t) \\ = \ &\mathrm{Proj}_{T_{\dots}\widehat{\mathcal{M}}_{x,z}}\left(D\#_\rho(u,v)\cdot(\xi,\zeta) + D\#_\rho(u,v)\cdot(\dot{u}, -\dot{v})\cdot t\right) \ .\end{aligned} \tag{2.82}$$

Thus, assuming that $D\widehat{\#}(\hat{u}, \hat{v}, \rho)$ is not injective, we find non-vanishing tangent vectors $(\xi, \zeta, t) \neq 0$ satisfying

$$D\#_\rho(u,v)\cdot(\xi,\zeta) + D\#(u,v)\cdot(\dot{u}, -\dot{v})\cdot t \in \mathbb{R}\cdot\frac{\partial}{\partial t}(u\#_\rho v) \ .$$

Without loss of generality we can carry out the normalization

$$\|\xi\|^2 \ + \ \|\zeta\|^2 \ + t^2 = 1 \ .$$

We now accomplish the proof by contradiction. Let us assume the existence of sequences

$$\rho_n \to \infty, \ \ \big((\xi_n, \zeta_n)\big)_{n\in\mathbb{N}} \text{ bounded and } (t_n)_{n\in\mathbb{N}} \subset \mathbb{R}$$

satisfying the identity

$$D\#_{\rho_n}\cdot(\xi_n, \zeta_n) + D\#_{\rho_n}\cdot(\dot{u}, -\dot{v}) = t_n\cdot\frac{\partial}{\partial t}(u\#_{\rho_n} v) \tag{2.83}$$

for all $n \in \mathbb{N}$. Due to the construction of the gluing operation in the last section we compute

$$\begin{aligned} & D\#_{\rho_n}(u,v)\cdot(\xi,\zeta) \\ = \;& \nabla_1 \exp\big(\chi_n,\gamma(\chi_n)\big)\cdot(\xi\#_{\rho_n}\zeta) + \nabla_2\exp(\ldots)\cdot\big(D\gamma(\chi_n)\cdot(\xi,\zeta)\big)\ , \end{aligned}$$

so that the sequence $(\,\|D\#_{\rho_n}\|_{\mathrm{Op}}\,)_{n\in\mathbb{N}} \subset \mathbb{R}$ turns out to be bounded. Since the sequence $\frac{\partial}{\partial t}(u\#_{\rho_n}v)_{\tau(\rho_n)}$ is also bounded uniformly as we see from the equation

$$\left\|\tfrac{\partial}{\partial t}(u\#_{\rho_n}v)_{\tau(\rho_n)}\right\|_{L^2} = \left\|\tfrac{\partial}{\partial t}(u\#_{\rho_n}v)\right\|_{L^2} = \sqrt{f(x)-f(z)}\ ,$$

we deduce the same for the sequence $(t_n)_{n\in\mathbb{N}}$. A detailed calculation involving the formulas

$$\begin{aligned} & D\#_{\rho_n}\cdot(\xi_n,\zeta_n) + D\#_{\rho_n}(\dot u,-\dot v) \\ = \;& \nabla_1\exp\big(\chi_n,\gamma(\chi_n)\big)\cdot\big(\xi_n\#_{\rho_n}\zeta_n + \dot u\#_{\rho_n}(-\dot v)\big) \\ & + \big[\nabla_2\exp\big(\chi_n,\gamma(\chi_n)\big)\circ D\gamma(\chi_n)\big]\cdot(\ \ldots\ ) \end{aligned}$$

and

$$\begin{aligned} & \frac{\partial}{\partial t}(u\#_{\rho_n}v) = \frac{\partial}{\partial t}\big(\exp_{\chi_n}\gamma(\chi_n)\big) \\ = \;& \nabla_1\exp\big(\chi_n,\gamma(\chi_n)\big)\cdot\dot\chi_n + \nabla_2\exp\big(\chi_n,\gamma(\chi_n)\big)\cdot\nabla_t\gamma(\chi_n) \end{aligned}$$

yields the identity (2.83) as

$$\begin{aligned} & \nabla_1\exp\big(\chi_n,\gamma(\chi_n)\big)\cdot\big(\xi_n\#_{\rho_n}\zeta_n + \dot u\#_{\rho_n}(-\dot v) + t_n\tfrac{\partial}{\partial t}(u\#^o_{\rho_n}v)\big) \\ + \;& \nabla_2\exp(\ldots)\cdot\big(D\gamma(\chi_n)\cdot(\xi_n+\dot u,\ \zeta_n-\dot v) + t_n\nabla_t\gamma(\chi_n)\big) = 0\ . \end{aligned} \tag{2.84}$$

Exploiting the well-known identity[18]

$$(\nabla_2\exp^{-1}\circ\nabla_1\exp)(p,0) = \mathrm{id}_{T_pM}$$

and the exponential decrease

$$\|\gamma(\chi_n)\|,\ \|D\gamma(\chi_n)\| \overset{n\to\infty}{\longrightarrow} 0\ ,$$

which was analysed above in the proof of Lemma 2.54, we are led to the decisive convergence

$$\left\|x_n\#_{\rho_n}y_n + \dot u\#_{\rho_n}(-\dot v) + t_n\tfrac{\partial}{\partial t}(u\#^o_{\rho_n}v)\right\|_{1,2} \to 0\ . \tag{2.85}$$

18See Appendix A.2.

According to the definition of the linear gluing version and of the pre-gluing operation, this implies convergence toward zero

$$\begin{aligned} &\lim_{n\to\infty} \|x_n + \dot{u}\,(1+t_n)\|_{H^{1,2}((-\infty,\rho_n-1])} \\ = &\lim_{n\to\infty} \|y_n + \dot{v}\,(t_n-1)\|_{H^{1,2}([1-\rho_n,\infty))} \quad = 0 \ , \end{aligned}$$

that is

$$x_n + (1+t_n)\cdot\dot{u} \overset{H^{1,2}}{\longrightarrow} 0 \quad \text{and} \quad y_n + (t_n-1)\cdot\dot{v} \overset{H^{1,2}}{\longrightarrow} 0 \ . \tag{2.86}$$

Taking into account that, due to the assumption and the decomposition (2.81), the relations

$$\begin{aligned} x_n \in T_u\widehat{\mathcal{M}}_{x,y},\ T_u\widehat{\mathcal{M}}_{x,y} \cap \mathbb{R}\dot{u} = \{0\}, \\ y_n \in T_v\widehat{\mathcal{M}}_{y,z},\ T_v\widehat{\mathcal{M}}_{y,z} \cap \mathbb{R}\dot{v} = \{0\} \end{aligned}$$

hold and that $T_u\widehat{\mathcal{M}}_{x,y}$ and $T_v\widehat{\mathcal{M}}_{y,z}$ are finite-dimensional, and thus locally compact, we observe that the zero-convergence in (2.86) leads to

$$x_n \to 0 \quad \text{and} \quad t_n \to -1$$

and

$$y_n \to 0 \quad \text{and} \quad t_n \to +1 \ ,$$

respectively. Hence, we deduce a contradiction.

(b): We now have to verify the injectivity of $\widehat{\#} : \widehat{K} \times [\rho_{\widehat{K}}, \infty) \to \widehat{\mathcal{M}}_{x,z}$. Let us now assume the existence of sequences

$$\rho_n \to \infty \quad \text{and} \quad (\hat{u}_{1,n}, \hat{v}_{1,n}) \neq (\hat{u}_{2,n}, \hat{v}_{2,n}) \in \widehat{K} \ ,$$

which satisfy the identity

$$[\hat{u}_{1,n}\#_{\rho_n}\hat{v}_{1,n}] = [\hat{u}_{2,n}\#_{\rho_n}\hat{v}_{2,n}] \quad \text{for all } n \in \mathbb{N} \ ,$$

where we refer to Definition 2.55, that is,

$$[\hat{u}_{i,n}\#_{\rho_n}\hat{v}_{i,n}] = (\hat{u}_{i,n}\#_{\rho_n}\hat{v}_{i,n})_{\tau_{i,n}} \ , \quad i = 1,2$$

with $f\left((\ldots)_{\tau_{i,n}}(0)\right) = a$. Resorting again to the compactness of $\widehat{K}$ we can consider without loss of generality the convergent sequences

$$\hat{u}_{i,n} \to \hat{u}_i,\ \ \hat{v}_{i,n} \to \hat{v}_i;\ \ i = 1,2 \ .$$

Then the representation

$$\hat{u}_{i,n}\#_{\rho_n}\hat{v}_{i,n} = \exp_{\chi_{i,n}} \gamma(\chi_{i,n}),\ \ \chi_{i,n} = \hat{u}_{i,n}\#^{o}_{\rho_n}\hat{v}_{i,n}$$

together with the exponential decrease

$$\gamma(\chi_{i,n}) \overset{n\to\infty}{\longrightarrow} 0$$

yields the following $H^{1,2}$-convergence with respect to local coordinates:

$$\left\| (\hat{u}_1 \#^o_{\rho_n} \hat{v}_1)_{\tau_{1,n}} - (\hat{u}_2 \#^o_{\rho_n} \hat{v}_2)_{\tau_{2,n}} \right\|_{1,2} \overset{n\to\infty}{\longrightarrow} 0 \ . \tag{2.87}$$

According to the construction of the pre-gluing map $\#^o$, the fixing of the time-parametrization

$$f\big(\hat{u}_1(0)\big) = f\big(\hat{u}_2(0)\big)$$

gives rise to the asymptotic synchronization

$$\tau_{1,n} - \tau_{2,n} \to 0$$

and thus to the identities

$$\hat{u}_1 = \hat{u}_2 = \hat{u} \quad \text{and} \quad \hat{v}_1 = \hat{v}_2 = \hat{v} \ .$$

However, knowing from part (a) that $\widehat{\#}$ is a local diffeomorphism, in particular at $(\hat{u}, \hat{v}) \in \widehat{K}$, we find an initial index $n_0 \in \mathbb{N}$, such that the identity

$$\big(\hat{u}_{1,n}, \hat{v}_{1,n}\big) = \big(\hat{u}_{2,n}, \hat{v}_{2,n}\big)$$

holds for all $n \geqslant n_0$ in contradiction to the assumption. □

At the end of this analysis of the gluing operation for unparametrized trajectories of time-independent gradient flow, we now discuss the relation with weak convergence toward broken trajectories as was defined in the last chapter.

Proposition 2.57 *Given a broken trajectory* $(\hat{u}, \hat{v})$ *from* $\widehat{\mathcal{M}}_{x,y} \times \widehat{\mathcal{M}}_{y,z}$, *each arbitrary increasing sequence of parameters* $\rho_n \to \infty$ *together with the gluing operation for unparametrized trajectories* $\widehat{\#}$ *induces the geometrical convergence*[19]

$$\hat{u} \widehat{\#}_{\rho_n} \hat{v} \overset{C^\infty_{\text{loc}}}{\longrightarrow} (\hat{u}, \hat{v}) \ .$$

The converse is also true: Any sequence of unparametrized trajectories converging to a simply broken trajectory finally lies within the range of such a gluing map $\widehat{\#}$.

Proof. The C^∞_{loc}-convergence toward $\hat{u}$ and $\hat{v}$ arises from a respectively appropriate choice of the value a with regard to the identification $\widehat{\mathcal{M}}_{x,z} = \mathcal{M}^a_{x,z}$. Firstly, let us choose a value satisfying

$$f(z) < a < f(y) \ ,$$

19i.e.: weak convergence with respect to a suitable sequence τ_n of reparametrization

that is

$$\big((\hat{u}\#_{\rho_n}\hat{v})(0)\big)_{n\in\mathbb{N}} \subset M^a \ .$$

Due to the compactness of M^a we can find without loss of generality a limit point $x_0 \in M^a$,

$$\widehat{\hat{u}\#_{\rho_n}\hat{v}}(0) \overset{n\to\infty}{\longrightarrow} x_0 \ .$$

Considering the unique trajectory $s \in C^\infty(\mathbb{R}, M)$ satisfying

$$\dot{s} = -\nabla f \circ s \quad \text{and} \quad s(0) = x_0 \ ,$$

we are led to the weak convergence

$$\widehat{\hat{u}\#_{\rho_n}\hat{v}} \overset{C^\infty_{\text{loc}}}{\longrightarrow} s \ ,$$

where the reparametrization sequence τ_n for

$$\widehat{\hat{u}\#_{\rho_n}\hat{v}} = (\hat{u}\#_{\rho_n}\hat{v})_{\tau_n} = \exp_{(\hat{u}\#^o_{\rho_n}\hat{v})_{\tau_n}}[\gamma(\chi_n)]_{\tau_n}$$

is determined by

$$f\big((\hat{u}\#_{\rho_n}\hat{v})(\tau_n)\big) = a \ .$$

The exponential decrease of $\gamma(\chi_n)$ analysed above implies the estimate

$$|\gamma(\chi_n)(\tau_n)| \leqslant \text{const} \cdot \mathrm{e}^{-\alpha\rho_n} \to 0$$

and thus the convergence

$$(\hat{u}\#^o_{\rho_n}\hat{v})(\tau_n) \to x_0 \ .$$

Hence, due to the construction of the pre-gluing operation, the condition $f(x_0) < f(y)$ gives rise to the limit point

$$\lim_{n\to\infty} \hat{v}(\tau_n - \rho_n) = x_0 \ ,$$

that is, $\hat{s} = \hat{v}$.

An entirely analogous argument now starting from the assumption $f(y) < a < f(x)$ yields weak convergence to the other part of the broken trajectory,

$$\widehat{\hat{u}\#_{\rho_n}\hat{v}} \overset{C^\infty_{\text{loc}}}{\longrightarrow} \hat{u} \ .$$

The converse statement is an immediate result from the uniqueness provided by the Banach fixed point principle by means of which we developed the gluing maps. If we are given any sequence of trajectories with fixed endpoints which converge weakly toward a specified simply broken trajectory, the elements finally lie within a $H^{1,2}$-neighbourhood in which the above contraction mapping principle (see Lemma 2.52) works. Hence, these elements have to lie within the range of the gluing map belonging to this specified simply broken trajectory. □

Thus, we have accomplished the proof of the central gluing Theorem 3.

Conclusions

At the end of this section, regarding the application within our homology theory in question, we wish to discuss some consequences of the compactness-gluing-complementarity which has been proven now. Actually, we shall consider the unparametrized trajectory spaces of second order and the simply broken trajectories. This means that we restrict ourselves to the relative Morse index $\mu(x)-\mu(z)=2$. Then we are able to classify the connected components $\widehat{\mathcal{M}}^o_{x,z}$ of the one-dimensional manifold without boundary $\widehat{\mathcal{M}}_{x,z}$ by the diffeomorphism type of S^1 and $(-1,1)$, respectively. We obtain the following

Corollary 2.58 *Let* $\phi : \widehat{\mathcal{M}}^o_{x,z} \xrightarrow{\approx} (-1,1)$ *be a diffeomorphism. Then there are exactly two different broken trajectories*

$$(\hat{u}_1,\hat{v}_1) \neq (\hat{u}_2,\hat{v}_2), \quad (\hat{u}_i,\hat{v}_i) \in \widehat{\mathcal{M}}_{x,y_i} \times \widehat{\mathcal{M}}_{y_i,z}; \;\; i=1,2$$

satisfying

$$\mu(x)-1=\mu(y_1)=\mu(y_2)=\mu(z)+1 \;\; ,$$

and, given any small $\epsilon>0$, *we find gluing parameters* $\rho_1(\epsilon),\rho_2(\epsilon)\in\mathbb{R}$, *such that we obtain the identities*

$$(\phi\circ\widehat{\#})\big(\{\hat{u}_1\}\times\{\hat{v}_1\}\times(\rho_1,\infty)\big)=(-1,-1+\epsilon)$$

and

$$(\phi\circ\widehat{\#})\big(\{\hat{u}_2\}\times\{\hat{v}_2\}\times(\rho_2,\infty)\big)=(1-\epsilon,1) \;\; .$$

Proof. The boundary property is a consequence of the gluing theorem. The fact that the broken trajectories cannot be identical arises from the uniqueness of the correction term $\gamma(\hat{u}_i\#^o_\rho\hat{v}_i)$ with respect to the uniquely supplied bundle $L^\perp$ with regard to large gluing parameters. This is the same argument which was used for the last statement of Proposition 2.57. □

From this corollary we gain an equivalence relation for simply broken trajectories with relative Morse index 2. This equivalence relation will prove crucial for the construction of Morse homology. In fact, it is the kernel of what we called in the introduction the cobordism relation for trajectory spaces.

Definition 2.59 *Let us define the set of simply broken trajectories with fixed endpoints* $\widetilde{\mathcal{M}}_{x,z}=\{(\hat{u},\hat{v})\in\widehat{\mathcal{M}}_{x,y}\times\widehat{\mathcal{M}}_{y,z}\mid \mu(x)-1=\mu(y)=\mu(z)+1\}$. *Then, the gluing cobordism from Corollary 2.58 induces an equivalence relation on* $\widetilde{\mathcal{M}}_{x,z}$ *by*

$$(\hat{u}_1,\hat{v}_1)\sim(\hat{u}_2,\hat{v}_2) \overset{\text{def.}}{\Longleftrightarrow} (\hat{u}_1\#_{\rho_1}\hat{v}_1)\simeq(\hat{u}_2\#_{\rho_2}\hat{v}_2) \;\; in \;\; \widehat{\mathcal{M}}_{x,z} \;\; .$$

This means that these respective broken trajectories correspond to the ends of the same pathwise connected component of $\widehat{\mathcal{M}}_{x,z}$, so that they are cobordant in the sense of 'compactness' and 'gluing'. It is obvious from the possible diffeomorphism types of one-dimensional manifolds without boundary that it holds that

$$\#[(\hat{u}, \hat{v})] = 2 \quad \textit{for all } (\hat{u}, \hat{v}) \in \widetilde{\mathcal{M}}_{x,z} \ .$$

We can thus decompose the set $\widetilde{\mathcal{M}}_{x,z}$ into pairs of different equivalent broken trajectories.

2.5.2 Gluing of Trajectories of the Time-Dependent Gradient Flow

The above defined equivalence relation for simply broken trajectories of the time-independent gradient flow yields the crucial argument in the proof of the chain complex property $\partial^2 = 0$. We now require the corresponding gluing results for the time-dependent and the λ-parametrized situation, in order to prove the homotopy invariance of the Morse homology in an analogous way. The essential difference between the earlier gluing operation and the gluing operation which is to be constructed later, is a lack of invariance under time-shifting

$$u \bullet \tau = u_\tau$$

with regard to the actual homotopy trajectories. However, we notice that we merely admitted Morse homotopies for the time-dependent gradient fields which were asymptotically time-independent, that is,

$$\frac{\partial}{\partial t}(\nabla h_t) = 0 \quad \text{for} \quad |t| \geqslant R \ .$$

Therefore, we must distinguish three different types of gluing operations:

- gluing of shifting-invariant trajectories, as analysed above,
- gluing of broken trajectories, which consist of shifting-invariant pieces as well as of trajectory pieces corresponding to the time-dependent gradient field, called mixed broken trajectories, and
- gluing of broken trajectories which do not contain any shifting-invariant pieces at all.

We shall obtain the result for the mixed broken trajectories analogously to the first gluing theorem. Let h_t be any regular Morse homotopy

$$f^\alpha \overset{h_t}{\simeq} f^\beta \ .$$

Theorem 4 *Given a compact set of mixed simply broken trajectories* $\widehat{K} \subset \mathcal{M}^h_{x_\alpha,x_\beta} \times \widehat{\mathcal{M}}^{f^\beta}_{x_\beta,y_\beta}$*, there is a lower parameter bound* $\rho_{\widehat{K}}$ *and a smooth embedding*

$$\begin{aligned} \#: \quad \widehat{K} \times [\rho_{\widehat{K}}, \infty) &\hookrightarrow \mathcal{M}^h_{x_\alpha,y_\beta} \\ (u_h, \hat{v}, \rho) &\mapsto u_h \#_\rho \hat{v} \ , \end{aligned}$$

such that, given any arbitrary sequence of gluing parameters $\rho_n \to \infty$*, we are provided with the weak convergence*

$$u_h \#_{\rho_n} \hat{v} \xrightarrow{C^\infty_{\mathrm{loc}}} (u_h, \hat{v}) \ .$$

Proof. Essentially, the construction of this gluing operation for mixed broken trajectories differs from that in Theorem 3 merely by the definition of the pre-gluing map. Instead of shifting both parts of the trajectories by the value ρ in opposite directions regarding the time-parametrization, we now define the pre-gluing map $\#^o$ for the mixed broken trajectory

$$(u_h, v) \in \mathcal{M}^h_{x_\alpha,x_\beta} \times \mathcal{M}^{f^\beta}_{x_\beta,y_\beta}$$

by

$$(2.88) \quad (u_h \#^o_\rho v)(t) = \begin{cases} u_h(t), & t \leqslant \rho - 1 \\ \exp_{x_\beta} \begin{pmatrix} \beta^-_{-\rho}[\exp^{-1}_{x_\beta}(u_h)] \\ +\beta^+_{-\rho}[\exp^{-1}_{x_\beta}(v)]_{-2\rho} \end{pmatrix}(t), & |t-\rho| \leqslant 1 \\ v_{-2\rho}(t), & t \geqslant \rho + 1 \ . \end{cases}$$

In this framework, the lower bound ρ_0 for the gluing parameters is now additionally bounded below via the time interval during which the Morse homotopy

$$f^\alpha \overset{h_t}{\simeq} f^\beta$$

is active, that is $\rho_0 \geqslant R(h_t)$.

The construction of a smooth gluing operation for these trajectories is accomplished by steps entirely analogous to that of the former gluing operation. What proves decisive for transferring the essential stages of the construction, is the invariance under time-shifting of the $H^{1,2}$- and L^2-products along the trajectories. It is worth mentioning, that this time we can consider equivalently the gluing operation as an embedding

$$\#_\rho : K \hookrightarrow \mathcal{M}^h_{x_\alpha,y_\beta}$$

of compact sets $K \subset \mathcal{M}^h_{x_\alpha,x_\beta} \times \mathcal{M}_{x_\beta,y_\beta}$ for fixed gluing parameters $\rho \geqslant \rho_K$ as well as the embedding stated above,

$$\# : \widehat{K} \times [\rho_{\widehat{K}}, \infty) \hookrightarrow \mathcal{M}^h_{x_\alpha,y_\beta} \ .$$

Once again, it is immediate from the representation

$$u_h \#_{\rho_n} \hat{v} = \exp_{u_h \#^o_{\rho_n} \hat{v}} \gamma(\chi_n)$$

together with $|\gamma(\chi_n)(t)| \leqslant \text{const}\, e^{-\alpha \rho_n} \to 0$, that the original trajectory $(u_h,\, \hat{v})$ may be reproduced from the weakly convergent sequence

$$u_h \#_{\rho_n} \hat{v} \xrightarrow{C^\infty_{\text{loc}}} (u_h,\, \hat{v}) \ .$$

□

The respective result for mixed broken trajectories of the form

$$(\hat{u},\, v_h) \in \widehat{\mathcal{M}}_{x_\alpha, y_\alpha} \times \mathcal{M}^h_{y_\alpha, y_\beta}$$

is obtained likewise.

Broken Pure h-Trajectories

For the case of the gluing of broken pure homotopy trajectories, that is a similar operation which glues trajectories $(u_{\alpha\beta},\, u_{\beta\gamma}) \in \mathcal{M}^{h^{\alpha\beta}}_{x_\alpha, x_\beta} \times \mathcal{M}^{h^{\beta\gamma}}_{x_\beta, x_\gamma}$ and maps them into the trajectory space $\mathcal{M}^{h^{\alpha\gamma}}_{x_\alpha, x_\gamma}$, we should present a separate approach. The crucial feature in this situation is that the gluing parameter ρ, for which the above gluing construction gives rise to an embedding into a trajectory space, if ρ is large enough, now also determines the target space $\mathcal{M}^{h^{\alpha\gamma}}_{x_\alpha, x_\gamma}$ itself. The result is stated as

Proposition 2.60 *Given a compact set of broken h-trajectories*

$$K \subset \mathcal{M}^{h^{\alpha\beta}}_{x_\alpha, x_\beta} \times \mathcal{M}^{h^{\beta\gamma}}_{x_\beta, x_\gamma} \ ,$$

there is a lower parameter bound R_0, such that for any $R \geqslant R_0$ a homotopy exists,

$$f^\alpha \overset{h^{\alpha\gamma}(R)}{\simeq} f^\gamma \ ,$$

together with a smooth mapping

$$\#_R : K \hookrightarrow \mathcal{M}^{h^{\alpha\gamma}(R)}_{x_\alpha, x_\gamma} \ ,$$

which represents an embedding.

Proof. For this case of the broken trajectories $(u_{\alpha\beta},\, u_{\beta\gamma})$ we use again the original, 'symmetric' pre-gluing:

$$(2.89) \quad (u_{\alpha\beta} \#^o_R v_{\beta\gamma})(t) = \begin{cases} u_{\alpha\beta}(t+R), & t \leqslant -1 \\ \exp_{x_\beta} \begin{pmatrix} \beta^-[\exp^{-1}_{x_\beta} u_{\alpha\beta}]_R \\ +\beta^+[\exp_{x_\beta} v_{\beta\gamma}]_{-R} \end{pmatrix}(t), & |t| \leqslant 1 \\ v_{\beta\gamma}(t-R), & t \geqslant 1 \ . \end{cases}$$

In order to obtain the correction term for an h-trajectory by means of a contraction map, we choose a suitable homotopy between f^α and f^β, namely

$$h_R^{\alpha\gamma}(t) = \begin{cases} h^{\alpha\beta}(t+R,\cdot), & t \leqslant 0 \\ h^{\beta\gamma}(t-R,\cdot), & t \geqslant 0 \end{cases} . \tag{2.90}$$

Provided that the gluing parameters are large enough, $R \geqslant R_0$, this homotopy is well-defined. We are now able to construct the required gluing operation $\#_R$ using the same methods of Banach calculus as above.

Essentially, the difference between this gluing operation and the gluing in the first theorem of this chapter consists of the fact that the invariance under time-shifting for the trajectories of a given time-independent gradient field has now been replaced by the appropriate choice of the time-dependent gradient field in the case of explicit time-dependence on both trajectories. □

Concerning this type of gluing operation we wish to discuss a special case as given in the following

Corollary 2.61 *Given the situation of isolated h-trajectories, that is*

$$\mu(x_\alpha) = \mu(x_\beta) = \mu(x_\gamma) \ ,$$

there is an $R > 0$ *such that the associated gluing map*

$$\#_R : \mathcal{M}^{h^{\alpha\beta}}_{x_\alpha,x_\beta} \times \mathcal{M}^{h^{\beta\gamma}}_{x_\beta,x_\gamma} \stackrel{\cong}{\to} \mathcal{M}^{h^{\alpha\gamma}(R)}_{x_\alpha,x_\gamma}$$

is a one-to-one correspondence of finite sets.

Proof. According to the above result it is sufficient to verify surjectivity. Let $u_{\alpha\gamma}$ be an h-trajectory from $\mathcal{M}^{h^{\alpha\gamma}(R)}_{x_\alpha,x_\gamma}$. Then, consideration of the solutions

$$u_{\alpha\beta} \in \mathcal{M}^{h^{\alpha\beta}}_{x_\alpha,x_\beta} \quad \text{and} \quad u_{\beta\gamma} \in \mathcal{M}^{h^{\beta\gamma}}_{x_\beta,x_\gamma}$$

of the respective time-dependent gradient flow, which are determined uniquely by

$$u_{\alpha\beta}(0) = u_{\alpha\gamma}(-R) \quad \text{and} \quad u_{\beta\gamma}(0) = u_{\alpha\gamma}(R) \ ,$$

leads us immediately to the broken trajectory forming the pre-image with respect to $\#_R$. □

2.5.3 Gluing for λ-Parametrized Trajectories

Let us now consider the manifold $\mathcal{M}^H_{x_\alpha,y_\beta} \subset [0,1] \times \mathcal{P}^{1,2}_{x_\alpha,y_\beta}$ of λ-parametrized trajectories as a zero set with respect to the mapping

$$\begin{aligned} G^{\alpha\beta} : [0,1] \times \mathcal{P}^{1,2}_{x_\alpha,y_\beta} &\to L^2\left(\mathcal{P}^{1,2}_{x_\alpha,y_\beta}{}^*TM\right) \\ (\lambda, u) &\mapsto \dot{u} + \frac{1}{\sqrt{\cdots}} \nabla H^{\alpha\beta}(\lambda,\cdot) \circ u \ . \end{aligned}$$

Here, $H^{\alpha\beta}$ describes a λ-homotopy $h_0 \overset{H^{\alpha\beta}}{\underset{\lambda}{\simeq}} h_1$ of regular Morse homotopies between f^α and f^β. In this framework, too, we wish to discuss the problem of a gluing operation combining such λ-parametrized trajectories (λ, u) with shifting-invariant trajectories of the gradient flow with respect to f^α and f^β, respectively, such that we again obtain λ-trajectories (λ', v) with λ' generally differing from λ.

Theorem 5 *Let $\widehat{K} \subset \mathcal{M}^H_{x_\alpha, y_\beta} \times \widehat{\mathcal{M}}^{f^\beta}_{y_\beta, z_\beta}$ be a compact set of mixed simply broken, λ-parametrized trajectories $(\lambda, u_\lambda, \hat{v})$. Then there is a lower parameter bound $\rho_{\widehat{K}}$ and a smooth embedding*

$$\begin{aligned} \#^H : \widehat{K} \times [\rho_{\widehat{K}}, \infty) &\hookrightarrow \mathcal{M}^H_{x_\alpha, z_\beta} \\ (\lambda, u_\lambda, \hat{v}, \rho) &\mapsto (\lambda, u_\lambda)\#^H_\rho \hat{v} = (\tilde{\lambda}, w_{\tilde{\lambda}}) \ , \end{aligned}$$

such that, provided any increasing sequence of parameters $\rho_n \to \infty$, we obtain weak convergence of the form

$$w_{\tilde{\lambda}}(\lambda, u_\lambda, \hat{v}, \rho_n) \overset{C^\infty_{\mathrm{loc}}}{\longrightarrow} (u_\lambda, \hat{v}) \quad \text{and} \quad \tilde{\lambda}(\lambda, u_\lambda, \hat{v}, \rho_n) \to \lambda \ .$$

As to the underlying analysis, we do not come upon anything new. We merely expand the Banach manifold, within which we are searching for the zeroes of a Fredholm map by means of the contraction principle, by the compact interval $[0, 1]$ of λ-parameters. Therefore, we will not provide a detailed proof at this stage. The strategy for proving this theorem is organized in two stages in a way which is absolutely analogous to the above investigations: At first, we have to define an appropriate pre-gluing map, in this situation similar to the time-dependent case in Section 2.5.2, together with a suitable normal bundle $\widetilde{L}^\perp$, which now lies within $\mathbb{R} \times H^{1,2}_{\mathbb{R}}(\mathcal{P}^{1,2}_{x_\alpha, z_\beta}{}^*TM)$. Finally, we must derive a unique smooth correction section within this normal bundle by means of a suitable contraction map.

Definition 2.62 *Referring to the corresponding pre-gluing map for mixed broken trajectories from Section 2.5.2 for an orientation, we now define the pre-gluing operation*

$$\#^{H,o} : K \times [\rho_0, \infty) \to [0,1] \times \mathcal{P}^{1,2}_{x_\alpha, z_\beta}$$

by

$$(\lambda, u_\lambda) \, \#^{H,o}_\rho v = (\lambda, u_\lambda \#^o_\rho v) \ , \tag{2.91}$$

where we denote by $\#^o_\rho$ the operation from (2.88), that is

$$(u_\lambda \#^o_\rho v)(t) = \begin{cases} u_\lambda(t), & t \leqslant \rho - 1 \\ v_{-2\rho}(t), & t \geqslant \rho + 1 \end{cases} \ , \quad \text{etc.}$$

In strict analogy with Definitions 2.48 and 2.49 we find a construction of a linear version of a pre-gluing map $\#_\chi$, this time for $\chi = (\lambda, u_\lambda) \#^{H,\circ}_\rho v \in [0,1] \times \mathcal{P}^{1,2}_{x_\alpha, z_\beta}$. Thus, we obtain a normal bundle $\widetilde{L}^\perp$ from

Proposition 2.63 *There is a lower parameter bound $\rho_1 \geqslant \rho_0$ such that the Fredholm operator $DG^{\alpha\beta}(\lambda, u_\lambda \#^\circ_\rho v) : \mathbb{R} \times H^{1,2}_{u_\lambda \#^\circ_\rho v} \to L^2_{u_\lambda \#^\circ_\rho v}$ is onto for all gluing parameters $\rho \geqslant \rho_1$ and mixed broken trajectories $\big((\lambda, u_\lambda), v\big) \in \widehat{K}$. Additionally, the composition of the linear version $\#_\chi$ with the $(\mathbb{R} \times L^2)$-projection onto $\ker DG^{\alpha\beta}(\chi)$ then induces an isomorphism:*

$$\phi_\chi = \mathrm{Proj}^{\mathbb{R} \times L^2_\chi}_{\ker DG^{\alpha\beta}(\chi)} \circ \#_\chi :$$
$$\ker DG^{\alpha\beta}(\lambda, u_\lambda) \times \ker D_v \xrightarrow{\cong} \ker DG^{\alpha\beta}(\lambda, u_\lambda \#^\circ_\rho v) \ .$$

As to the proof. Essentially, the proof of 2.50 remains the same up to expansion by $\mathbb{R}$. At this stage we wish to point to the trivialized version, which shall be mentioned again in the next chapter in relation to the discussion of induced coherent orientations. This will be treated in Proposition 3.14. Regarding a detailed proof in the trivial case together with the additional set of real parameters, the reader is also referred to [F-H]. This linear version of gluing in the trivial case is of the type

$$\mathbb{R} \times H^{1,2} \times H^{1,2} \ni (\tau, \xi, \zeta) \mapsto (\tau,\ \beta^-_{-\rho} \cdot \xi + \beta^+_{-\rho} \cdot \zeta_{-2\rho}) \in \mathbb{R} \times H^{1,2} \ .$$

□

Finally, considering the normal bundle $\widetilde{L}^\perp$ over $\widehat{K} \times [\rho_1, \infty)$, which is induced by this proposition in analogy with the former bundle $L^\perp$, and given a large enough lower parameter bound $\rho_2 \geqslant \rho_1$, we are again able to find a unique smooth correction section of

$$\begin{aligned} &\gamma : \widehat{K} \times [\rho_2, \infty) \to \widetilde{L}^\perp \\ &\gamma(\lambda, u_\lambda, v, \rho) = (\tau, \xi) \in \mathbb{R} \times H^{1,2}_{u_\lambda \#^\circ_\rho v} \ , \end{aligned} \tag{2.92}$$

such that

$$(\lambda, u_\lambda) \#^H_\rho v = (\lambda + \tau, \exp_{u_\lambda \#^\circ_\rho v} \xi) \in \mathcal{M}^H_{x_\alpha, z_\beta}$$

represents the required gluing operation. Once again we observe exponential convergence $(\tau, \xi) \to (0,0)$ as the gluing parameter tends to infinity, $\rho \to \infty$.

Chapter 3

Orientation

Summing up the analytical foundational results we have developed up to this stage, we notice that this knowledge about the trajectory spaces of the time-independent and time-dependent negative gradient flow enables us already to build a Morse homology theory with coefficients in the field $\mathbb{Z}_2$. However, in order to admit arbitrary coefficient groups, i.e. coefficients in $\mathbb{Z}$, we still have to accomplish more elaborate results concerning the characteristic intersection numbers for the unparametrized trajectories. Referring to the introduction, we may deduce these intersection numbers from a comparison of the canonical orientation of the intersection manifold

$$W^u(x) \pitchfork W^s(y) \approx \mathcal{M}_{x,y}$$

by the negative gradient field with some coherent orientation related to the critical points.

Considering the framework of our analytical methods we need a concept for such a coherent orientation of the trajectory spaces, which are treated as zero sets of the fundamental Fredholm map with respective endpoints. Actually, within this analytical context, coherence denotes the compatibility of the respective orientations with the cobordism relations from the compactness-gluing-complementarity in the last chapter. In fact, the purpose of this chapter is to extend the former cobordism concept for trajectory spaces to a concept of oriented cobordisms. The fundamental feature of this concept will be the notion of orientations for Fredholm operators arising from the already well-known determinant bundle (see for instance [Don]). The reason for using this determinant bundle for Fredholm operators may be explained as a generalization of the orientation of a manifold. Knowing that in our framework the tangent space of a trajectory manifold can be identified with the kernel of the surjective linearization of the fundamental Fredholm map, we wish to drop this regularity assumption and define a substitute for the maximal exterior product of the tangent space, in which a non-vanishing element represents an orientation. This substitute concerning a non-surjective Fredholm operator consists of an appropriate combination of kernel and cokernel, so that a continuous variation of a thus oriented Fredholm operator preserves the orientation, whether the operator is onto or not. Hence we shall be provided with an orientation concept for trajectory spaces including the regular trajectory manifolds as special cases.

Another point which seems worth mentioning concerns the topology of the underlying manifold M in the sense of the axiomatic homology theory in question. It appears to be founded in an abstract way within the existence and construction problem of such a coherent orientation for the Fredholm operators $D_u \in \Sigma_{u^*TM}$ as u describes the trajectories of the negative gradient flow.

Throughout the development of this analytical concept of a coherent orientation we keep in close analogy to the Floer homology in the symplectic case. The corresponding results can be found in [F-H]. At first, we shall review the construction of an orientation bundle for Fredholm operators in general. Then we shall verify the compatibility with the gluing calculus for Fredholm operators of the type $F \in \Sigma_{\text{triv}}$. In the second half of the chapter we shall transfer the orientation concept from the trivialized framework to the manifold M, so that the topology enters when we ask for the existence of a coherent orientation. In contrast to the symplectic case in [F-H] we shall also consider non-orientable manifolds, whereby we shall need a slightly modified concept.

Finally, we still have to verify that this method for obtaining a coherent orientation on the trajectory manifolds based on the Fredholm calculus is equivalent to the classical process of induced orientations from the differential topological viewpoint. This shall be postponed to Chapter B of the appendix.

3.1 Orientation and Gluing in the Trivial Case

3.1.1 The Determinant Bundle

In order to define the orientation bundle for Fredholm operators, we require the following preparations:

Definition 3.1 *Let E and F be finite-dimensional $\mathbb{R}$-vector spaces. Then, by means of the short notations for the one-dimensional vector spaces*

$$\Lambda^{\max} E = \Lambda^{\dim E} E \quad \textit{and} \quad \Lambda^0 E = \mathbb{R} \ ,$$

we define the one-dimensional space of determinants with respect to E and F by

$$\mathrm{Det}\,(E,F) = (\Lambda^{\max} E) \otimes (\Lambda^{\max} F)^* \ .$$

Now let H_0 and H_1 be fixed Banach spaces and let us denote by

$$\mathcal{F}(H_0, H_1) \subset \mathcal{L}(H_0, H_1)$$

the open subset of Fredholm operators. Then we define the space of determinants associated to such a Fredholm operator $F \in \mathcal{F}(H_0, H_1)$ as

$$\mathrm{Det}(F) = \mathrm{Det}(\ker F, \mathrm{coker}\, F) \ .$$

If we further consider the continuous mapping of an arbitrary topological space X into the space of Fredholm operators

$$f : X \to \mathcal{F}(H_0, H_1) \ ,$$

we may define the space

$$\mathrm{Det}\, f = \bigcup_{x \in X} \{x\} \times \mathrm{Det}\big(f(x)\big) \ .$$

Thus, the canonical projection map is endowed with the one-dimensional $\mathbb{R}$-vector spaces $\mathrm{Det}\,\big(f(x)\big) = \pi^{-1}(x)$ as fibres.

The aim of this section is to show that the space $\mathrm{Det}\, f$ is a real line bundle on the topological space X. Then, the local trivializations uniquely determine the topology of this vector bundle. If we assumed $\dim \ker f$ or $\dim \mathrm{coker}\, f$ to be locally constant functions on X,[1] this result could be immediately concluded. By the following fundamental algebraic lemma we will derive the existence of the determinant bundle for general Fredholm operators without this assumption.

Lemma 3.2 *Let*

$$0 \to E_1 \overset{d_1}{\to} \dots \overset{d_{k-1}}{\to} E_k \to 0$$

be a sequence of finite-imensional $\mathbb{R}$-vector spaces. Then there is a canonical isomorphism

$$\phi : \bigotimes_{i \text{ even}} (\Lambda^{\max} E_i) \overset{\cong}{\longrightarrow} \bigotimes_{i \text{ odd}} (\Lambda^{\max} E_i) \ .$$

1Note that due to the continuity of the integer valued Fredholm index the continuity of one of the two functions implies that of the other.

Proof. Let $e_{11}, \ldots, e_{1n_1}$ be a basis of E_1. Since d_1 is injective we may extend the linear independent vectors $d_1(e_{11}), \ldots, d_1(e_{1n_1})$ by $e_{21}, \ldots, e_{2n_2}$ to a basis of E_2. Corresponding to this first step and due to exactness, we can find successively at each stage of the sequence an extension to a basis of the form

$$d_i(e_{i1}), \ldots, d_i(e_{in_i}), e_{i+1\,1}, \ldots, e_{i+1\,n_{i+1}}, \quad \text{for } i = 1, \ldots, k-1 \ ,$$

with $n_k = 0$. Then we define the isomorphism ϕ by

$$\begin{pmatrix} d_1(e_{11}) \wedge \ldots \wedge d_1(e_{1n_1}) \wedge e_{21} \wedge \ldots \wedge e_{2n_2} \\ \otimes d_3(e_{31}) \wedge \ldots \wedge e_{4n_4} \otimes \ldots \otimes \ldots \wedge e_{2j\,n_{2j}} \end{pmatrix}$$
$$\mapsto \quad \begin{pmatrix} e_{11} \wedge \ldots \wedge e_{1n_1} \otimes d_2(e_{21}) \wedge \ldots \wedge d_2(e_{2n_2}) \\ \wedge e_{31} \wedge \ldots \wedge e_{3n_3} \otimes \ldots \otimes \ldots \wedge e_{2j'+1\,n_{2j'+1}} \end{pmatrix} .$$

Obviously, this isomorphism is independent of the special choices of the vectors of the bases, because any change in the bases operates by multiplication with the same determinant of the transformation on either side of the homomorphism ϕ. □

This lemma enables us to provide local trivializations of the determinant bundle $\pi : \operatorname{Det} f \to X$. (See the careful exposition in [F-H] for details.) Given $x \in X$, we associate to the Fredholm operator $f(x)$ a linear map $\psi : \mathbb{R}^n \to H_1$, such that

$$\text{(3.1)} \qquad \begin{array}{rll} \widehat{f}_\psi(x) : & \mathbb{R}^n \times H_0 & \to H_1 \\ & (h,k) & \mapsto \psi(h) + f(x) \cdot k \end{array}$$

describes a surjective Fredholm operator. Since surjectivity is a regular property, it must be satisfied on an open set in X. Thus there is a neighbourhood $U(x)$ such that the surjectivity of

$$\widehat{f}_\psi(y) \in \mathcal{F}(\mathbb{R}^n \times H_0, H_1)$$

holds throughout $U(x)$ together with the identity

$$\operatorname{ind} \widehat{f}_\psi(y) = \operatorname{ind} \widehat{f}_\psi(x) \text{ for all } y \in U(x) \ .$$

Thus, the mapping

$$\begin{array}{rll} f_\psi(y) : & \mathbb{R}^n \times H_0 & \to \mathbb{R}^n \times H_1 \\ & (h,k) & \mapsto \left(0, \widehat{f}_\psi(y)(h,k)\right) \end{array}$$

gives rise to a real line bundle $\operatorname{Det} f_\psi$ on $U(x)$, because we now deal with the constant functions

$$\dim \ker f_\psi(y) \quad \text{and} \quad \dim \operatorname{coker} f_\psi(y)$$

on $U(x)$. By the fundamental Lemma 3.2, we can verify that these local determinant bundles associated to the covering

$$\bigcup_{x \in X} U(x) = X$$

yield a local trivialization of $\mathrm{Det}\, f$ on X. Actually, given $\big(x, \psi, U(x)\big)$ and $y \in U(x)$, we have an exact sequence

$$0 \longrightarrow \ker f(y) \xrightarrow{d_1} \ker f_\psi(y) \xrightarrow{d_2} \mathbb{R}^n \xrightarrow{d_3} \operatorname{coker} f(y) \longrightarrow 0 \ ,$$

where

$$\begin{aligned} d_1(k) &= (0,k) \\ d_2(h,k) &= h \\ d_3(h) &= [\psi(h)]_{R\left(f(y)\right)} \ . \end{aligned}$$

This induces a natural isomorphism

$$\mathrm{Det}\, f(y) \xrightarrow{\cong} (\Lambda^{\max} \ker f_\psi(y)) \otimes (\Lambda^{\max}\mathbb{R}^n)^* \ ,$$

i.e. a natural isomorphism

(3.2) $$\mathrm{Det}\, f(y) \xrightarrow{\cong} \mathrm{Det}\, f_\psi(y)$$

due to the surjectivity of $\widehat{f}_\psi(y)$. The corresponding calculation of the transition maps therefore provides the result that

$$\pi : \mathrm{Det}(f) \to X$$

is a real line bundle.

Definition 3.3 *The bundle* $\mathrm{Det}\, f$ *is called the determinant bundle of* f *and any non-vanishing section in this line bundle induces an orientation for the family*

$$f : X \to \mathcal{F}(H_0, H_1)$$

of Fredholm operators. We remark that such a section does not necessarily exist.

3.1.2 Gluing and Orientation for Fredholm Operators

Throughout this section, we shall use the results of Fredholm theory developed in the last chapter. In order to define a gluing operation for Fredholm operators of the type[2]

$$F_A \in \Sigma_\xi = \Sigma_{\mathrm{triv}}, \ \ \xi = \overline{\mathbb{R}} \times \mathbb{R}^n$$

2See Definition 2.14.

analogous to the construction in the last chapter, we must restrict ourselves to asymptotically constant operators, that is, to $K = \frac{\partial}{\partial t} + A_K \in \Sigma_{\text{triv}}$ satisfying

$$A_K(t) = \text{const} \quad \text{as} \quad |t| \geqslant T \text{ for a } T \geqslant 0 \ .$$

Actually, referring to the equivalence relation

$$K_A \sim L_B \Leftrightarrow A^{\pm} = B^{\pm}$$

and recalling that according to Lemma 2.15 these equivalence classes are topologically contractible, we may choose an asymptotically constant representative from each equivalence class Θ_K.

Definition 3.4 *Considering the isometric* $\mathbb{R}$*-action*

$$\begin{aligned} \mathbb{R} \times \Sigma \quad &\to \Sigma \\ (\rho, F_A) \quad &\mapsto F_\rho = F_{A(\cdot+\rho)} \ , \end{aligned}$$

we are able to define the gluing operation for asymptotically constant operators $K, L \in \Sigma$ *with matching ends* $K^+ = L^-$ *by*

$$\begin{aligned} K\#_\rho L &= F_{A_\rho} \in \Theta(K^-, L^+) \\ A_\rho(t) &= \begin{cases} A_K(t+\rho), & t \leqslant 0 \\ A_L(t-\rho), & t \geqslant 0 \end{cases} \end{aligned}$$

for all $\rho \geqslant \rho_0(K, L)$*. Here, the lower parameter bound is determined by the asymptotical behaviour*

$$A_K(\cdot + \rho)_{|\,[-1,\infty]} \equiv A_K^+ \equiv A_L^- \equiv A_L(\cdot - \rho)_{|\,[-\infty,1]} \quad \textit{for all } \rho \geqslant \rho_0 \ .$$

Remark According to the index theorem from Section 2.2, the Fredholm index operates additively under this gluing operation, i.e.

$$\text{ind}\,(K\#_\rho L) = \text{ind}\, K + \text{ind}\, L$$

holds for all asymptotically constant operators $K, L \in \Sigma$ with matching ends $K^+ = L^-$. Moreover, the equivalence class $\Theta_{K\#L}$ does not depend on the actual gluing parameter ρ and the respective choice of the representatives $K \in \Theta_K$, $L \in \Theta_L$. In a nutshell, if we consider the set

$$\widehat{\Sigma} = \Sigma / \sim$$

of equivalence classes $[K] = \Theta_K$, $K \in \Sigma$ and the map

$$\text{ind} : \widehat{\Sigma} \to \mathbb{Z} \ ,$$

we can state the following relation:

$$\operatorname{ind} \circ \# = + \circ (\operatorname{ind}, \operatorname{ind}) \ .$$

The crucial question now is whether there is a similar orientation concept for Fredholm operators as defined above, which is likewise naturally compatible with the gluing operation. At this stage it is important to state that, due to the contractibility of the equivalence classes $[K] = \Theta(K^+, K^-)$, the associated determinant bundles

$$\operatorname{Det}[K] \to \Theta(K^+, K^-)$$

are in fact orientable.

First, we require an isomorphism between the spaces of determinants associated to the respective Fredholm operators

$$\operatorname{Det}(K) \otimes \operatorname{Det}(L) \xrightarrow{\cong} \operatorname{Det}(K \#_\rho L) \ .$$

Second, regarding the equivalence classes Θ_K, Θ_L and $\Theta_{K\#_\rho L}$, we have to verify that the orientation on $\Theta_{K\#_\rho L}$ induced by this isomorphism is independent of the choice of the actual representatives. Once again, the fundamental Lemma 3.2, by which we may transform in a natural way the Fredholm operators K and L into the surjective operators $\widehat{K}_\psi$ and $\widehat{L}_\psi$, turns out to be the decisive utility. Thus let K and L be asymptotically constant operators from

$$\Sigma_{\text{triv}} \subset \mathcal{L}\big(H^{1,2}(\mathbb{R}, \mathbb{R}^n), L^2(\mathbb{R}, \mathbb{R}^n)\big)$$

with matching ends $K^+ = L^-$. We find a linear mapping

$$\psi : \mathbb{R}^k \to L^2(\mathbb{R}, \mathbb{R}^n) \ ,$$

such that the Fredholm operators

$$\begin{aligned} \widehat{K}_\psi,\ \widehat{L}_\psi : \quad \mathbb{R}^k \times H^{1,2} &\to L^2 \\ (h, u) &\mapsto L \cdot u + \psi \cdot h \\ (h, u) &\mapsto K \cdot u + \psi \cdot h \end{aligned}$$

are both onto. By analogy with the derivation of the determinant bundle, Lemma 3.2 supplies us with natural isomorphisms

$$\operatorname{Det}(K_\psi) \xrightarrow{\cong} \operatorname{Det}(K) \quad \text{and} \quad \operatorname{Det}(L_\psi) \xrightarrow{\cong} \operatorname{Det}(L)$$

concerning the operators

$$\begin{aligned} K_\psi,\ L_\psi : \quad \mathbb{R}^k \times H^{1,2} &\to \mathbb{R}^k \times L^2 \\ (a, u) &\mapsto (0, L \cdot u + \psi \cdot a) \\ (a, u) &\mapsto (0, K \cdot u + \psi \cdot a) \ . \end{aligned}$$

We note that without loss of generality we may assume an $R \geqslant 0$ satisfying

$$\operatorname{supp}\big(\psi(a)\big) \subset [-R, R] \quad \text{for all } a \in \mathbb{R}^k \ ,$$

because the bounded linear mapping

$$\begin{aligned} \beta_R : \quad L^2 &\to L^2 \\ u &\mapsto \beta_R \cdot u \end{aligned}$$

associated to a smooth cut-off function β_R with compact support in $[-R, R]$ gives rise to a continuous deformation satisfying

$$\widehat{L}_{\beta_R \cdot \psi} \overset{\mathcal{L}(H^{1,2}, L^2)}{\longrightarrow} \widehat{L}_{\psi} \quad \text{as } R \to \infty \ ,$$

and the surjective Fredholm operators form an open subset of the space of operators $\mathcal{L}(H^{1,2}, L^2)$.

Given these surjectively extended Fredholm operators $\widehat{K}_\psi$ and $\widehat{L}_\psi$, we now define a gluing operation which is adapted to the situation:

Definition 3.5 *Let $\rho_0 = \rho_0(K, L)$ be the lower parameter bound associated to the asymptotically constant operators K and L. In order to take into consideration the extension by ψ, we lift this lower bound to $\rho_1 = R + 1 + \rho_0$, and we consequently define the mapping*

$$\widehat{K}_\psi \#_\rho \widehat{L}_\psi : \mathbb{R}^k \times \mathbb{R}^k \times H^{1,2} \to L^2$$

for $\rho \geqslant \rho_1$ by

$$(\widehat{K}_\psi \#_\rho \widehat{L}_\psi)(a, b, u)(t) = \big((K \#_\rho L) \cdot u\big)(t) + \psi(a)(t + \rho) + \psi(b)(t - \rho) \ .$$

The central proposition with respect to this gluing of surjective Fredholm operators may be stated now as

Proposition 3.6 *Let P_ρ be the orthogonal projection in the Hilbert space $\mathbb{R}^k \times \mathbb{R}^k \times L^2$ onto the finite-dimensional subspace $\ker(\widehat{K}_\psi \#_\rho \widehat{L}_\psi)$. Then there is a lower parameter bound $\rho_2 \geqslant \rho_1$ such that the following holds for all gluing parameters $\rho \geqslant \rho_2$:*

The Fredholm operator $\widehat{K}_\psi \#_\rho \widehat{L}_\psi$ is surjective and the linear mapping

$$\begin{aligned} \phi_\rho : \quad \ker \widehat{K}_\psi \times \ker \widehat{L}_\psi &\overset{\cong}{\longrightarrow} \ker(\widehat{K}_\psi \#_\rho \widehat{L}_\psi) \\ \big((a, u), (b, v)\big) &\mapsto P_\rho(a, b, u_\rho + v_{-\rho}) \end{aligned}$$

is an isomorphism.

Proof. We observe that this proposition corresponds to Proposition 2.50 about the linear version of the pre-gluing of trajectories up to the finite-dimensional surjectivity extensions $(\mathbb{R}^k, \psi)$. Regarding these extensions with compact support $\operatorname{supp} \psi(a) \subset [-R, R]$, we may accomplish the proof by strict analogy. The reader is also referred to [F-H]. □

Moreover, this 'surjectivation' by means of $(\mathbb{R}^k, \psi)$ comes to the core of the discussion of the trajectory gluing for the λ-parametrized gradient flow with respect to the orientation problem. There, too, the Fredholm operators D_u become surjective generally at the earliest after extension by the partial derivative for the variational parameter λ.[3]

Now considering the induced mapping

$$\begin{aligned} &K_\psi \#_\rho L_\psi : \mathbb{R}^k \times \mathbb{R}^k \times H^{1,2} \to \mathbb{R}^k \times \mathbb{R}^k \times L^2 \\ &(K_\psi \#_\rho L_\psi)(a, b, u) = \left(0, 0, (\widehat{K}_\psi \#_\rho \widehat{L}_\psi)(a, b, u)\right) , \end{aligned}$$

which is equivalent to

$$K_\psi \#_\rho L_\psi \equiv (K \#_\rho L)_{\psi_\rho \oplus \psi_{-\rho}} ,$$

according to the above derivation of the determinant bundle, we are provided with the natural ismomorphism

$$\operatorname{Det}(K_\psi \#_\rho L_\psi) \xrightarrow{\cong} \operatorname{Det}(K \#_\rho L) \tag{3.3}$$

by Lemma 3.2. Noting that the gluing Proposition 3.6 induces an isomorphism

$$\left(\Lambda^{\max} \ker \widehat{K}_\psi\right) \otimes \left(\Lambda^{\max} \ker \widehat{L}_\psi\right) \xrightarrow{\cong} \Lambda^{\max} \ker (\widehat{K}_\psi \#_\rho \widehat{L}_\psi)$$

and resorting to the natural isomorphism

$$\left(\Lambda^{\max} \mathbb{R}^k\right)^* \otimes \left(\Lambda^{\max} \mathbb{R}^k\right)^* \xrightarrow{\cong} \left(\Lambda^{\max} (\mathbb{R}^k \times \mathbb{R}^k)\right)^*$$

we obtain an isomorphism for the associated determinant spaces of

$$\operatorname{Det}(K_\psi) \otimes \operatorname{Det}(L_\psi) \xrightarrow{\cong} \operatorname{Det}(K_\psi \#_\rho L_\psi) , \tag{3.4}$$

provided that the gluing parameter ρ is large enough. Finally, the natural isomorphisms yielded by the algebraic Lemma 3.2, which describe the transformation of the determinant spaces with respect to the surjective extension, i.e.

$$\operatorname{Det}(K_\psi) \xrightarrow{\cong} \operatorname{Det} K, \text{ etc.,}$$

lead us to the isomorphism for the original determinant spaces

$$\operatorname{Det} K \otimes \operatorname{Det} L \xrightarrow{\cong} \operatorname{Det}(K \#_\rho L) . \tag{3.5}$$

3See also Theorem 2 in Section 2.3.2.

The next step is to verify that this isomorphism may be transferred to the determinant bundles of the associated equivalence classes Θ_K, Θ_L and $\Theta_{K\#L}$. In other words, we have to verify the coherence with respect to different representatives. Since the equivalence classes as topological spaces are contractible within Σ_{triv}, we may choose connecting arcs K^τ and L^τ with $\tau \in [0,1]$ for each pair of representatives (K^0, K^1) and (L^0, L^1), respectively. We are able to repeat the above analysis starting from these arcs instead of fixed operators together with the additional argument that the set of parameters $[0,1]$ is compact and the arcs are continuous. Therefore, we once again find surjective extensions and corresponding isomorphisms uniformly with respect to τ or dependent on τ. These isomorphisms give rise to a vector bundle isomorphism on $[0,1]$, i.e.

$$
\begin{array}{ccc}
\operatorname{Det} K^\tau \otimes \operatorname{Det} L^\tau & \xrightarrow{\;\cong\;} & (K^\tau \#_\rho L^\tau) \\
& \searrow \quad \oplus \quad \swarrow & \\
& [0,1] &
\end{array}
\tag{3.6}
$$

for $\rho \geqslant \rho_0(K^{[0,1]}, L^{[0,1]})$. Hence we deduce that our orientation concept for the equivalence classes Θ_K and Θ_L is compatible with the gluing which yields $\Theta_{K\#L}$.

Before we finish this section by stating the result in a theorem, we wish to mention an important property. As we already noticed in the section on trajectory gluing, due to our construction, this gluing operation itself does not satisfy an associativity rule. However the mere orientation problem can actually be solved. Let K, L, $M \in \Sigma_{\text{triv}}$ be asymptotically constant operators with matching ends $K^+ = L^-$ and $L^+ = M^-$. Then, provided suitable gluing parameters $\rho_1, \ldots, \rho_4$, the operators $(K\#_{\rho_1}L)\#_{\rho_2}M$ as well as $K\#_{\rho_4}(L\#_{\rho_3}M)$ are well-defined representatives of the same equivalence class $\Theta(K^-, M^+)$. Thus we can construct a smooth line bundle E on $[0,1]$ with the boundary fibers

$$E_0 = \operatorname{Det}((K\#_{\rho_1}L)\#_{\rho_2}M) \quad \text{and} \quad E_1 = \operatorname{Det}\big(K\#_{\rho_4}(L\#_{\rho_3}M)\big)$$

and a vector bundle isomorphism

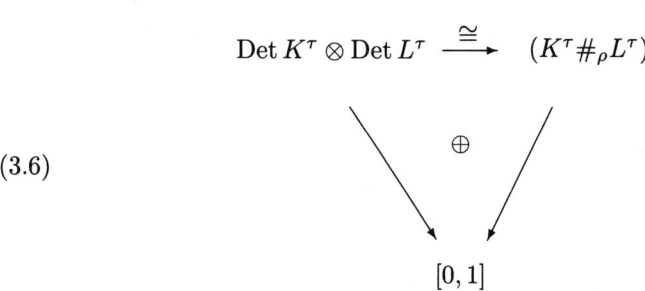

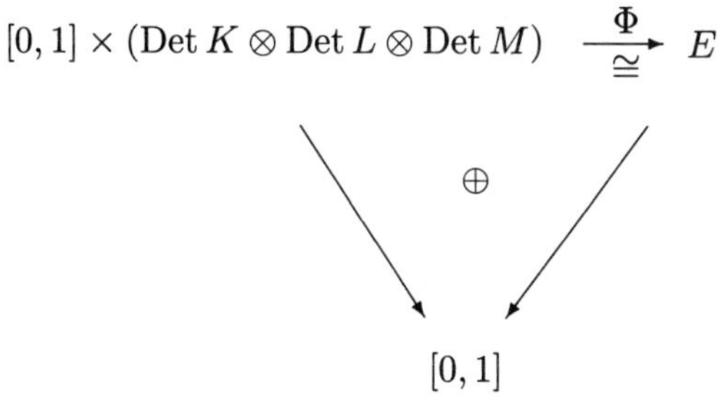

$$\begin{array}{ccc} [0,1] \times (\operatorname{Det} K \otimes \operatorname{Det} L \otimes \operatorname{Det} M) & \xrightarrow[\cong]{\Phi} & E \\ & \searrow \quad \oplus \quad \swarrow & \\ & [0,1] & \end{array}$$

such that Φ_0 and Φ_1 represent exactly the compositions of the gluing isomorphisms from Proposition 3.6 in the respective order.
We sketch the proof. In principle, it resorts to the same analytical construction elements which we have used in our discussion about gluing up to this stage. At first, we define a gluing operation for three operators simultaneously with two free, independent gluing parameters. Then we have to prove the associated isomorphism property with respect to the surjectively extended Fredholm operators. Secondly, we have to show that, given gluing parameters that are chosen appropriately, we are able to homotope this isomorphism for three operators to the respective composition. The crucial step consists of finding a uniform estimate together with a homotopy to 0 for the difference term between the respective isomorphisms as the gluing parameters become large enough. Finally we are able to sum up this section in the following theorem:

Theorem 6 *Let K and L be asymptotically constant operators from Σ_{triv} with matching ends $K^+ = L^-$ and let o_K and o_L be orientations of the canonical determinant bundles on Θ_K and Θ_L, respectively. Then for all $\rho \geqslant \rho_0(K,L)$, the gluing operation*

$$\#_\rho : (K,L) \mapsto K\#_\rho L$$

induces an orientation $o_K \# o_L$ on $\Theta_{K\#_\rho L} = \Theta_{K\#L}$ independently of ρ. This induced orientation does not depend on the actual choice of representatives $K \in \Theta_K$ and $L \in \Theta_L$. Moreover, the associativity rule of is fulfilled,

$$(o_K \# o_L) \# o_M = o_K \# (o_L \# o_M) \ .$$

3.2 Coherent Orientation

The outcome of the above analysis in the case of the trivial $\overline{\mathbb{R}}$-vector bundle $\overline{\mathbb{R}} \times \mathbb{R}^n$ is an orientation concept which is supplied uniformly for all asymptotically constant operators with coinciding ends. This means an orientation for the equivalence classes $\Theta_K \in \widehat{\Sigma}$, and it is compatible with the gluing operation. We now intend to transfer this result to the situation of the $\overline{\mathbb{R}}$-bundles u^*TM induced by the gradient trajectories u. Thus we principally have to consider the dependence on the curve u. Actually, this a priori issue touches on the topology of the underlying manifold. The crucial observation will be that mainly the knowledge about asymptotical behaviour suffices and that we are able to orient all Fredholm operators uniformly as one having identical end terms and associated to curves with coinciding fixed endpoints. The only delicate point is the question if two given smooth curves u and v with identical endpoints $u(\pm\infty) = v(\pm\infty)$ give rise to pull-back bundles u^*TM and v^*TM admitting trivializations coinciding appropriately at $T_{u(\pm\infty)}M = T_{v(\pm\infty)}M$. If this is guaranteed, we may transfer the results from the last section in a unique and canonical way.

It is at this stage that we come upon the problem that not all two curves with coinciding endpoints satisfy this trivialization condition on the pull-back bundles in an a priori way. The obstruction of a common trivialization in the above sense is hidden in the topology of the connection to a loop $u \cdot v^{-1}$ (provided a suitable reparametrization). Actually, the vanishing of the characteristic class $w_1((u \cdot v^{-1})^*TM)$ is the necessary and sufficient condition for the trivializability of the bundle $(u \cdot v^{-1})^*TM$. In other words, we find simultaneous canonical orientations except for curves with coinciding endpoints, which close up to n-dimensional 'Möbius bands'. This topological obstruction is in fact an expression for the orientability of the underlying manifold M.

Nevertheless, a more detailed analysis will show that already the coincidence of the trivializations at one end proves sufficient, if we only can specify in a uniform way which end is the right one. Then, the possible difference at the other end, which may be expressed as a fixed reflection, turns out to be irrelevant. The only essential feature is the uniform and necessarily noncanonical specification for all curves, at which end we wish to have the identical trivialization. Although it might look unexpected or even cryptic to the reader, it is worth mentioning at this stage that this specification is in one-to-one correspondence with the dimension axiom of the later homology theory and with the obstruction to the well-known Poincaré duality in the case of closed but non-orientable manifolds. We shall come back to this issue in the section 'topology and coherent orientation' within the next chapter which is central to this monograph.

Finally, having accomplished the transfer of the above notion of an orientation to the non-trivial framework of curves, we shall develop the concept of a coherent orientation by means of the compatibility with the gluing operation for trajectories from the last chapter. This coherent orientation is the main feature which allows us to set up a homology theory with coefficients in $\mathbb{Z}$.

3.2.1 Orientation and Gluing on the Manifold M

We shall accomplish the orientation process for the Fredholm operators from Σ_{u^*TM} not only for the trajectories of a special gradient field but in general for all smooth compact curves $u \in C^\infty(\overline{\mathbb{R}}, M)$. Therefore, we consider this space of curves to be equipped with the Whitney C^∞-topology.[4] Since we merely deal with pull-back bundles u^*TM on $\overline{\mathbb{R}}$, Lemma 2.19 guarantees that the induced covariant derivation from TM is Fredholm admissible. Hence, we drop ∇ from the notation.

Definition 3.7 *We call two Fredholm operators*

$$K \in \Sigma_{u^*TM}, \quad L \in \Sigma_{v^*TM}$$

along supporting curves $u, v \in C^\infty(\overline{\mathbb{R}}, M)$ *equivalent if the asymptotical identities*

$$u(\pm\infty) = v(\pm\infty) \quad \text{and} \quad K^\pm = L^\pm$$

hold. Any pair of trivializations

$$\begin{aligned} \phi_u : \quad u^*TM &\xrightarrow{\cong} \overline{\mathbb{R}} \times \mathbb{R}^n \\ \psi_v : \quad v^*TM &\xrightarrow{\cong} \overline{\mathbb{R}} \times \mathbb{R}^n \end{aligned}$$

is called admissible for such equivalent operators $(u, K) \sim (v, L)$, *if the identity* $\phi_u(-\infty) = \psi_v(-\infty)$ *holds at the identical lower ends of the curves, the relation*

$$\phi_u(+\infty) \cdot \psi_v(+\infty)^{-1} = \begin{pmatrix} \pm 1 & & & \\ & 1 & & \\ & & \ddots & \\ & & & 1 \end{pmatrix} \in \mathrm{GL}(n, \mathbb{R})$$

at the identical upper ends and if the equivalence defined above for the trivial framework

$$\phi_u(K) = \phi_u K \phi_u^{-1} \sim_{\text{triv}} \psi_v L \psi_v^{-1} = \psi_v(L)$$

is given. Let us denote the equivalence class of (u, K) *by* $\big[(u, K)\big]$ *or merely by* $[K^-, K^+]$, *as it is already determined uniquely by the endpoint operators* $K^\pm \in \mathrm{End}(TM)$.

Remark Given two curves u and v, which are connected to a trivializable loop with respect to a suitable reparametrization, i.e. $w_1\big((u \cdot v^{-1})^*TM\big) = 0$, we always find an admissible pair (ϕ_u, ψ_v) associated to $(u, K) \sim (v, L)$, so that

---

4Relating to the differentiable structure on $\overline{\mathbb{R}}$ introduced in the first chapter, we note that weak and strong Whitney topology on $C^\infty(\overline{\mathbb{R}}, M)$ fall together.

the identity $\phi_u(\pm\infty) = \psi_v(\pm\infty)$ holds. Otherwise, if $u \cdot v^{-1}$ yields an n-dimensional 'Möbius band', we choose asymptotical trivializations $C(\pm\infty) : T_{u(\pm\infty)}M \xrightarrow{\cong} \mathbb{R}^n$, such that

$$C(\pm\infty)K^{\pm}C(\pm\infty)^{-1} = C(\pm\infty)L^{\pm}C(\pm\infty)^{-1}$$

has a diagonal shape. Then, we compose $C(+\infty)$ with the reflection

$$S = \begin{pmatrix} -1 & & & \\ & +1 & & \\ & & \ddots & \\ & & & +1 \end{pmatrix}$$

thus yielding $C^S(+\infty) = S \cdot C(+\infty)$. Hence, there are trivializations ϕ_u, ψ_v satisfying

$$\phi_u(-\infty) = \psi_v(-\infty) = C(-\infty)$$

and

$$\phi_u(+\infty) = C(+\infty), \;\; \psi_v(+\infty) = C^S(+\infty)$$

or vice versa $\phi_u(+\infty) = C^S(+\infty)$, $\psi_v(+\infty) = C(+\infty)$, and which form an admissible pair by construction. So, to sum up, there is always an admissible pair of trivializations associated to $(u, K) \sim (v, L)$.

The next lemma shows how to orient such equivalent operators simultaneously in a unique way by means of these trivializations.

Lemma 3.8 *Let $(u, K) \sim (v, L)$ be equivalent Fredholm operators and let (ϕ, ψ) and (ϕ', ψ') be respectively admissible pairs of trivializations. Moreover, let $\operatorname{Det} K$ and $\operatorname{Det} L$ be oriented by o_K and o_L. Then the pair (ϕ, ψ) induces compatible orientations*

$$\phi(o_K) \simeq \psi(o_L)$$

on the determinant bundle of the trivialized class $\Theta_{\phi K \phi^{-1}} = \Theta_{\psi L \psi^{-1}}$ if and only if the same is true for the pair (ϕ', ψ') on $\Theta_{\phi' K \phi'^{-1}} = \Theta_{\psi' L \psi'^{-1}}$, that is

$$\phi'(o_K) \simeq \psi'(o_L) \; .$$

Let us state separately the decisive step of the proof of this lemma as

Auxiliary Proposition 3.9 *Let ξ be a smooth $\overline{\mathbb{R}}$-vector bundle and o_K a fixed orientation for $K \in \Sigma_\xi$. We denote the set of sections $\psi \in C^\infty(\operatorname{End}(\xi))$, which are pointwise invertible, by GL_ξ. Further, let us choose any section ψ in GL_ξ satisfying the additional asymptotical condition*

$$\psi(\pm\infty) = \mathrm{id}_{\xi(\pm\infty)} \; .$$

Considering the determinant bundle $\operatorname{Det}\Theta_K$ *on* Θ_K *together with the orientation induced by* o_K*, we are provided with an orientation* $\psi(o_K)$ *for*

$$\psi(K) = \psi \cdot K \cdot \psi^{-1} \in \Theta_K$$

via the Hilbert space isomorphisms

$$H^{1,2}(\xi) \xrightarrow{\cong} H^{1,2}(\xi) \text{ and resp. } L^2(\xi) \xrightarrow{\cong} L^2(\xi)$$

which arise from ψ *and which are again denoted by* ψ*.*

Then this new orientation of the determinant bundle is compatible with the original one, that is

$$\psi(o_K) \simeq o_K \ .$$

Proof. Resorting to the structure group of ξ, we reduce the given operator $K \in \Sigma_\xi$ and the transformation ψ successively to a form which can be analysed easily and explicitly. At first, we observe that we may start without loss of generality from the trivial bundle $\xi = \overline{\mathbb{R}} \times \mathbb{R}^n$. Namely, if we choose any trivialization

$$\phi : \xi \to \overline{\mathbb{R}} \times \mathbb{R}^n \ ,$$

ψ and K transform as

$$\psi_{\text{triv}} = \phi \cdot \psi \cdot \phi^{-1} \in \mathrm{GL}_{\text{triv}} \quad \text{and} \quad \psi_{\text{triv}}(K_{\text{triv}}) = \psi(K)_{\text{triv}} \ .$$

Provided that the assertion is true for the trivial bundle, that is

$$(\psi\,(o_K))_{\text{triv}} = \psi_{\text{triv}}\big(o_{K_{\text{triv}}}\big) \simeq o_{K_{\text{triv}}} \ ,$$

we can transfer this orientation equivalence to the original bundle by means of ϕ, because it holds that

$$o_K \simeq o_L \quad \text{if and only if} \quad \phi(o_K) \simeq \phi(o_L) \ .$$

The next step of the proof consists of showing that it is sufficient to verify the equivalence

$$\psi(o_L) \simeq o_L$$

for any Fredholm operator $L \in \mathcal{F}(H^{1,2}, L^2)$ which is homotopical with K, i.e. there is a continuous arc

$$L_\tau : [0,1] \to \mathcal{F}(H^{1,2}, L^2) \quad \text{satisfying} \quad L_0 = K,\ L_1 = L \ .$$

Such an arc would then yield the equivalence

$$\psi(o_K) \simeq \psi(o_L) \simeq o_L \simeq o_K$$

due to the vector bundle isomorphism

$$\begin{array}{ccc}
\mathrm{Det}(L_\tau) & \xrightarrow{\cong} & \mathrm{Det}\big(\psi(L_\tau)\big) \\
& \searrow \quad \oplus \quad \swarrow & \\
& [0,1] &
\end{array}$$

induced by ψ. Hence, we may assume without loss of generality that the operator $K \in \Sigma_{\text{triv}}$ is asymptotically constant and that it is admissible with respect to the gluing operation introduced in the last section.

We shall now simplify the section

$$\psi \in \mathrm{GL}_{\text{triv}} = C^\infty\big(\overline{\mathbb{R}}, \mathrm{GL}(n, \mathbb{R})\big)$$

appropriately. Since we know from the assumption that ψ is endowed with the end operators

$$\psi(\pm\infty) = \mathbb{1} \in \mathrm{GL}^+(n, \mathbb{R})$$

and that $SO(n, \mathbb{R})$ is a deformation retract of $\mathrm{GL}^+(n, \mathbb{R})$, we are able to find a continuous path ψ_τ from ψ to a $\psi_0 \in C^\infty\big(\overline{\mathbb{R}}, SO(n, \mathbb{R})\big)$. Moreover, by means of a continuous asymptotical reparametrization, we may find a continuous arc ending in an asymptotically constant ψ. Regarding such a $\psi \in \mathrm{GL}_{\text{triv}}$, we can construct a similar gluing operation, so that, given

$$\psi, \widetilde{\psi} \quad \text{with} \quad \psi(+\infty) = \widetilde{\psi}(-\infty)$$

and

$$K, \widetilde{K} \quad \text{with} \quad K(+\infty) = \widetilde{K}(-\infty) \ ,$$

we obtain the identity

$$(\psi \# \widetilde{\psi})(K \# \widetilde{K}) = \psi(K) \# \widetilde{\psi}(\widetilde{K}) \ . \tag{3.7}$$

Since our orientation concept for Σ_{triv} is compatible with gluing, we now choose asymptotically constant operators L, $M \in \Sigma_{\text{triv}}$ satisfying

$$A_L^- = \mathbb{1}, \ A_L^+ = A_K^-, \ A_K^+ = A_M^- \quad \text{and} \quad A_M^+ = \mathbb{1} \ .$$

Let us assume for the moment that we have already proven the assertion for the special case

$$N \in \Sigma_{\text{triv}} \quad \text{with} \quad A_N^\pm = \mathbb{1} \ . \tag{3.8}$$

Then, due to the gluing operation, the general assertion is concluded from

$$\begin{aligned}
& o(L) \# o(K) \# o(M) \\
\simeq \ & o(L \# K \# M) \overset{(3.8)}{\simeq} o\big((\mathbb{1} \# \psi \# \mathbb{1})(L \# K \# M)\big) \\
\overset{(3.7)}{\simeq} \ & o(L) \# o\big(\psi(K)\big) \# o(M) \ .
\end{aligned}$$

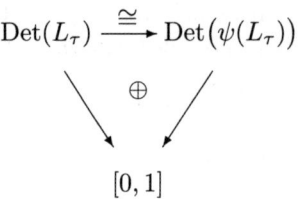

Considering all these reduction steps, we may start without loss of generality from the special situation

$$K = \frac{\partial}{\partial t} + \mathbb{1} \quad \text{and} \quad \psi(t) = \begin{pmatrix} \cos\varphi(t) & \sin\varphi(t) & & & \\ -\sin\varphi(t) & \cos\varphi(t) & & 0 & \\ & & 1 & & \\ & 0 & & \ddots & \\ & & & & 1 \end{pmatrix} ,$$

where $\varphi \in C^\infty(\overline{\mathbb{R}}, S^1)$ is asymptotically constant. We obtain the representation

$$\begin{aligned} \psi(K) &= \psi \cdot K \cdot \psi^{-1} = \frac{\partial}{\partial t} + \left(\psi(\dot{\psi^{-1}}) + \mathbb{1}\right) \\ &= \frac{\partial}{\partial t} + \mathbb{1} + \dot{\varphi} \cdot \begin{pmatrix} 0 & -1 & & & \\ 1 & 0 & & 0 & \\ & & 1 & & \\ & 0 & & \ddots & \\ & & & & 1 \end{pmatrix} \\ &= K + \Delta_\psi \ . \end{aligned}$$

Regarding the difference term Δ_ψ representing a compact perturbation of K, we calculate the estimate

$$\|\Delta_\psi\|_{\mathcal{L}(H^{1,2},L^2)} \leqslant \|\dot{\varphi}\|_{0,2} < \infty \ . \tag{3.9}$$

Given any $r > 0$, we define

$$\psi_r(t) = \psi(\frac{t}{r}) \ ,$$

so that the estimate (3.9) amounts to

$$\|\Delta_{\psi_r}\| \leqslant \frac{1}{r} \cdot \text{const} \ . \tag{3.10}$$

Therefore, provided that $r_0 > 0$ has been chosen large enough, the continuous arc

$$\begin{aligned} K_\tau^{r_0} : \ [0,1] &\to \mathcal{F}(H^{1,2}, L^2) \\ \tau &\mapsto K + \tau\, \Delta_{\psi_{r_0}} \end{aligned}$$

runs merely through isomorphism operators $H^{1,2} \xrightarrow{\cong} L^2$, because we already know that

$$K = \frac{\partial}{\partial t} + \mathbb{1} : H^{1,2} \xrightarrow{\cong} L^2$$

describes an isomorphism. Finally, we see that on one side the transformation ψ_{r_0} maps the orientation $[1\otimes 1^*]$ of K onto the orientation $[1\otimes 1^*]$ of $\psi_{r_0} K \psi_{r_0}^{-1}$. On the other side, we can deduce from the homotopy $K_\tau^{r_0}$ that both orientations correspond to each other by continuation. Hence, we obtain the relation

$$o_K \simeq o_{\psi_{r_0}(K)} = \psi_{r_0}(o_K) \ . \tag{3.11}$$

Regarding the continuous homotopy with respect to the variation of $r \in [1, r_0]$, we are led similarly to the correspondence of the orientations $\psi_{r_0}(o_K)$ and $\psi(o_K)$ in the determinant bundle $\operatorname{Det}\psi_r(K)$ on $[1, r_0]$. From this we conclude the assertion

$$\psi(o_K) \simeq o_K \ .$$

□

This result enables us to prove Lemma 3.8.

Proof of Lemma 3.8. Let us start from the orientation equivalence

$$o(\phi K \phi^{-1}) = \phi(o_K) \simeq \psi(o_L) = o(\psi L \psi^{-1}) \ . \tag{3.12}$$

Then we have to verify the equivalence of the orientations

$$\phi'(o_K) = \phi'\phi^{-1}\big(o(\phi K \phi^{-1})\big) \quad \text{and} \quad \psi'(o_L) = \psi'\psi^{-1}\big(o(\psi L \psi^{-1})\big) \ . \tag{3.13}$$

Moreover, on the one hand, we assume the asymptotical identities $\phi(-\infty) = \psi(-\infty)$ and $\phi'(-\infty) = \psi'(-\infty)$, so that

$$(\psi\psi'^{-1}\phi'\phi^{-1})(-\infty) = \mathbb{1} \in \mathrm{GL}(n, \mathbb{R}) \ .$$

Since on the other hand $\phi'\phi^{-1}$ and $\psi\psi'^{-1}$ describe curves within $\mathrm{GL}(n, \mathbb{R})$, the inequality $\det\big(\psi\psi'^{-1}\phi'\phi^{-1}\big)(+\infty) > 0$ follows. Thus, we are led to the identity

$$(\psi\psi'^{-1}\phi'\phi^{-1})(+\infty) = \mathbb{1} \in \mathrm{GL}(n, \mathbb{R}) \ .$$

Altogether, we are able to apply the auxiliary proposition and hence derive the equivalence

$$\phi(o_K) \simeq (\psi\psi'^{-1}\phi'\phi^{-1})\big(\phi(o_K)\big) \ . \tag{3.14}$$

To sum up, the assumption (3.12) implies the assertion (3.13). □

By means of this last lemma, we now are able to transfer the orientation concept from the trivial bundle as treated in the first section to the Fredholm operators

$$\{\, (u, K) \mid K \in \Sigma_{u^*TM} \,\}$$

in relation with the above equivalence relation:

Definition 3.10 *Any orientation of an* $L \in [(u,K)]$ *induces a unique orientation of the equivalence class* $[(u,K)]$, *that is to say that given* $K \sim L$ *the equivalence*

$$o_K \simeq o_L \quad \text{holds exactly if} \quad \phi(o_K) \simeq \psi(o_L)$$

holds for any admissible pair of trivializations (ϕ, ψ).

Remark If we defined the equivalence classes to be of smaller a priori extent by the condition that trivializations with coincidence at both ends should exist, we obviously would be provided with more classes to be oriented coherently in the case of a non-orientable manifold. As we shall see from later features within the construction of a Morse homology, however, we do not have that much freedom regarding the choice of orientations for operators on curves which may be linked together yielding a 'Möbius band'. Actually, the dimension axiom will give rise to a condition on coupling between the orientations of two such classes. The crucial point worth mentioning is that we have already built in this condition by demanding the coincidence of trivializations at the negative end within Definition 3.7 of equivalence which might have seemed rather unmotivated to the reader. Assuming that we had fixed the coincidence at the positive end of the curve, we would be led to the 'wrong' dimension axiom, as shall be illustrated below in Section 4.1.4 on the example of $\mathbb{P}^2(\mathbb{R})$.

Having accomplished the transfer to the non-trivial framework of varying curves, we now introduce a similar corresponding gluing operation by the trivialization method. We have to verify that this gluing is compatible with the trajectory gluing as regards the orientation induction by gluing.

Definition 3.11 *First, we once again define the pre-gluing operation from the last chapter for arbitrary curves* $u, v \in C^\infty(\overline{\mathbb{R}}, M)$ *satisfying* $u(+\infty) = v(-\infty)$. *Given* $\rho \geqslant \rho_0(u,v)$ *we define*

$$u\#^o_\rho v(t) = \begin{cases} u(t+\rho), & t \leqslant -1 \\ \exp_y \begin{pmatrix} \beta^-(t) \cdot \exp_y^{-1}\left(u(t+\rho)\right) \\ +\beta^+(t) \cdot \exp_y^{-1}\left(v(t-\rho)\right) \end{pmatrix}, & |t| \leqslant 1 \\ v(t-\rho), & t \geqslant 1\ . \end{cases}$$

Here, the lower parameter bound $\rho_0(u,v)$ *is again derived from the asymptotical condition*

$$u(t+\rho_0),\ v(t-\rho_0) \in U(y) \quad \text{for } t > -1 \text{ and } t < 1 \text{ respectively}$$

in relation with the normal neighbourhood $U(y)$ *of the exponential map.*

Let us now consider trivializations

$$\begin{aligned} \phi: \quad & u^*TM \xrightarrow{\cong} \overline{\mathbb{R}} \times \mathbb{R}^n\ , \\ \psi: \quad & v^*TM \xrightarrow{\cong} \overline{\mathbb{R}} \times \mathbb{R}^n\ , \end{aligned}$$

by starting first from a smooth trivialization

$$\Gamma : TM_{|\,U(y)} \xrightarrow{\cong} U(y) \times \mathbb{R}^n$$

on the neighbourhood $U(y)$. *This induces immediately* $\overline{\mathbb{R}}$*-smooth trivializations on the pull-back bundles* $u_{|\,(-1,\infty]}{}^*TM$ *and* $v_{|\,[-\infty,1)}{}^*TM$*, which may be extended to trivializations* ϕ *and* ψ*, such that the identities*

$$\Gamma_{|\,u((-1,\infty])} \equiv \phi_{|\,(-1,\infty]}, \quad \Gamma_{|\,v([-\infty,1))} \equiv \psi_{|\,[-\infty,1)}$$

are satisfied. Thus we are able to define a gluing operation for trivializations in a compatible way:

$$\phi\#^o_\rho\psi : (u\#^o_\rho v)^*TM \xrightarrow{\cong} \overline{\mathbb{R}} \times \mathbb{R}^n$$

$$\phi\#^o_\rho\psi(t) = \begin{cases} \phi(t+\rho), & t \leqslant -1 \\ \Gamma_{u\#^o_\rho v(t)}, & |t| \leqslant 1 \\ \psi(t-\rho), & t \geqslant 1 \ . \end{cases}$$

By means of these trivializations we shall transfer the gluing operation for Σ_{triv} to the pull-back $\overline{\mathbb{R}}$-bundles on the curves, but first, we have to introduce the condition of asymptotical constance.

Definition 3.12 *Given* $K \in \Sigma_{u^*TM}$ *and* $L \in \Sigma_{v^*TM}$*, we obtain*

$$K_\phi = \phi K \phi^{-1} \quad \text{and} \quad L_\psi = \psi L \psi^{-1} \in \Sigma_{\text{triv}} \ .$$

With regard to the time-shifting we notice the identity

$$\left(K_\phi\right)_\rho = \left(K_\rho\right)_{\phi_\rho} \ ,$$

because the $\mathbb{R}$*-action commutes with the partial derivative* $\frac{\partial}{\partial t}$*,*

$$\left(\dot{\phi}\right)_\rho = \left(\dot{\phi_\rho}\right) \ .$$

The next step is to define the halfway asymptotically constant operator associated to K_ϕ

$$K_{\phi\,\text{as}} = \frac{\partial}{\partial t} + \beta^- \cdot A_{K_\phi} + \beta^+ \cdot A^+_{K_\phi}$$

and $L_{\psi\,\text{as}}$ *analogously, having in mind the identity* $A^-_{L_\psi} \equiv A^+_{K_\phi}$*. Obviously, the following relations are true:*

$$K_\phi \simeq K_{\phi\,\text{as}} \quad \text{and} \quad L_\psi \simeq L_{\psi\,\text{as}} \quad \text{in} \quad \Theta_{K_\phi} \ \text{and} \ \Theta_{L_\psi}, \ \text{respectively.}$$

Hence, the gluing operation already known on the trivial bundle for the operator $K_{\phi\,\text{as}}\,\#_\rho L_{\psi\,\text{as}}$ *induces the isomorphism*

$$\operatorname{Det} K_{\phi\,\text{as}} \otimes \operatorname{Det} L_{\psi\,\text{as}} \xrightarrow{\cong} \operatorname{Det}\left(K_{\phi\,\text{as}} \# L_{\psi\,\text{as}}\right) \ .$$

Thus, defining

$$K_\phi \#_\rho^o L_\psi = K_{\phi\,\mathrm{as}} \#_\rho L_{\psi\,\mathrm{as}} \quad \textit{for} \quad \rho \geqslant \rho_0 = 1 \ ,$$

we obtain an orientation of $K_\phi \#_\rho^o L_\psi$ *induced by* $\#_\rho^o$ *in a way which is independent of the construction of* $K_{\phi\,\mathrm{as}}$ *and* $L_{\psi\,\mathrm{as}}$ *based respectively on* K_ϕ *and* L_ψ*, due to the homotopy invariance described in the last section. Moreover, since there is a* $T \geqslant 0$ *such that*

$$K_\phi \#_\rho^o L_\psi(t) = \begin{cases} (K_\phi)_\rho(t), & t \leqslant -T \\ (L_\psi)_{-\rho}(t), & t \geqslant T \end{cases}$$

holds for all $\rho \geqslant \rho_0$*, the class* $\Theta_{K_\phi \#_\rho^o L_\psi}$ *is also independent of the noncanonical construction of* $K_{\phi\,\mathrm{as}}$ *and* $L_{\psi\,\mathrm{as}}$*. Finally, we may transform this gluing construction back to the non-trivial situation by means of the analogously introduced trivialization* $\phi \#_\rho^o \psi$*,*

$$K \#_\rho^o L = (\phi \#_\rho^o \psi)^{-1} (K_\phi \#_\rho^o L_\psi)(\phi \#_\rho^o \psi) \in \Sigma_{(u \#_\rho^o v)^* TM} \ .$$

We notice again that, due to homotopy invariance, the class $[u \#_\rho^o v, K \#_\rho^o L]$ and the associated orientation induced by $\#_\rho^o$ from $o([u, K])$ and $o([v, L])$ are independent of all the above noncanonical construction elements. Summing up, we consider the noncanonical gluing operation

$$\begin{aligned} \#^o \quad : \quad & [u, K^-, K^+] \times [v, L^-, L^+] \to [u \#_\rho^o v, K^-, L^+] = [u, K^\pm] \# [v, L^\pm] \\ & (u, K) \#_\rho^o (v, L) = (u \#_\rho^o v, K \#_\rho^o L) \ , \end{aligned}$$

which induces a unique and canonical mapping of orientations

$$\big(o[u, K],\, o[v, L]\big) \mapsto o[u, K] \# o[v, L]$$

with regard to the determinant bundles

$$\mathrm{Det}\big([u, K]\big) \to [u, K], \quad \text{etc.}$$

Due to the construction, the homotopy invariance and particularly the associativity may be deduced from the former trivial framework:

$$(3.15) \quad \big(\, o[u, K] \# o[v, L] \,\big) \# o[w, M] \;=\; o[u, K] \# \big(\, o[v, L] \# o[w, M] \,\big) \ .$$

We shall round off this section by verifying that this orientation concept in relation to the gluing of arbitrary smooth curves $u \in C^\infty(\overline{\mathbb{R}}, M)$ and Fredholm operators $K \in \Sigma_{u^*TM}$ includes our former trajectories as a special case. Along these trajectories we consider the surjective linearizations in local coordinates

$$D_u = DF_u(0) : \, H^{1,2}(u^*TM) \to L^2(u^*TM), \; u \in \mathcal{M}_{x,y} \subset C^\infty_{x,y} \ ,$$

that is, the differentials of

$$F_u = \nabla_2 \exp_u^{-1} \circ \left(\frac{\partial}{\partial t} + \nabla f\right) \circ \exp_u : H^{1,2}(u^*\mathcal{D}) \to L^2(u^*TM) \ .$$

In other words, the gluing $\#^o$ defined above is compatible with trajectory gluing as far as the orientation induction is concerned.

Lemma 3.13 *Regarding the gluing operation for time-independent trajectories*

$$\mathcal{M}_{x,y} \times \mathcal{M}_{y,z} \ni (u,\, v) \mapsto u\#_\rho v \in \mathcal{M}_{x,z}$$

endowed with fixed orientations $o[u, D_u]$ *and* $o[v, D_v]$, *we observe that the isomorphism*

$$D\#_\rho : \ker D_u \times \ker D_v \xrightarrow{\cong} \ker D_{u\#_\rho v}$$

induces the same orientation $o[u\#_\rho v, D_{u\#_\rho v}]$ *as we obtain from*

$$o[u, D_u]\#o[v, D_v] \ .$$

Additionally, the identity of the classes

$$[u\#_\rho v, D_{u\#_\rho v}] = [u\#_\rho^o v, D_u\#_\rho^o D_v]$$

holds.

Proof. Let us assume any fixed pair of trajectories together with compact neighbourhoods $N(u) \subset \mathcal{M}_{x,y}$ and $N(v) \subset \mathcal{M}_{y,z}$. Given the compact set $K = N(u) \times N(v)$ we choose a fixed gluing parameter $\rho \geqslant \rho(K)$. Then the glued trajectory is represented as

$$u\#_\rho v = \exp_{u\#_\rho^o v} \gamma_K(u, v, \rho) \ .$$

Let us now consider

$$\begin{aligned} H: \quad K \times [0,1] \quad &\to C^\infty_{x,z}(\overline{\mathbb{R}}, M) \subset \mathcal{P}^{1,2}_{x,z} \\ (u, v, \tau) \quad &\mapsto \exp_{u\#_\rho^o v}\big(\tau \cdot \gamma_K(u, v, \rho)\big) \end{aligned}$$

yielding the homotopy $h_\tau = H(\,\cdot\,, \tau)$ from $\#_\rho^o$ to $\#_\rho$. Provided that the gluing parameter ρ has been chosen large enough, the differential

$$Dh_\tau(u, v) : \ker D_u \times \ker D_v \to H^{1,2}(h_\tau(u, v)^*TM)$$

proves injective for all $\tau \in [0,1]$. This follows from the deductions in the last chapter in the same way as we obtain the surjectivity of $D_{h_\tau(u,v)}$ and the isomorphism relation

$$\text{(3.16)} \qquad \mathrm{Proj}^{L^2}_{\ker D_{h_\tau}} \circ Dh_\tau : \ker D_u \times \ker D_v \xrightarrow{\cong} \ker D_{h_\tau}$$

for all $\tau \in [0,1]$. As a consequence, the isomorphism

$$D\#_\rho = Dh_1 = \mathrm{Proj}^{L^2}_{\ker Dh_1} \circ Dh_1$$

induces the same orientation on

$$[\,u\#_\rho v,\, D_{u\#_\rho v}\,] = [\,u\#^o_\rho v,\, D_{u\#^o_\rho v}\,]$$

as

$$\mathrm{Proj}^{L^2}_{\ker Dh_0} \circ Dh_0 = \mathrm{Proj}^{L^2}_{\ker D\#^o_\rho} \circ D\#^o_\rho \ .$$

According to the construction of the pre-gluing map in the last chapter we can find admissible trivializations of the pull-back bundles u^*TM and v^*TM, such that we are led to the linear mapping

$$\begin{array}{llll}\Phi: & \ker D_{u,\mathrm{triv}} \times \ker D_{v,\mathrm{triv}} & \xrightarrow{\cong} & \ker D_{u\#^o_\rho v,\mathrm{triv}} \\ & (\,\xi,\,\zeta\,) & \mapsto & \mathrm{Proj}^{L^2}_{\ker D_{u\#^o_\rho v,\mathrm{triv}}}(\beta^-\xi_\rho + \beta^+\zeta_{-\rho})\end{array}$$

on $\overline{\mathbb{R}} \times \mathbb{R}^n$. We denote the operators from Σ_{triv} by

$$K = D_{u,\mathrm{triv}} \text{ and } L = D_{v,\mathrm{triv}}$$

throughout the remainder of the proof. It remains to show that we can homotope this trivialized version Φ continuously to the gluing version used in Definition 3.12 above. Actually, we may construct suitable homotopies from K and L to respective halfway asymptotically constant operators K_{as} and L_{as}, which exclusively go through surjective operators.[5] Hence, we find a homotopy from the map Φ to the isomorphism

$$\begin{array}{lll}\ker K_{\mathrm{as}} \times \ker L_{\mathrm{as}} & \xrightarrow{\cong} & \ker K_{\mathrm{as}}\#_\rho L_{\mathrm{as}} \\ (\,\xi,\,\zeta\,) & \mapsto & \mathrm{Proj}^{L^2}_{\ker}\,(\xi_\rho + \zeta_{-\rho})\ ,\end{array}$$

which similarly goes merely through isomorphisms. This is in fact possible, because the compact interval of time around 0, on which the curves $A_{K_{\mathrm{as}}\#_\rho L_{\mathrm{as}}}$ and $A_{D_{u\#^o_\rho v},\mathrm{triv}}$ from $C^\infty\big(\overline{\mathbb{R}}, \mathrm{End}(\mathbb{R}^n)\big)$ differ, decreases as the gluing parameter increases. The central element of this homotopy through isomorphisms is given by a continuation of the form

$$\tau \mapsto \big[(1-\tau)\,\beta^- + \tau\big]\ \xi_\rho\ +\ \big[(1-\tau)\,\beta^+ + \tau\big]\ \zeta_{-\rho}\ .$$

Actually, the proof that the corresponding linear gluing version is an isomorphism may be accomplished according to the scheme of Proposition 2.50 for each parameter τ. Hence we conclude the proof of the lemma. □

5Note that the surjective operators form an open subset such that suitable asymptotically constant operators can be found.

Remark Similarly, we may state a corresponding result for the gluing operation for mixed broken trajectories, now based on a linear gluing version of the type

$$(\xi, \zeta, \rho) \mapsto \beta^-_{-\rho} \cdot \xi + \beta^+_{-\rho} \cdot \zeta_{-2\rho} \quad \text{and} \quad \beta^-_\rho \cdot \xi_{2\rho} + \beta^+_\rho \cdot \zeta, \quad \text{respectively} \ .$$

We can reduce this result to Lemma 3.13 by means of a continuous variation of the form

$$(\xi, \zeta, \rho) \mapsto \beta^-_{\tau\rho} \cdot \xi_{(1+\tau)\rho} + \beta^+_{\tau\rho} \cdot \zeta_{(-1+\tau)\rho}, \quad \tau \in [-1, 1] \ .$$

Orientation and Gluing for λ-Parametrized Trajectories

Let $G^{\alpha\beta} : [0,1] \times \mathcal{P}^{1,2}_{x_\alpha, y_\beta} \to L^2(\mathcal{P}^{1,2}_{x_\alpha, y_\beta} {}^*TM)$ be the mapping

$$G^{\alpha\beta}(\lambda, \gamma) = \dot\gamma + \frac{1}{\sqrt{\cdots}} \nabla H^{\alpha\beta} \circ \gamma$$

associated to the λ-homotopy $H^{\alpha\beta} : [0,1] \times M \to \mathbb{R}$. Given the local coordinates $(\exp_u, \nabla_2 \exp_u)$ at $(\lambda, u_\lambda) \in \mathcal{M}^H_{x_\alpha, y_\beta} = {G^{\alpha\beta}}^{-1}(0)$, we consider the corresponding linearization

$$\begin{aligned} DG^{\alpha\beta}(\lambda, u_\lambda) &: T_\lambda[0,1] \times H^{1,2}(u^*TM) \to L^2(u^*TM) \\ (\tau, \xi) &\mapsto D_1 G^{\alpha\beta}(\lambda, u_\lambda) \cdot \tau + D_2 G^{\alpha\beta}(\lambda, u_\lambda) \cdot \xi \ . \end{aligned}$$

As we know from the transversality results, $DG^{\alpha\beta}(\lambda, u_\lambda)$ proves surjective for a generic metric. The Fredholm operator $D_{u_\lambda} = D_2 G^{\alpha\beta}(\lambda, u_\lambda) \in \Sigma_{u^*TM}$ with the index $\mu(x_\alpha) - \mu(y_\beta)$, however, does not need to be surjective at all. This can merely be provided as an a priori statement for the boundary parameters $\lambda = 0, 1$. Hence we conclude that $DG^{\alpha\beta}(\lambda, u_\lambda)$ is a Fredholm operator with index $\mu(x_\alpha) - \mu(y_\beta) + 1$ for all $(\lambda, u_\lambda) \in \mathcal{M}^H_{x_\alpha, y_\beta}$.

We now wish to compare the gluing embedding

$$\#^H_\rho : \mathcal{M}^H_{x_\alpha, y_\beta} \times \mathcal{M}^{f^\beta}_{y_\beta, z_\beta}, \ \mathcal{M}^{f^\alpha}_{x_\alpha, y_\alpha} \times \mathcal{M}^H_{y_\alpha, z_\beta} \supset \widehat{K} \hookrightarrow \mathcal{M}^H_{x_\alpha, z_\beta} \ \text{, resp.,}$$

for λ-parametrized trajectories with the more general gluing of Fredholm classes with regard to the orientation induction. At this point, we once again have to resort to the fundamental sequence Lemma 3.2. Dealing with the orientation problem is crucially simplified by the determinant bundle and the gluing with respect to classes of Fredholm operators $[K]$ always being led back to the respective surjectively extended operator $\widehat{K}_\psi : \mathbb{R}^k \times H^{1,2} \to L^2$ as far as the concrete construction is concerned. But this exactly concerns the relation between the operators $DG^{\alpha\beta} : \mathbb{R} \times H^{1,2} \to L^2$ and $D_2 G^{\alpha\beta} : H^{1,2} \to L^2$. Let us consider the following short exact sequence:

$$(3.17) \quad 0 \longrightarrow \ker D_2 G^{\alpha\beta} \xrightarrow{d_1} \ker DG^{\alpha\beta} \xrightarrow{d_2} T_\lambda[0,1] \xrightarrow{d_3} \operatorname{coker} D_2 G^{\alpha\beta} \longrightarrow 0$$

where

$$d_1(\xi) = (0, \xi)$$
$$d_2(\tau, \xi) = \tau$$
$$d_3(\tau) = \left[D_1 G^{\alpha\beta} \cdot \tau\right]_{\mathrm{R}(D_2 G^{\alpha\beta})} .$$

According to the fundamental Lemma 3.2, it induces a canonical isomorphism

$$\text{(3.18)} \qquad \begin{aligned} & \Lambda^{\max}\ker D_2 G^{\alpha\beta}(\lambda, u_\lambda) \otimes T_\lambda[0,1] \\ \xrightarrow{\cong} \ & \Lambda^{\max}\ker DG^{\alpha\beta}(\lambda, u_\lambda) \otimes \Lambda^{\max}\mathrm{coker}\, D_2 G^{\alpha\beta}(\lambda, u_\lambda) \end{aligned}$$

for all $(\lambda, u_\lambda) \in \mathcal{M}^H_{\dots}$. This amounts to a canonical bundle isomorphism

$$\mathrm{Det}\, D_2 G^{\alpha\beta} \otimes T_*[0,1] \xrightarrow{\cong} \Lambda^{\max}\ker DG^{\alpha\beta}$$

on the components of the parameter-trajectory spaces $\mathcal{M}^H_{\dots}$. Thus, the choice of the fixed orientation $\frac{\partial}{\partial\lambda} \equiv 1$ on $T_*[0,1] \cong \mathbb{R}$ gives rise to natural isomorphisms of the above line bundles on these manifolds of λ-parametrized trajectories

$$\text{(3.19)} \qquad \Omega : \mathrm{Det}\, D_2 G^{\alpha\beta} \xrightarrow{\cong} \Lambda^{\max}\ker DG^{\alpha\beta} \ .$$

Given any pair $(\lambda, u_\lambda) \in \mathcal{M}^H_{x_\alpha, y_\beta}$, Ω uniformly induces an identification of an orientation of the operator $D_{u_\lambda} \in \Sigma_{u_\lambda^* TM}$ with an orientation of the tangent space $T_{(\lambda, u_\lambda)}\mathcal{M}^H_{x_\alpha, y_\beta}$.

Proposition 3.14 *Let us consider the gluing operation for λ-parametrized trajectories*

$$\mathcal{M}^H_{x_\alpha, y_\beta} \times \mathcal{M}^{f^\beta}_{y_\beta, z_\beta} \ni \big((\lambda, u_\lambda), v\big) \mapsto (\lambda, u_\lambda) \#^H_\rho v = (\tilde\lambda, w_{\tilde\lambda}) \in \mathcal{M}^H_{x_\alpha, z_\beta}$$

together with given orientations $o[u_\lambda, D_{u_\lambda}]$ and $o[v, D_v]$. Then, the isomorphism

$$D\#^H_\rho : T_{(\lambda, u_\lambda)}\mathcal{M}^H_{x_\alpha, y_\beta} \times T_v \mathcal{M}^{f^\beta}_{y_\beta, z_\beta} \xrightarrow{\cong} T_{(\lambda, u_\lambda)\,\#^H_\rho v}\mathcal{M}^H_{x_\alpha, z_\beta}$$

in connection with the isomorphisms

$$\begin{aligned} \Omega(\lambda, u_\lambda) \ &: \ \mathrm{Det}\, D_{u_\lambda} \xrightarrow{\cong} \Lambda^{\max} T_{(\lambda, u_\lambda)}\mathcal{M}^H_{x_\alpha, y_\beta}, \\ \Omega\big((\lambda, u_\lambda)\,\#^H_\rho v\big) \ &: \ \mathrm{Det}\, D_{w_{\tilde\lambda}} \xrightarrow{\cong} \Lambda^{\max} T_{(\lambda, u_\lambda)\,\#^H_\rho v}\mathcal{M}^H_{x_\alpha, z_\beta} \end{aligned}$$

induces the same orientation $o[w_{\tilde\lambda}, D_{w_{\tilde\lambda}}]$ as we obtain from the gluing $o[u_\lambda, D_{u_\lambda}] \# o[v, D_v]$. Here we use the identity of the classes of glued operators

$$[w_{\tilde\lambda}, D_{w_{\tilde\lambda}}] = [u_\lambda \#^o_\rho v, D_{u_\lambda} \#^o_\rho D_v] \ .$$

Proof. By analogy with the proof of Lemma 3.13 we regard homotopies and trivializations at suitably large gluing parameters, so that we can reduce the assertion to the following special case.

Let $K, L \in \Sigma_{\text{triv}}$ be halfway asymptotically constant with matching ends $K^+ = L^-$. We further assume that L is already surjective and that $\psi : \mathbb{R} \to L^2$ is a linear mapping, such that

$$\begin{aligned} \widehat{K}_\psi : \mathbb{R} \times H^{1,2} &\to L^2 \\ (\tau, \xi) &\mapsto \psi \cdot \tau + K \cdot \xi \end{aligned}$$

is surjective, too. Then, defining $\widehat{K}_\psi \#^H_\rho L : \mathbb{R} \times H^{1,2} \to L^2$ by

$$\begin{aligned} (\widehat{K}_\psi \#^H_\rho L)(a, \xi) &= \psi \cdot a + K \#^h_\rho L \cdot \xi \\ \text{with} \quad A_{K \#^h_\rho L}(t) &= \begin{cases} A_K(t), & t \leqslant \rho \\ A_L(t - 2\rho), & t \geqslant \rho \end{cases} \end{aligned}$$

and choosing ρ large enough, we deduce that

$$\begin{aligned} \vartheta_\rho : \ker \widehat{K}_\psi \times \ker L &\xrightarrow{\cong} \ker (\widehat{K}_\psi \#^H_\rho L) \\ (a, \xi, \zeta) &\mapsto \operatorname{Proj}^{\mathbb{R} \times L^2}_{\widehat{K}_\psi \#^H_\rho L} (a, \xi + \zeta_{-2\rho}) \end{aligned} \tag{3.20}$$

is an isomorphism. The proof is accomplished by exactly the same steps as are presented in [F-H] for the more general version of the gluing of asymptotically constant Fredholm operators. By means of the homotopy of the form

$$(\xi, \zeta) \mapsto \xi_{(1-\tau)\rho} + \zeta_{(-1-\tau)\rho}, \quad \tau \in [0, 1] \ ,$$

we are able to transform the actual linear version of gluing into the shape used in Proposition 3.6 for any lower parameter bound ρ_1 that is large enough. Therefore, without loss of generality, we may start from the version

$$\begin{aligned} \vartheta_\rho : \ker \widehat{K}_\psi \times \ker L &\xrightarrow{\cong} \ker (\widehat{K}_\psi \#_\rho L) \\ ((a, u), v) &\mapsto P_\rho (a, u_\rho + v_{-\rho}) \end{aligned} \tag{3.21}$$

and compare it to the linear gluing version

$$\phi_\rho : \ker \widehat{K}_\psi \times \ker \widehat{L}_\psi \xrightarrow{\cong} \ker (\widehat{K}_\psi \#_\rho \widehat{L}_\psi) \tag{3.22}$$

used in Proposition 3.6. Considering the isomorphism

$$\Lambda^{\max} \ker \widehat{K}_\psi \otimes \Lambda^{\max} \ker L \xrightarrow{\cong} \Lambda^{\max} \ker \widehat{K}_\psi \#_\rho L$$

induced by ϑ_ρ, we observe that due to the identities $\ker \widehat{K}_\psi = \ker K_\psi$ for

$$\begin{aligned} K_\psi : \mathbb{R} \times H^{1,2} &\to \mathbb{R} \times L^2 \\ (a, u) &\mapsto (0, \widehat{K}_\psi \cdot (a, u)) \end{aligned}$$

and $\ker \widehat{K}_\psi \#_\rho L = \ker (K\#_\rho L)_{\psi_\rho \oplus 0}$ we may also state it as

$$\Lambda^{\max}\ker K_\psi \otimes \Lambda^{\max}\ker L \xrightarrow{\cong} \Lambda^{\max}\ker (K\#_\rho L)_{\psi_\rho\oplus 0} \ .$$

Thus on the one hand, the choice of the canonical basis $1^* \in \mathbb{R}^*$ gives rise to the isomorphism

$$\text{(3.23)} \qquad \operatorname{Det}\vartheta_\rho : \operatorname{Det} K_\psi \otimes \operatorname{Det} L \xrightarrow{\cong} \operatorname{Det}(K\#_\rho L)_{\psi_\rho\oplus 0}$$

induced by ϑ_ρ by analogy with the deductions in Section 3.1.2. On the other hand, ϕ_ρ yields the isomorphism

$$\text{(3.24)} \qquad \operatorname{Det}\phi_\rho : \operatorname{Det} K_\psi \otimes \operatorname{Det} L_\psi \xrightarrow{\cong} \operatorname{Det}(K\#_\rho L)_{\psi_\rho\oplus\psi_{-\rho}} \ .$$

Using the natural isomorphisms $\operatorname{Det} K \xrightarrow{\cong} \operatorname{Det} K_\psi$, $\operatorname{Det} L \xrightarrow{\cong} \operatorname{Det} L_\psi$, etc., we are led to the result that $\operatorname{Det}\vartheta_\rho$ as well as $\operatorname{Det}\phi_\rho$ induces the same isomorphism $\operatorname{Det} K \otimes \operatorname{Det} L \xrightarrow{\cong} \operatorname{Det}(K\#_\rho L)$. Since the composition

$$\Lambda^{\max}\ker \widehat{K}_\psi = \Lambda^{\max}\ker K_\psi \overset{\mathrm{id}\otimes 1^*}{\longrightarrow} \Lambda^{\max}\ker K_\psi \otimes \mathbb{R}^* = \operatorname{Det} K_\psi \longrightarrow \operatorname{Det} K$$

is expressed exactly by the natural isomorphisms defined above,

$$\Omega^{-1} : \Lambda^{\max}\ker \widehat{K}_\psi \xrightarrow{\cong} \operatorname{Det} K \ ,$$

the proof is accomplished. □

Coherent Orientation

Up to now we have developed a concept which enables us to join Fredholm operators $K \in \Sigma_{u^*TM}$ along compact curves $u \in C^\infty(\overline{\mathbb{R}}, M)$ into equivalence classes, which may be oriented as one with consideration of the gluing operation. The last step toward the key feature of our intended homology theory consists of uniformly orienting all those operator classes in a way which is compatible with gluing, i.e. in a coherent way.

Definition 3.15 *Let*

$$\Lambda = \{\, [u,K] \mid u \in C^\infty(\overline{\mathbb{R}}, M),\ K \in \Sigma_{u^*TM} \,\}$$

be the set of the equivalence classes of Fredholm operators along smooth compact curves. Then any map σ, which associates an orientation $\sigma[u,K]$ of the corresponding determinant bundle $\operatorname{Det}([u,K]) \to [u,K]$ *to each class $[u,K] \in \Lambda$, is called a coherent orientation, if and only if it is compatible with the gluing operation*

$$\sigma[u,K]\#\sigma[v,L] = \sigma[u\#v, K\#L]$$

for all pairs $[u,K]$ *and* $[v,L]$, *which are admissible for gluing, that is, which are endowed with matching ends* $u(+\infty)=v(-\infty)$ *and* $K^+=L^-$. *Let us denote the set of all such coherent orientations by* $\mathcal{C}_\Lambda$. *Moreover, we are interested in the group*

$$\Gamma=\Big\{f\in\{+1,-1\}^\Lambda \mid f([u,K]\#[v,L])=f([u,K])\cdot f([v,L])\Big\}$$

with respect to pointwise multiplication.

By the following proposition we verify that this definition makes sense.

Proposition 3.16 *Coherent orientations exist, i.e.* $\mathcal{C}_\Lambda\neq\emptyset$, *and the group* Γ *acts freely and transitively on* $\mathcal{C}_\Lambda$ *by*

$$\begin{aligned}\Gamma\times\mathcal{C}_\Lambda &\to \mathcal{C}_\Lambda\\ (f,\sigma) &\mapsto f\,\sigma\\ (f,\sigma)\ \big([u,K]\big) &= f\big([u,K]\big)\cdot\sigma[u,K]\ .\end{aligned}$$

Proof. In the first part of the proof, we present an algorithm for constructing such an orientation. It is crucially related to the associativity verified above (see also [F-H]). Let $x_0\in M$ be an arbitrary point in M and let $A\in\operatorname{End}(T_{x_0}M)$ be non-degenerate and conjugated self-adjoint, for instance $x_0\in\operatorname{Crit}f$ and $A=H^2f(x_0)$ for a Morse function f. Then, we consider $u_0\in C^\infty(\overline{\mathbb{R}},M)$ to be the constant curve

$$u_0\equiv x_0=\text{const}$$

together with the operator

$$K_0=\frac{\partial}{\partial t}+A\in\Sigma_{u_0^*TM}\ ,$$

which is an isomorphism as $A\in C^\infty\big(\operatorname{End}(u_0^*TM)\big)$ is constant.[6] Thus, we are able to fix the orientation $1\otimes1^*$ for the determinant space $\operatorname{Det}K_0=\mathbb{R}\otimes\mathbb{R}^*$. Briefly, the class $[u_0,K_0]\in\Lambda$, called the anchoring class, is oriented by $[1\otimes1^*]$.

The next step is to consider the subsets

$$\begin{aligned} &S^-=\{\,[u,K]\in\Lambda\mid u(-\infty)=x_0\,\}\\ \text{and}\quad &S^+=\{\,[u,K]\in\Lambda\mid u(+\infty)=x_0\,\}\end{aligned}$$

of Λ and to choose an arbitrary orientation $\sigma[u,K]$ for each class $[u,K]\neq[u_0,K_0]=[u_0]$ from S^-. Considering the one-to-one correspondence $S^-\cong S^+$ induced by the reversal of the time-parametrization, that is,

$$[v^{-1},L^{-1}]\in S^-\quad\text{for}\quad[v,L]\in S^+$$

[6] $u_0=\text{const}$ also implies the identity $\nabla_t=\frac{\partial}{\partial t}$.

with

$$v^{-1}(t) = v(-t), \quad A_{L^{-1}}(t) = A_L(t) \ ,$$

we consequently obtain unique orientations on the classes $[v, L] \in S^+$ already fixed by the condition

$$\sigma[v^{-1}, L^{-1}] \# \sigma[v, L] = \sigma[u_0] \ .$$

Obviously, this proves consistent in relation to the anchoring class

$$\begin{aligned} &[u_0] \in S^- \cap S^+ \\ &\sigma[u_0] \# \sigma[u_0] = \sigma[u_0] \ . \end{aligned}$$

By means of these two types of classes from S^- and S^+ anchored at x_0 together with the associativity of gluing as far as the orientations are concerned, we are now able to accomplish the last step toward a coherent orientation. Considering an arbitrary $[u, K] \in \Lambda$, we find unique classes

$$[v_u, K_0, K^-] \in S^- \quad \text{with} \quad v_u(+\infty) = u(-\infty)$$

and

$$[w_u, K^+, K_0] \in S^+ \quad \text{with} \quad w_u(-\infty) = u(+\infty) \ .$$

Then, an orientation $\sigma[u, K]$ is determined uniquely by the condition

$$\sigma[v_u, K_0, K^-] \# \sigma[u, K] \# \sigma[w_u, K^+, K_0] = \sigma[u_0] \ ,$$

which is well-defined due to associativity. Moreover, associativity guarantees that this construction in fact provides a coherent orientation.

We now analyse the Γ-action: Given two arbitrary coherent orientations σ_1 and σ_2, we define

$$f : \Lambda \to \{\pm 1\} \quad \text{by} \quad \sigma_1 = f \cdot \sigma_2 \ .$$

Obviously, this is equivalent to

$$\sigma_2 = f \cdot \sigma_1 \ .$$

Hence, given two classes $[u, K]$ und $[v, L]$ admitted for gluing, we are led to the identity

$$\begin{aligned} & \sigma_1[u, K] \# \sigma_1[v, L] = \sigma_1[u \# v, K \# L] \\ = \ & (f\,\sigma_2)[u \# v, K \# L] = f[\ldots \# \ldots] \cdot \sigma_2[\ldots \# \ldots] \\ = \ & f([\ldots \# \ldots]) \cdot \big(\sigma_2[u, K] \# \sigma_2[v, L]\big) \\ = \ & f([\ldots \# \ldots]) \Big(f[u, K] \sigma_1[u, K] \# f[v, L] \sigma_1[v, L] \Big) \ . \end{aligned}$$

This amounts to

$$1 = f[u\#v, K\#L] \cdot f[u, K]\, f[v, L] \ .$$

Applied in both conclusion directions, this means that such a mapping $f : \Lambda \to \{\pm 1\}$ represents an element of the group Γ, exactly if it transfers coherent orientations again to coherent orientations. Hence we conclude that the asserted group action is in fact well-defined and transitive. Finally, it is straightforward to verify that this action is free. □

Chapter 4

Morse Homology Theory

4.1 The Main Theorems of Morse Homology

We introduce Morse homology theory by two main theorems about the existence of a canonical boundary operator associated to a given Morse function f and about the existence of canonical isomorphisms between each pair of such Morse complexes. These theorems appear as the essence from the theory on compactness, gluing and orientation which has been developed throughout the last three chapters. It is the isolated trajectories of the time-independent, time-dependent and the λ-parametrized gradient flow which form the crucial features in the core of these main theorems. The following preparatory section describes how we associate respective characteristic signs to these isolated trajectories.

4.1.1 Canonical Orientations

Concerning the Fredholm classes $[u, K] \in \Lambda$ analysed in the last chapter, we henceforth denote for all trajectories u of the time-independent, time-dependent or λ-parametrized gradient flow by $[u]$ the class $[u, D_u]$, where D_u stands for the surjective[1] Fredholm operator yielded by the linearization of

$$\frac{\partial}{\partial t} + \nabla f, \quad \text{etc.,}$$

in local coordinates at u. Due to the surjectivity of these operators together with the representation

$$T_u \mathcal{M}_{x,y} \equiv \ker D_u \subset H^{1,2}(u^*TM)$$

[1] except for the λ-parametrized gradient flow; see below

for tangent spaces of the trajectory manifolds, an orientation of the class $[u]$ gives unique rise to a differential-topological orientation of the connected component of $\mathcal{M}_{x,y}$ containing u. Now, in contrast with the notion of a coherent orientation, we find the following canonical orientations for the classes $[u] \in \Lambda$ in the cases of isolated trajectories u:

- As to the *time-dependent* gradient flow, i.e. for isolated trajectories in $\mathcal{M}^{h^{\alpha\beta}}_{x_\alpha,x_\beta}$, $\ker D_u$ vanishes, so that we may orient the trivial line bundle $\operatorname{Det}[u] = \mathbb{R} \otimes \mathbb{R}^*$ canonically by $1 \otimes 1^*$.
- In the case of the *time-independent* gradient flow, i.e. $u \in \mathcal{M}^f_{x,y}$, $\ker D_u$ turns out to be of dimension 1. We find a canonical orientation by
$$0 \neq \dot{u} = -\nabla f \circ u \in \ker D_u \ ,$$
because these solutions of an autonomous differential equation are endowed with a one-dimensional shifting-invariance. This orientation $o([u])$ is briefly denoted by $[u_t]$.
- As to the λ-*parametrized* gradient flow, we consider the natural bundle isomorphism from the last chapter
$$\Omega : \operatorname{Det} D_2 G^{\alpha\beta} \xrightarrow{\cong} \Lambda^{\max} \ker DG^{\alpha\beta} \ , \tag{4.1}$$
which was introduced on components of the parameter-trajectory spaces $\mathcal{M}^H_{x_\alpha,y_\beta}$, where
$$D_2 G^{\alpha\beta}(\lambda, u_\lambda) = D_{u_\lambda} \in \Sigma_{u_\lambda^* TM}, \ \ u \in \mathcal{M}^{H_\lambda}_{x_\alpha,y_\beta}$$
is generally not a surjective operator. Restricting the analysis to a relative Morse index $\mu(x_\alpha) - \mu(y_\beta) = -1$, this means $\dim\ker D_{u_\lambda} = 0$ and $\dim\operatorname{coker} D_{u_\lambda} = 1$ and therefore isolated λ-trajectories, we obtain by Ω the orientation $1 \otimes \big(D_1 G^{\alpha\beta} \cdot \frac{\partial}{\partial\lambda}\big)^*$ on $\operatorname{Det} D_2 G^{\alpha\beta} = \operatorname{Det} D_{u_\lambda}$ from the canonical orientation 1 on $\Lambda^{\max}\ker DG^{\alpha\beta} = \mathbb{R}$. Here, we apply the canonical identification
$$\operatorname{coker} D_2 G^{\alpha\beta}(\lambda, u_\lambda) \cong \mathrm{R}\big(D_1 G^{\alpha\beta}(\lambda, u_\lambda)\big) \ .$$
We deal equivalently with the canonical orientations
  - ▷ $1 \otimes 1^*$ on $T_{(\lambda,u_\lambda)}\mathcal{M}^H_{x_\alpha,y_\beta}$ and
  - ▷ $1 \otimes \big(D_1 G^{\alpha\beta}(\lambda, u_\lambda) \cdot \frac{\partial}{\partial\lambda}\big)^*$ on $[u_\lambda, D_{u_\lambda}]$.

  If we consider the 1-dimensional components of $\mathcal{M}^H_{x_\alpha,y_\beta}$, i.e. $\mu(x_\alpha) = \mu(y_\beta)$, and regular λ-values, which means regular with respect to the projection map
$$\pi : \mathcal{M}^H_{x_\alpha,y_\beta} \to [0,1] \ ,$$

so that $D_2 G^{\alpha\beta}(\lambda, u_\lambda) = D_{u_\lambda}$ is onto, the natural isomorphism

$$\Omega(\lambda, u_\lambda) : \operatorname{Det} D_{u_\lambda} \xrightarrow{\cong} T_{(\lambda,u_\lambda)}\mathcal{M}^H_{x_\alpha,y_\beta}$$

identifies the canonical orientation $1 \otimes 1^*$ on $[D_{u_\lambda}]$ with that orientation on $\mathcal{M}^H_{x_\alpha,y_\beta}$ which is mapped onto the canonical orientation $\frac{\partial}{\partial\lambda}$ of $[0,1]$ by the projection map π. Deciding whether two orientations $[1 \otimes 1^*]_{u_{\lambda_1}}$ and $[1 \otimes 1^*]_{u_{\lambda_2}}$ of operator classes $[D_{u_{\lambda_i}}]$ at regular λ-values and within the same component of $\mathcal{M}^H_{x_\alpha,y_\beta}$ are equivalent amounts to analysing the pre-image of the projection map on the interval of λ-parameters; see also Figure 4.4 on page 146.

Definition 4.1 *Given any fixed coherent orientation σ as developed in the last chapter, we are able to associate characteristic signs $\tau_\sigma(u)$ to the isolated trajectories by comparing the above canonical orientations for $[u, D_u]$ with σ, with respect to the three different types of isolated trajectories, that is*

1. *for* $u \in \mathcal{M}^f_{x,y}$: $\quad \tau_\sigma(u)\,\sigma[u] = [u_t]$
2. *for* $u \in \mathcal{M}^{h^{\alpha\beta}}_{x_\alpha,x_\beta}$: $\quad \tau_\sigma(u)\,\sigma[u] = [1 \otimes 1^*]$
3. *for* $u \in \mathcal{M}^{H^{\alpha\beta}_\lambda}_{x_\alpha,y_\beta}$: $\quad \tau_\sigma(u)\,\sigma[u] = [1 \otimes (D_1 G^{\alpha\beta}(\lambda, u) \cdot \frac{\partial}{\partial\lambda})^*]$.

4.1.2 The Morse Complex

Definition 4.2 *Let $f \in C^\infty(\mathbb{R}, M)$ be a fixed Morse function and*

$$\operatorname{Crit}_k f = \{\, x \in \operatorname{Crit} f \mid \mu(x) = k \,\} \quad \textit{for} \quad 0 \leqslant k \leqslant n$$

the discrete subset of M containing the critical points of f with Morse index k. We define

$$C_k(f) = \operatorname{Crit}_k f \otimes \mathbb{Z}$$

as the free abelian group generated by the critical points with Morse index k, i.e. in particular $C_k(f) = 0$ for $k < 0$ or $k > n$.

Given a fixed coherent orientation $\sigma \in \mathcal{C}_\Lambda$ on M we now define the mapping

$$\langle \cdot, \cdot \rangle : \operatorname{Crit} f \times \operatorname{Crit} f \to \mathbb{Z}$$

by

$$\langle x, y \rangle^\sigma = \begin{cases} \sum\limits_{\hat{u} \in \widehat{\mathcal{M}}^f_{x,y}} \tau_\sigma(\hat{u}), & \textit{for} \quad \mu(x) - \mu(y) = 1\,, \\ 0, & \textit{otherwise}\,. \end{cases} \tag{4.2}$$

The pairing $\langle x, y \rangle^\sigma$ *amounts to counting the connecting orbits with relative Morse index 1 between* x *and* y *equipped with their characteristic signs from the above definition. Here it is worth mentioning that the characteristic number* $\langle x, y \rangle^\sigma$ *corresponds exactly to the geometrical intersection number of*

$$W^u(x) \pitchfork W^s(y)/\mathbb{R}$$

as described in the introduction, up to a suitable change of the noncanonical orientation of the respective critical points. This noncanonical ingredient is equivalent to the noncanonical choice of a coherent orientation. We shall illustrate this interrelation between the present approach to Morse homology and the classical approach by Thom, Smale, Milnor and Witten in more details in Appendix B.

Since $\mathrm{Crit}_k f$ *is the generating set of the group* $C_k(f)$ *for* $k = 0,\dots,n$, *we extend* $\langle \cdot, \cdot \rangle^\sigma$ *to*

$$\langle \cdot, \cdot \rangle^\sigma : C_*(f) \otimes C_*(f) \to \mathbb{Z} \tag{4.3}$$

and we define a homomorphism sequence $(\partial_k^\sigma)_{k=0,\dots,n}$ *associated to* $(C_k)_{k=0,\dots,n}$ *by*

$$\partial_k^\sigma : C_k(f) \to C_{k-1}(f)$$
$$\partial_k^\sigma x = \sum_{y \in \mathrm{Crit}_{k-1} f} \langle x, y \rangle^\sigma y \ .$$

This definition leads us to the first main theorem within this work on Morse homology:

Theorem 7 *Given any Morse function* f *and any coherent orientation* σ, *the family of homomorphisms*

$$\left(C_k(f), \partial_k^\sigma\right)_{k\in\mathbb{Z}} = \left(C_*(f), \partial^\sigma\right)$$

represents a chain complex, i.e. it holds that

$$\partial_{k-1}^\sigma \circ \partial_k^\sigma = 0 \ \text{for all}\ k = 0,\dots,n \ .$$

Within this monograph, we call this chain complex the Morse complex , in contrast to classical theory, which denotes by Morse complex the associated cellular complex[2].

Proof. Given any $x \in C_k(f)$, i.e. without loss of generality $x \in \mathrm{Crit}_k f$ and

2associated to f by homotopical equivalence with M, see [M1]

$k \geqslant 1$, straightforward computation according to the above definition leads to the equation

$$\begin{aligned}\partial^2 x &= \sum_{z\in\mathrm{Crit}_{k-2}f}\ \sum_{y\in\mathrm{Crit}_{k-1}f} \langle x, y\rangle^\sigma \langle y, z\rangle^\sigma z \\ &= \sum_{z\in\mathrm{Crit}_{k-2}f}\ \sum_{(\hat u,\hat v)\in\widetilde{\mathcal{M}}_{x,z}} \tau_\sigma(\hat u)\cdot\tau_\sigma(\hat v)\, z\ ,\end{aligned}$$

where we denote by $\widetilde{\mathcal{M}}_{x,z}$ the set of simply broken trajectories between x and z, as was defined in the gluing Section 2.5. Now, resorting to the cobordism equivalence, we may reorder the double sum on the right hand side, obtaining

$$\partial^2 x = \sum_{z\in\mathrm{Crit}_{k-2}f}\ \sum_{[(\hat u,\hat v)]\in\widetilde{\mathcal{M}}_{x,z}/\sim}\ \sum_{(\tilde u,\tilde v)\in[(\hat u,\hat v)]} \tau_\sigma(\tilde u)\cdot\tau_\sigma(\tilde v)\, z\ . \tag{4.4}$$

Since we already know that each of these equivalence classes contains exactly two different broken trajectories, we could deduce the assertion in the special case of $\mathbb{Z}_2$ coefficients from the identity

$$\sum_{(\tilde u,\tilde v)\in[(\hat u,\hat v)]} 1 \equiv 0 \bmod 2\ .$$

The analogous general step concluding the proof,

$$\sum_{(\tilde u,\tilde v)\in[(\hat u,\hat v)]} \tau_\sigma(\tilde u)\cdot\tau_\sigma(\tilde v) = 0\ , \tag{4.5}$$

will be proven in the following lemma by means of the interrelations between gluing, coherent and canonical orientation, i.e. briefly by oriented cobordism theory for simply broken trajectories. Thus, the theorem is concluded from (4.5). □

Lemma 4.3 *Let $x, y, y', z \in \mathrm{Crit}\, f$ be critical points satisfying*

$$\mu(x) - 1 = \mu(y) = \mu(y') = \mu(z) + 1\ ,$$

and let

$$(\hat u_1, \hat v_1) \in \widehat{\mathcal{M}}^f_{x,y} \times \widehat{\mathcal{M}}^f_{y,z},\ (\hat u_2, \hat v_2) \in \widehat{\mathcal{M}}^f_{x,y'} \times \widehat{\mathcal{M}}^f_{y',z}$$

be equivalent broken trajectories, i.e.

$$(\hat u_1, \hat v_1) \sim (\hat u_2, \hat v_2) \quad \textit{within } \widetilde{\mathcal{M}}^f_{x,z}\ .$$

Then, with respect to any arbitrary coherent orientation σ on M, the identity

$$\tau_\sigma(\hat u_1)\cdot\tau_\sigma(\hat v_1) = -\tau_\sigma(\hat u_2)\cdot\tau_\sigma(\hat v_2)$$

holds.

Proof. The cobordism-equivalence signifies per definition that both broken trajectories are mapped by the respective gluing operation $\#$ into the same component of the unparametrized trajectory space $\widehat{\mathcal{M}}_{x,z}$, which is diffeomorphic to $(-1,1)$. The strategy of proving this lemma is to compare the orientations which are induced by the gluing operation on this component from the respective canonical orientations. The following relation between the canonical orientations defined above has to be verified:

$$[\hat{u}_{1\,t}]\,\#\,[\hat{v}_{1\,t}] \;=\; -[\hat{u}_{2\,t}]\,\#\,[\hat{v}_{2\,t}] \;. \tag{4.6}$$

Then, according to the definition of $\tau_\sigma(\hat{u}_1)$, etc., the assertion is concluded from the coherence condition and the cobordism-equivalence:

$$\begin{array}{rcl} \sigma[\hat{u}_1]\,\#\,\sigma[\hat{v}_1] & \overset{\sigma\in\mathcal{C}_\Lambda}{=} & \sigma[\hat{u}_1\#\hat{v}_1] \\ & \overset{(\hat{u}_1,\hat{v}_1)\simeq(\hat{u}_2,\hat{v}_2)}{=} & \sigma[\hat{u}_2\#\hat{v}_2] \\ & \overset{\sigma\in\mathcal{C}_\Lambda}{=} & \sigma[\hat{u}_2]\,\#\,\sigma[\hat{v}_2] \;. \end{array}$$

Hence, we have to analyse how the canonical orientations behave with respect to simply broken trajectories in relation to the gluing of trajectories. We consider the diffeomorphism from the time-shifting action

$$\begin{array}{ccc} \widehat{\mathcal{M}}_{x,y}\times\mathbb{R} & \overset{\approx}{\longrightarrow} & \mathcal{M}_{x,y} \\ (\hat{u},\,\tau) & \mapsto & \hat{u}_\tau \end{array} \tag{4.7}$$

in relation to both versions of the trajectory gluing, for parametrized as well as for associated unparametrized trajectories. Let $(\hat{u},\hat{v})$ be a fixed unparametrized broken trajectory from the connected component $\widehat{\mathcal{M}}_{x,z}{}^o$, ρ_0 an appropriate gluing parameter and $\epsilon>0$ fixed, so that we obtain the following commutative diagram,

$$\begin{array}{ccc} \{\hat{u}\}_{(-\epsilon,\epsilon)}\times\{\hat{v}\}_{(-\epsilon,\epsilon)} & \xrightarrow{\;\#_{\rho_0}\;} & \mathcal{M}_{x,z}{}^o \\ \varphi\Big\uparrow & & \Big\downarrow\cdot/\mathbb{R} \\ (\rho_0-\epsilon,\rho_0+\epsilon) & \xrightarrow[\;\psi\;]{} & \widehat{\mathcal{M}}_{x,z}{}^o \end{array}$$

together with the embeddings

$$\begin{array}{rl} \varphi & :\rho\mapsto\left(\hat{u}_{(\rho-\rho_0)},\,\hat{v}_{(\rho_0-\rho)}\right) \\ \text{and}\quad\psi & :\rho\mapsto\widehat{\hat{u}\#_\rho\hat{v}}\;. \end{array}$$

Note that the appropriate choice of ρ_0 depends on ϵ. Now, let

$$\phi:\widehat{\mathcal{M}}_{x,z}{}^o\overset{\approx}{\to}(-1,\,1)$$

denote the diffeomorphism which endows the one-dimensional connected component with a fixed orientation vector, $e \in T_{\psi(\rho_0)}\widehat{\mathcal{M}_{x,z}}^o$, i.e.

$$D\phi \cdot e = \frac{\partial}{\partial s} \in T_{s_0}(-1,\, 1)\ .$$

As a consequence, time-shifting (4.7) and the orientation $[e]$ on $\widehat{\mathcal{M}_{x,z}}^o = \mathcal{M}_{x,z}{}^o/\mathbb{R}$ induce a fixed orientation $(\frac{\partial}{\partial \tau_\sigma},\, e)$ on $\mathcal{M}_{x,z}{}^o$. On the other hand, we may also consider the canonical orientations on

$$\{\hat{u}\}_{(-\epsilon,\epsilon)} \times \{\hat{v}\}_{(-\epsilon,\epsilon)}$$

by

$$\left((\dot{\hat{u}},\, \dot{\hat{v}}),\, (\dot{\hat{u}},\, -\dot{\hat{v}}) \right)\ ,$$

where the embedding φ identifies the pair of tangent vectors

$$(\dot{\hat{u}},\, -\dot{\hat{v}}) \in T_{(\hat{u}_0,\, \hat{v}_0)}\left(\mathcal{M}_{x,y} \times \mathcal{M}_{y,z}\right)$$

exactly with $\frac{\partial}{\partial \rho} \in T_{\rho_0}(\rho_0 - \epsilon,\, \rho_0 + \epsilon)$.

Referring to these fixed orientations, we now have to determine whether the embedding $\#_{\rho_0}$ acts in a preserving or in a reversing manner. According to the above commutative diagram, it identifies the pair of tangent vectors $(\dot{\hat{u}},\, -\dot{\hat{v}})$ with the pre-image of the vector $D\psi \cdot \frac{\partial}{\partial \rho}$ with regard to the differential of the projection map $\cdot/\mathbb{R}$. It likewise identifies the pair $(\dot{\hat{u}},\, \dot{\hat{v}})$ with the vector $\frac{\partial}{\partial \tau_\sigma} \in T_{\hat{u}\#_{\rho_0}\hat{v}}\mathcal{M}_{x,z}{}^o$ canonically induced by time-shifting. Hence, the question of orientation preserving or reversing by the gluing operation $\#_{\rho_0}$ is reduced to the comparison of the fixed vector $\frac{\partial}{\partial s}$ with the tangent vector on $(-1,1)$ which is induced by $\phi \circ \psi$ together with $\frac{\partial}{\partial \rho} \in T_{\rho_0}(\rho_0, -\epsilon,\, \rho_0 + \epsilon)$. Exactly at this stage we derive different characteristic signs for the two equivalent broken trajectories $(\hat{u}_1,\, \hat{v}_1)$ and $(\hat{u}_2,\, \hat{v}_2)$.

Assuming the convergence

$$\lim_{\rho \to \infty} \phi(\widehat{\hat{u}_1 \#_\rho \hat{v}_1}) = 1$$

and respectively

$$\lim_{\rho \to \infty} \phi(\widehat{\hat{u}_2 \#_\rho \hat{v}_2}) = -1$$

for a given fixed diffeomorphism ϕ, we observe that $\frac{\partial}{\partial \rho}$ is identified with

- $\frac{\partial}{\partial s}$ by $\psi_1(\cdot) = \widehat{\hat{u}_1 \#. \hat{v}_1}$ and ϕ and with
- $-\frac{\partial}{\partial s}$ by $\psi_2(\cdot) = \widehat{\hat{u}_2 \#. \hat{v}_2}$ and ϕ .

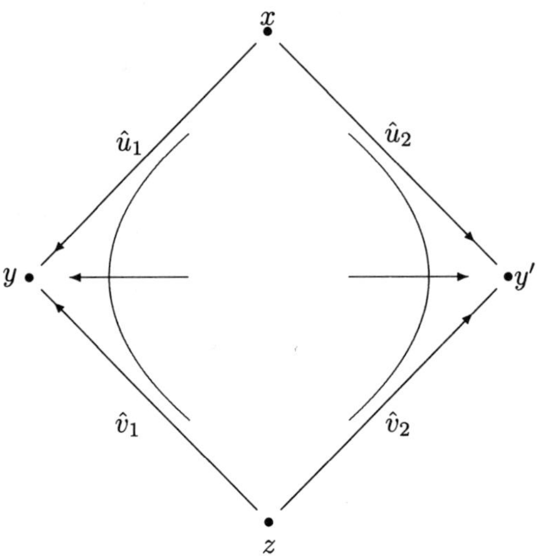

Figure 4.1: Cobordism-equivalence for ∂-trajectories

Thus, we deduce the relation asserted in (4.6) between the orientations induced by gluing of equivalent broken trajectories. This adverse orientation induction is illustrated in Figure 4.1. The arrows on the broken trajectories mark the canonical orientations $\big([\hat{u}_t], -[\hat{v}_t]\big)$, which under the gluing operation correspond to the respective orientation of the one-dimensional connected component $\widehat{\mathcal{M}}_{x,z}{}^o$ according to the respectively indicated weak convergence of unparametrized trajectories. □

4.1.3 The Canonical Isomorphism

Given two Morse functions, we now intend to derive a chain homomorphism between the associated chain complexes $C_*(f^\alpha)$ and $C_*(f^\beta)$, which gives rise to a canonical isomorphism between the respective homology groups. Actually, we will obtain this chain homomorphism in the same way as we extracted the canonical boundary operator from the negative gradient flow of a fixed Morse function. As to the isomorphism, 'canonical' means the independence from any further assumptions apart from f^α and f^β.

Definition 4.4 *Let f^α and f^β be any Morse functions on the manifold M, let $h^{\alpha\beta}$ be an arbitrarily associated Morse homotopy and let us choose any coherent orientation σ. By strict analogy with Definition 4.2, we define*

$$\langle \cdot, \cdot \rangle^\sigma_{\alpha\beta} : \big(C_k(f^\alpha)\big)_{k=0,\ldots,n} \times \big(C_k(f^\beta)\big)_{k=0,\ldots,n} \to \mathbb{Z}$$

by

$$(4.8)\qquad \langle x_\alpha, x_\beta \rangle^\sigma_{\alpha\beta} = \begin{cases} \sum\limits_{u\in\mathcal{M}^{h^{\alpha\beta}}_{x_\alpha,x_\beta}} \tau_\sigma(u), & \text{for } \mu(x_\alpha)=\mu(x_\beta)\,, \\ 0, & \text{otherwise} \end{cases}$$

for the generators $(x_\alpha, x_\beta) \in \operatorname{Crit} f^\alpha \times \operatorname{Crit} f^\beta$. *Once again, the compactness result for the trajectory space* $\mathcal{M}^{h^{\alpha\beta}}_{x_\alpha,x_\beta}$ *consisting of isolated trajectories guarantees the finiteness of this sum. Consequently, we define the group homomorphisms by analogy with the above* ∂_k, *that is*

$$\Phi_k(h^{\alpha\beta}) = \Phi_k^{\beta\alpha} : C_k(f^\alpha) \to C_k(f^\beta)$$
$$\Phi_k^{\beta\alpha} x_\alpha = \sum_{x_\beta\in\operatorname{Crit}_k f^\beta} \langle x_\alpha, x_\beta \rangle^\sigma_{\alpha\beta}\, x_\beta \ .$$

Now, the fact that the latter sum is finite is derived from the compactness results for Morse homotopies as was stated in Corollary 2.46. Note that this relies essentially on the coerciveness of the considered Morse functions.

The first step toward the second main theorem of Morse homology consists of the verification of the fundamental functorial relation for chain complex morphisms:

Proposition 4.5 *The family of morphisms*

$$\Phi_\bullet^{\beta\alpha} = \left(\Phi_k^{\beta\alpha} : C_k(f^\alpha) \to C_k(f^\beta)\right)_{k=0,\dots,n}$$

forms a chain homomorphism

$$\Phi_\bullet^{\beta\alpha} : C_*(f^\alpha \to C_*(f^\beta)\ ,$$

i.e. it holds

$$\partial_k^\sigma(f^\beta)\circ\Phi_k^{\beta\alpha} = \Phi_{k-1}^{\beta\alpha}\circ\partial_k^\sigma(f^\alpha) \ \text{for all } k=0,\dots,n\ .$$

Proof. The proof follows that of Theorem 7. Using the short notations

$$\partial^\alpha = \partial_k^\sigma(f^\alpha),\ \ \partial^\beta = \partial_k^\sigma(f^\beta)$$

we compute

$$\begin{aligned} &\left(\partial^\beta\circ\Phi^{\beta\alpha} - \Phi^{\beta\alpha}\circ\partial^\alpha\right)(x_\alpha) \\ =\ &\partial^\beta\Big(\sum_{\mu(x_\beta)=k}\ \sum_{u_{\alpha\beta}\in\mathcal{M}_{x_\alpha,x_\beta}} \tau_\sigma(u_{\alpha\beta})\,x_\beta\Big) - \Phi^{\beta\alpha}\Big(\sum_{\mu(y_\alpha)=k-1}\ \sum_{\hat u_\alpha\in\widehat{\mathcal{M}}_{x_\alpha,y_\alpha}} \tau_\sigma(\hat u_\alpha)\,y_\alpha\Big) \\ =\ &\sum_{\mu(y_\beta)=k-1} n(x_\alpha, y_\beta)\,y_\beta\ , \end{aligned}$$

where

$$n(x_\alpha,\, y_\beta) = \sum_{\mu(x_\beta)=k} \sum_{u_{\alpha\beta}\in\mathcal{M}^{h^{\alpha\beta}}_{x_\alpha,x_\beta}} \sum_{\hat{u}_\beta\in\widehat{\mathcal{M}}^{f^\beta}_{x_\beta,y_\beta}} \tau_\sigma(u_{\alpha\beta})\cdot\tau_\sigma(\hat{u}_\beta) - \sum_{\mu(y_\alpha)=k-1} \sum_{\hat{u}_\alpha\in\widehat{\mathcal{M}}^{f^\alpha}_{x_\alpha,y_\alpha}} \sum_{v_{\alpha\beta}\in\mathcal{M}^{h^{\alpha\beta}}_{y_\alpha,y_\beta}} \tau_\sigma(\hat{u}_\alpha)\cdot\tau_\sigma(v_{\alpha\beta}) .$$

We now prove the identity

$$n(x_\alpha,\, y_\beta) = 0. \tag{4.9}$$

in a way that is comparable with Lemma 4.3. The analysis of the cobordism-equivalence for mixed broken trajectories with fixed endpoints

$$(x_\alpha,\, y_\beta) \in \mathrm{Crit}_k f^\alpha \times \mathrm{Crit}_{k-1} f^\beta ,$$

this time requires a separate discussion of two different cases related to the order of the trajectories:

(a) $(u_{\alpha\beta},\, \hat{u}_\beta) \in \mathcal{M}^{h^{\alpha\beta}}_{x_\alpha,x_\beta} \times \widehat{\mathcal{M}}^{f^\beta}_{x_\beta,y_\beta}$ and

(b) $(\hat{u}_\alpha,\, v_{\alpha\beta}) \in \widehat{\mathcal{M}}^{f^\alpha}_{x_\alpha,y_\alpha} \times \mathcal{M}^{h^{\alpha\beta}}_{y_\alpha,y_\beta}$.

Starting from the broken trajectory $(u_{\alpha\beta},\, \hat{u}_\beta)$, we observe two different possibilities for cobordism-equivalence with regard to this distinction of cases:

(a) $(u_{\alpha\beta},\, \hat{u}_\beta) \sim (w_{\alpha\beta}, \hat{v}_\beta) \in \mathcal{M}_{x_\alpha,x'_\beta} \times \widehat{\mathcal{M}}_{x'_\beta,y_\beta}$ for a $x'_\beta \in \mathrm{Crit}_k f^\beta$. Analysing the orientation induction by the gluing operation for mixed broken trajectories, we are led to an identification of the canonical orientation

$$\big([1\otimes 1^*],\, [\hat{u}_{\beta,t}]\big)$$

on the component of $(v_{\alpha\beta},\, \hat{u}_\beta)$ in $\mathcal{M}_{x_\alpha,x_\beta} \times \widehat{\mathcal{M}}_{x_\beta,y_\beta}$ with that orientation on the one-dimensional component ${\mathcal{M}_{x_\alpha,y_\beta}}^o$, which corresponds to a decreasing parameter ρ regarding the gluing of broken trajectories. Actually, we observe this orientation correspondence at both ends, because we deal both times with broken trajectories of type (a) according to the above specification (see Figure 4.2). Hence, this case leads to the identity

$$\tau_\sigma(v_{\alpha\beta})\cdot\tau_\sigma(\hat{u}_\beta) = -\tau_\sigma(w_{\alpha\beta})\cdot\tau_\sigma(\hat{v}_\beta) . \tag{4.10}$$

(b) $(u_{\alpha\beta},\, \hat{u}_\beta) \sim (\hat{u}_\alpha,\, v_{\alpha\beta}) \in \widehat{\mathcal{M}}_{x_\alpha,y_\alpha} \times \mathcal{M}_{y_\alpha,y_\beta}$ for a $y_\alpha \in \mathrm{Crit}_{k-1} f^\alpha$. In the case of an equivalence with a mixed broken trajectory of type (b), this trajectory is endowed with the canonical orientation

$$\big([\hat{u}_{\alpha,t}],\, [1\otimes 1^*]\big) ,$$

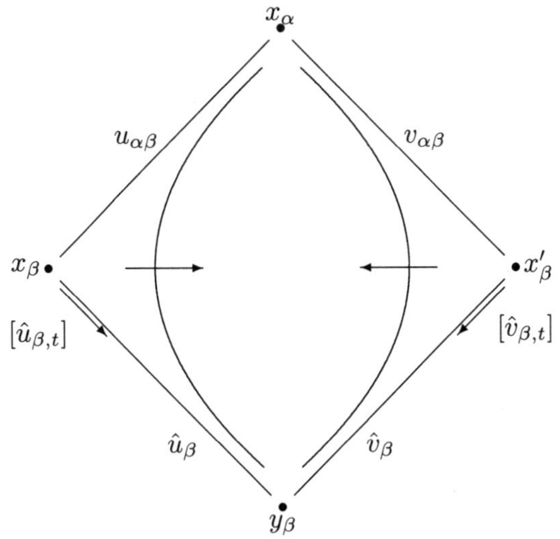

Figure 4.2: Cobordism-equivalence for mixed broken trajectories of type (a)-(a)

which now corresponds to an increasing gluing parameter (see Figure 4.3). Thus, the gluing operation induces at the cobordant broken trajectories of different type the same orientation on the associated one-dimensional component $\mathcal{M}_{x_\alpha,y_\beta}{}^o$. Hence, we verify the relation

$$\tau_\sigma(u_{\alpha\beta}) \cdot \tau_\sigma(\hat{u}_\beta) = \tau_\sigma(\hat{u}_\alpha) \cdot \tau_\sigma(v_{\alpha\beta}) \tag{4.11}$$

in this case.

As a consequence, we obtain products of characteristic signs with different signs in the case of mixed broken trajectories of the same type as in (4.10) and with equal signs in the case of different types as in (4.11). Since the cobordism-equivalence from the compactness-gluing-complementarity yields a one-to-one correspondence between mixed broken trajectories regardless of type, the above case distinction accomplishes the proof of the asserted identity

$$n(x_\alpha, y_\beta) = 0$$

and thus of the proposition, too. □

The next step now consists of verifying that the induced homomorphisms on the level of the homology groups,

$$\Phi_*^{\beta\alpha} : H_*(f^\alpha) \to H_*(f^\beta)$$

are in fact independent of the choice of the actual Morse homotopy $f^\alpha \overset{h^{\alpha\beta}}{\simeq} f^\beta$.

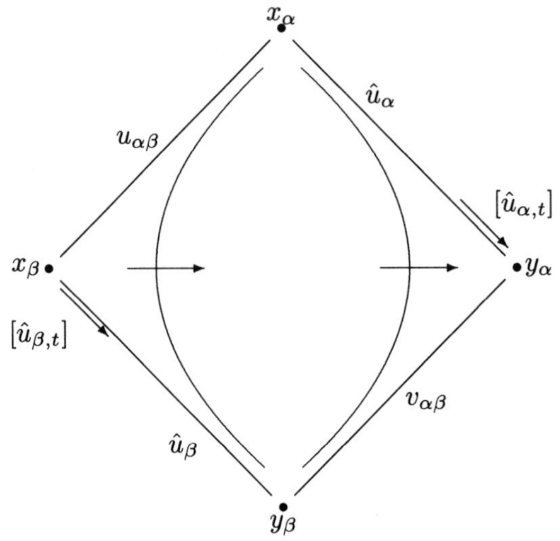

Figure 4.3: Cobordism-equivalence for mixed broken trajectories of type (a)-(b)

Proposition 4.6 *Let $h_0^{\alpha\beta}$ and $h_1^{\alpha\beta}$ be two regular Morse homotopies between f^α and f^β together with their associated chain morphisms $\Phi_0^{\beta\alpha}$ and $\Phi_1^{\beta\alpha}$. Then there is a family[3] of morphisms*

$$\Psi_k^{\beta\alpha} : C_k(f^\alpha) \to C_k(f^\beta),\ k \in \mathbb{Z}$$

satisfying the identities

$$\Phi_{1,k}^{\beta\alpha} - \Phi_{0,k}^{\beta\alpha} = \partial_{k+1}^{\beta} \circ \Psi_k^{\beta\alpha} - \Psi_{k-1}^{\beta\alpha} \circ \partial_k^{\alpha} \ . \tag{4.12}$$

In other words, $\Psi_*^{\alpha\beta}$ represents a chain homotopy between the chain complexes $C_*(f^\alpha)$ and $C_*(f^\beta)$.

Proof. At this stage we finally have to apply the results about transversality, compactness and orientation as far as the spaces of λ-parametrized trajectories are concerned. Let us choose a regular homotopy

$$h_0^{\alpha\beta} \overset{H_\lambda^{\alpha\beta}}{\simeq} h_1^{\alpha\beta}$$

with the homotopy parameter λ, so that, due to the regularity result in Theorem 2, we obtain the finite-dimensional manifold

$$\mathcal{M}_{x_\alpha,y_\beta}^{H^{\alpha\beta}} = \Big\{ (\lambda,\, u) \in [0,1] \times \mathcal{P}_{x_\alpha,y_\beta}^{1,2} \ \big|\ u \in \mathcal{M}_{x_\alpha,y_\beta}^{H^{\alpha\beta}}(\lambda) \Big\}$$

3We extend the finite family $C_*(f^\alpha)$ by $C_j(f^\alpha) = 0$ for $j < 0$ or $j > n$.

associated to the fixed endpoints $(x_\alpha, y_\beta) \in \operatorname{Crit} f^\alpha \times \operatorname{Crit} f^\beta$. Here, we use the brief notation $\mathcal{M}_{x_\alpha,y_\beta}(\lambda)$ for the trajectory space associated to the fixed parameter λ.[4] The ascending of the degree of the operator

$$\Psi_*^{\beta\alpha} : C_*(f^\alpha) \to C_{*+1}(f^\beta)$$

is due to the fact that, within this framework, we obtain isolated trajectories exactly if the relative Morse index is

$$\mu(x_\alpha) - \mu(y_\beta) = -1 \ .$$

Under this condition the finiteness of the set $\mathcal{M}_{x_\alpha,y_\beta}^{H^{\alpha\beta}}$ is provided by the respective compactness result. Again, these isolated trajectories can be counted with characteristic signs and we define

$$\Psi_k : C_k(f^\alpha) \to C_{k+1}(f^\beta)$$

for $k \in \mathbb{Z}$ on the generators $x_\alpha \in \operatorname{Crit} f^\alpha$ by

$$\begin{aligned} \Psi_k^{\alpha\beta} x_\alpha &= \sum_{\mu(z_\beta)=k+1} \langle x_\alpha, z_\beta \rangle z_\beta \\ \langle x_\alpha, z_\beta \rangle &= \sum_{(\lambda,u_\lambda)\in\mathcal{M}_{x_\alpha,z_\beta}^{H^{\alpha\beta}}} \tau_\sigma(u_\lambda) \ . \end{aligned}$$

Thus, we have to compute the difference of the terms

$$\partial^\beta \circ \Psi_k^{\beta\alpha}(x_\alpha) = \sum_{\mu(z_\beta)=k+1} \ \sum_{(\lambda,u_\lambda)\in\mathcal{M}_{x_\alpha,z_\beta}^{H^{\alpha\beta}}} \tau_\sigma(u_\lambda)\, \partial^\beta z_\beta$$

$$= \sum_{\mu(x_\beta)=k} \Big[\sum_{\mu(z_\beta)=k+1} \ \sum_{(\lambda,u_\lambda)\in\mathcal{M}_{x_\alpha,z_\beta}^{H^{\alpha\beta}}} \ \sum_{\hat{u}_\beta\in\widehat{\mathcal{M}}_{z_\beta,x_\beta}^{f^\beta}} \tau_\sigma(u_\lambda)\cdot\tau_\sigma(\hat{u}_\beta) \Big] x_\beta \tag{4.13}$$

and

$$\Psi_{k-1}^{\beta\alpha} \circ \partial^\alpha(x_\alpha) = \sum_{\mu(y_\alpha)=k-1} \ \sum_{\hat{u}_\alpha\in\widehat{\mathcal{M}}_{x_\alpha,y_\alpha}^{f^\alpha}} \tau_\sigma(\hat{u}_\alpha)\, \Psi^{\beta\alpha} y_\alpha$$

$$= \sum_{\mu(x_\beta)=k} \Big[\sum_{\mu(y_\alpha)=k-1} \ \sum_{\hat{u}_\alpha\in\widehat{\mathcal{M}}_{x_\alpha,y_\alpha}^{f^\alpha}} \ \sum_{(\lambda,v_\lambda)\in\mathcal{M}_{y_\alpha,x_\beta}^{H^{\alpha\beta}}} \tau_\sigma(\hat{u}_\alpha)\cdot\tau_\sigma(v_\lambda) \Big] x_\beta \tag{4.14}$$

with regard to the right hand side of equation (4.12) and

$$\left(\Phi_1^{\beta\alpha} - \Phi_0^{\beta\alpha}\right)(x_\alpha) = \sum_{\mu(x_\beta)=\mu(x_\alpha)} \Big[\sum_{u_1\in\mathcal{M}_{x_\alpha,x_\beta}^{h_1}} \tau_\sigma(u_1) - \sum_{u_0\in\mathcal{M}_{x_\alpha,x_\beta}^{h_0}} \tau_\sigma(u_0) \Big] x_\beta \tag{4.15}$$

4This is not necessarily a manifold!

with respect to the left hand side. In principle, all steps which have to be gone through are known from the above propositions as applications of cobordism-equivalence for trajectory spaces. In this framework, however, we have to apply our results on compactness, gluing and orientation to the 0- and 1-dimensional manifolds of the λ-parametrized trajectories. This leads to a slightly richer variety of possible cases to consider. This time, the connected components in question may also be manifolds with boundary. Hence, regarding the cobordism-equivalence, we have to take into consideration more separated cases. The sketch in Figure 4.4 suggests a geometrical visualization of the closed submanifold $\mathcal{M}^H_{x_\alpha,z_\beta} \subset [0,1] \times \mathcal{P}^{1,2}_{x_\alpha,z_\beta}$ for $\mu(x_\alpha) = \mu(z_\beta)$. There, the ends of the one-dimensional components, which have a boundary in the weak sense of convergence toward mixed broken trajectories, are indicated by open brackets ')'.

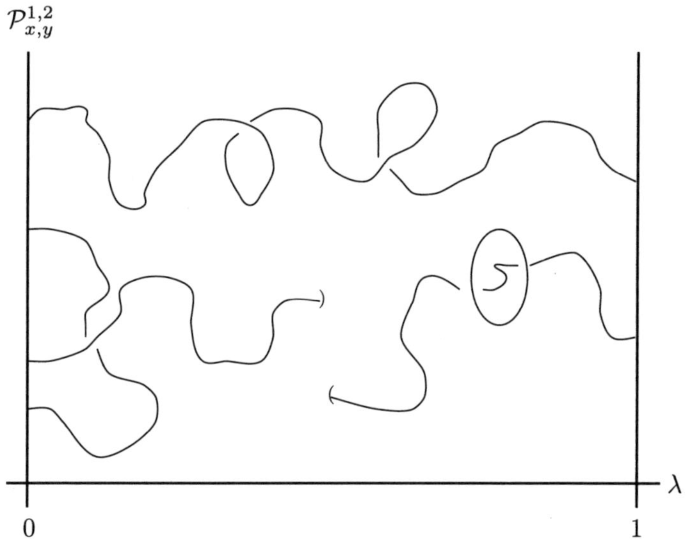

Figure 4.4: The 1-dimensional manifold with boundary of λ-parametrized trajectories

With regard to the components of the one-dimensional λ-parametrized trajectory spaces, we have to consider separately the following diffeomorphism types:

(a) $[0,1]$, with $\partial\mathcal{M}^H_{x_\alpha,x_\beta} \subset \mathcal{M}_{x_\alpha,x_\beta}(0) \cup \mathcal{M}_{x_\alpha,x_\beta}(1)$,

(b) S^1,

(c) $(-\infty,\infty)$,

(d) $[0,\infty)$ and $(-\infty,1]$.

Obviously, case (b) does not play a role in our cobordism-equivalence, because those closed components neither contain boundary points in the strong sense, that is, curves $u \in \mathcal{M}_{x_\alpha,x_\beta}(0) \cup \mathcal{M}_{x_\alpha,x_\beta}(1)$ as trajectories for the time-dependent gradient flows yielding $\mathcal{M}^{h_0}_{x_\alpha,x_\beta}$ and $\mathcal{M}^{h_1}_{x_\alpha,x_\beta}$, nor boundary points in the weak sense, that is, mixed broken trajectories, composed of trajectories $v_\lambda \in \mathcal{M}_{x',z'}(\lambda)$ satisfying $\mu(x') - \mu(z') = -1$ and respective ∂-trajectories. Concerning the oriented cobordism-equivalence for trajectory spaces, the gluing operation associates to each mixed broken trajectory

$$(u_\lambda,\, \hat{u}_\beta) \in \mathcal{M}_{x_\alpha,z_\beta}(\lambda) \times \mathcal{M}^{f^\beta}_{z_\beta,x_\beta} \quad \text{and} \quad (\hat{u}_\alpha,\, v_\lambda) \in \mathcal{M}^{f^\alpha}_{x_\alpha,y_\alpha} \times \mathcal{M}_{y_\alpha,x_\beta}(\lambda)$$

building up the terms in (4.13) and (4.14), a one-dimensional connected component of $\mathcal{M}^H_{x_\alpha,x_\beta}$ of type (c) or (d). Additionally, regarding the terms in (4.15), the trajectories $u_{\alpha\beta} \in \mathcal{M}_{x_\alpha,x_\beta}(0) \cup \mathcal{M}_{x_\alpha,x_\beta}(1)$ lead us to connected components with boundary in the strong sense, that is, of type (a) or (d). Hence, the discussion of the cobordism relations in the proof of Proposition 4.5 has to be extended by the cases (a) and (d) of manifolds with boundary.

As to (a): The connected component ${\mathcal{M}^H_{x_\alpha,x_\beta}}^o$ in question is diffeomorphic to the compact interval. As we analysed in the first section of this chapter, the natural isomorphism Ω induces the following orientations of the component ${\mathcal{M}^H_{x_\alpha,y_\beta}}^o$ from the canonical orientations $[1 \otimes 1^*]$ of the classes $[u_i]$, $u_i \in \mathcal{M}^{h_i}_{x,y}$, $i = 0, 1$, at the boundaries of the interval, $\lambda = 0, 1$:

- $\lambda = 0 \rightsquigarrow$ orientation toward the interior of the component,
- $\lambda = 1 \rightsquigarrow$ outward directed orientation.

Given the boundary curves

$$\partial\, {\mathcal{M}^H_{x_\alpha,x_\beta}}^o = \{\, u_{\alpha\beta},\, v_{\alpha\beta} \,\} \subset \mathcal{M}^{h_0}_{x_\alpha,x_\beta} \cup \mathcal{M}^{h_1}_{x_\alpha,x_\beta}\ ,$$

we consequently have to distinguish the following two cases:

1. Assume $\{\, u_{\alpha\beta},\, v_{\alpha\beta} \,\} \subset \mathcal{M}^{h_0}_{x_\alpha,x_\beta}$ or $\mathcal{M}^{h_1}_{x_\alpha,x_\beta}$. Then, in both cases, the canonical orientations of $u_{\alpha\beta}$ and $v_{\alpha\beta}$ give rise to opposite orientations on the one-dimensional connected component, such that the identity
$$\sigma[u_{\alpha\beta}] = \sigma[v_{\alpha\beta}]$$
implies the relation
$$\tau_\sigma(u_{\alpha\beta}) = -\tau_\sigma(v_{\alpha\beta})\ .$$
2. Considering the situation $u_{\alpha\beta} \in \mathcal{M}(\lambda = 0)$, $v_{\alpha\beta} \in \mathcal{M}(\lambda = 1)$, we obtain at both ends the same orientation on the component from the canonical orientations of the boundary points, so that we are led to the identity
$$\tau_\sigma(u_{\alpha\beta}) = \tau_\sigma(v_{\alpha\beta})\ .$$

Summing up these results, we may reduce the difference

$$\left(\Phi_1^{\beta\alpha} - \Phi_0^{\beta\alpha}\right)(x_\alpha)$$

from (4.15) to trajectories $u_{\alpha\beta} \in \mathcal{M}_{x_\alpha,x_\beta}(\lambda = 0, 1)$, which solely bound components ${\mathcal{M}^H_{x_\alpha,x_\beta}}^o$ of type (d).

As to (c): Considering connected components ${\mathcal{M}^H_{x_\alpha,x_\beta}}^o$ of the diffeomorphic type $(-\infty, \infty)$, we now have to deal with weak convergence toward mixed broken trajectories at both ends. The natural isomorphisms

$$\Omega(\lambda, u_\lambda) : \operatorname{Det} D_{u_\lambda} \xrightarrow{\cong} \Lambda^{\max} T_{(\lambda,u_\lambda)} \mathcal{M}^H_{\dots}$$

together with Proposition 3.14 enable us to reduce this case to the analysis in the proof of Proposition 4.5. Once again, we distinguish two types of cobordism-equivalence by the order in the mixed broken trajectories:

- $(u_\lambda,\, \hat{u}_\beta) \in \mathcal{M}_{x_\alpha,z_\beta}(\lambda) \times \widehat{\mathcal{M}}^{f^\beta}_{z_\beta,x_\beta}$ as in the right hand side of (4.13) and
- $(\hat{u}_\alpha,\, v_\lambda) \in \widehat{\mathcal{M}}^{f^\alpha}_{x_\alpha,y_\alpha} \times \mathcal{M}_{y_\alpha,x_\beta}(\lambda)$ as in (4.14).

The associated case distinction for the equivalence of broken trajectories is illustrated within Figures 4.5 and 4.6. We observe that the difference of both

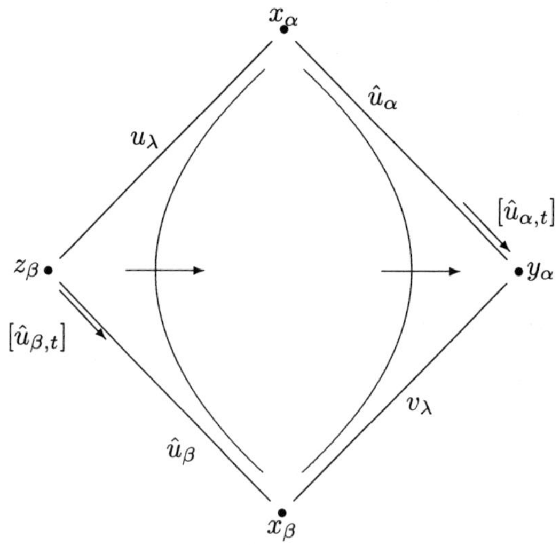

Figure 4.5: Equally directed induction of orientation

right hand sides (4.13) and (4.14) may also be reduced to the computation for

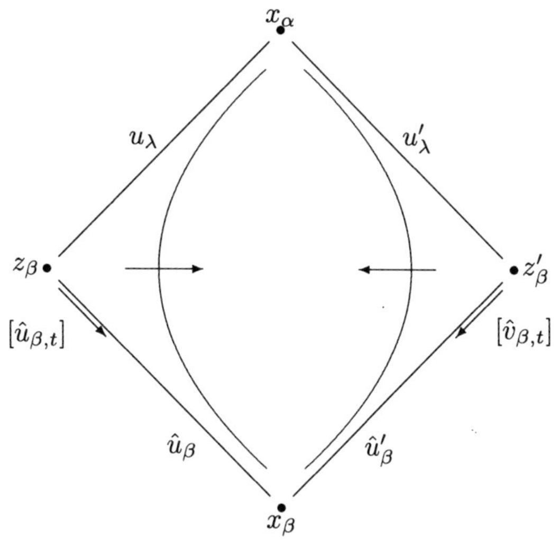

Figure 4.6: Induction of orientation in opposite directions

mixed broken trajectories, which the gluing operation associates to the one-dimensional components ${\mathcal{M}^H_{x_\alpha,x_\beta}}^o$ of type (d).

Thus, it remains to accomplish the analysis of the cobordism-equivalence for components of type (d), which, due to the discussions up to this point, establish the decisive link between the left hand side and the right hand side of the chain homotopy equation (4.12).

As to (d): On the one hand, as we know from the discussion of case (a), the canonical orientation $[1 \otimes 1^*]$ induces an inward pointing orientation on the entire component

$${\mathcal{M}^{H^{\alpha\beta}}_{x_\alpha,x_\beta}}^o \approx [0,\infty)$$

at boundary trajectories $u_0 \in \mathcal{M}_{x_\alpha,x_\beta}(\lambda = 0)$ and an outward pointing orientation at boundary trajectories $u_1 \in \mathcal{M}_{x_\alpha,x_\beta}(\lambda = 1)$. On the other hand, the case distinction for the orientation induction for mixed broken λ-trajectories referring to the canonical orientations and the natural isomorphism $\Omega(\lambda, u_\lambda)$ yields an inward pointing orientation for the order

$$(u_\lambda,\, \hat{u}_\beta) \in \mathcal{M}_{x_\alpha,z_\beta}(\lambda) \times \widehat{\mathcal{M}}^{f^\beta}_{z_\beta,x_\beta}$$

and an outward pointing orientation for the mixed broken trajectory with the order

$$(\hat{u}_\alpha,\, v_\lambda) \in \widehat{\mathcal{M}}^{f^\alpha}_{x_\alpha,y_\alpha} \times \mathcal{M}_{y_\alpha,x_\beta}(\lambda) \ .$$

An illustration is given by the sketches in Figure 4.7. Hence, in relation with

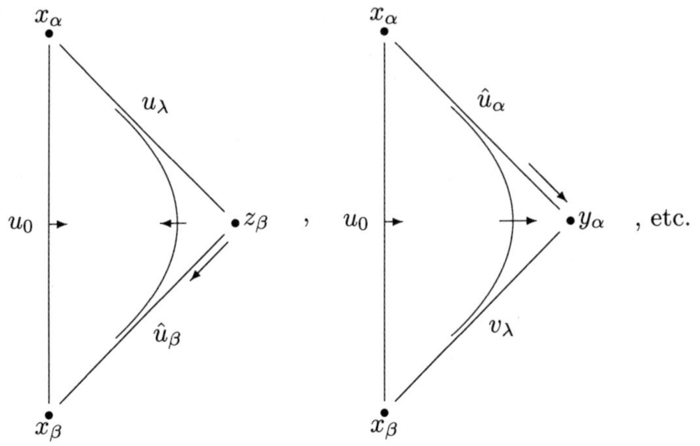

Figure 4.7: Cobordism-equivalence of type (d)

the boundary trajectories

$$u_0,\, u_1 \in \mathcal{M}_{x_\alpha, x_\beta}(\lambda = 0, 1)\ ,$$

we distinguish the following four combinations:

- $\tau_\sigma(u_\lambda) \cdot \tau_\sigma(\hat{u}_\beta) = -\tau_\sigma(u_0)$,
- $\tau_\sigma(\hat{u}_\alpha) \cdot \tau_\sigma(v_\lambda) = \tau_\sigma(u_0)$,
- $\tau_\sigma(u_\lambda) \cdot \tau_\sigma(\hat{u}_\beta) = \tau_\sigma(u_1)$ and
- $\tau_\sigma(\hat{u}_\alpha) \cdot \tau_\sigma(v_\lambda) = -\tau_\sigma(u_1)$.

We are finally able to deduce the asserted identity for $(x_\alpha,\, x_\beta) \in \mathrm{Crit}_k f^\alpha \times \mathrm{Crit}_k f^\beta$:

$$\begin{aligned}
& \sum_{\mu(z_\beta)=k+1} \ \sum_{(\lambda, u_\lambda) \in \mathcal{M}^H_{x_\alpha, z_\beta}} \ \sum_{\hat{u}_\beta \in \widehat{\mathcal{M}}^{f^\beta}_{z_\beta, x_\beta}} \tau_\sigma(u_\lambda) \cdot \tau_\sigma(\hat{u}_\beta) \\
- & \sum_{\mu(y_\alpha)=k-1} \ \sum_{\hat{u}_\alpha \in \widehat{\mathcal{M}}^{f^\alpha}_{x_\alpha, y_\alpha}} \ \sum_{(\lambda, v_\lambda) \in \mathcal{M}^H_{y_\alpha, x_\beta}} \tau_\sigma(\hat{u}_\alpha) \cdot \tau_\sigma(v_\lambda) \\
= & \sum_{u_1 \in \mathcal{M}_{x_\alpha, x_\beta}(1)} \tau_\sigma(u_1) \ - \sum_{u_0 \in \mathcal{M}_{x_\alpha, x_\beta}(0)} \tau_\sigma(u_0)\ .
\end{aligned}$$

□

Summing up this last analysis, we have obtained the result on the canonical homomorphism

$$\Phi_*^{\beta\alpha} : H_*^\sigma(f^\alpha) \to H_*^\sigma(f^\beta) \ ,$$

associated to given Morse functions f^α and f^β together with a fixed coherent orientation, by additionally taking into account the λ-parametrized trajectory spaces. The next proposition establishes the final independence of the homology groups from the choice of Morse function.

Proposition 4.7 *Each ordered triple of Morse functions f^α, f^β and f^γ together with the associated homomorphisms $\Phi_*^{\cdot\cdot}$ fulfills the composition rule*

$$\Phi_*^{\gamma\beta} \circ \Phi_*^{\beta\alpha} = \Phi_*^{\gamma\alpha} : H_*^\sigma(f^\alpha) \to H_*^\sigma(f^\gamma) \ .$$

Proof. At this stage we take advantage of the freedom of choosing an appropriate Morse homotopy, established by Proposition 4.6. Considering the homomorphisms $\Phi_*^{\gamma\beta}$ and $\Phi_*^{\beta\alpha}$ induced by $\Phi_\bullet^\sigma(h^{\beta\gamma})$ and $\Phi_\bullet^\sigma(h^{\alpha\beta})$, respectively, we now use the gluing construction for simply broken trajectories consisting exclusively of homotopy trajectories associated to $h^{\beta\gamma}$ and $h^{\alpha\beta}$. Due to Proposition 4.6, every Morse homotopy $h^{\alpha\gamma}(R)$ constructed in this way must yield the same homomorphism $\Phi_*^{\gamma\alpha}$. But in order to be able to apply the gluing result from corollary 2.61, we are only allowed to put a finite number of critical points x_α, x_β, x_γ into relation with each other by isolated h-trajectories. Hence, we generally cannot find a homotopy $h^{\alpha\gamma}$ such that the identity

$$\Phi_\bullet^\sigma(h^{\alpha\gamma}) \cdot c = \big(\Phi_\bullet^\sigma(h^{\beta\gamma}) \circ \Phi_\bullet^\sigma(h^{\alpha\beta})\big) \cdot c \tag{4.16}$$

is satisfied for all chains $c \in C_*(f^\alpha)$. Fortunately, this is not necessary with regard to the required result for the homology groups. It is sufficient to find a homotopy $h^{\alpha\gamma}\big(R(c_k)\big)$ according to the construction of the gluing for each fixed chain

$$c_k = \sum_{i=1}^{n} a_i x_{\alpha,i}^k \in C_k(f^\alpha), \ k \in \mathbb{Z} \ ,$$

so that relation (4.16) is satisfied. We consequently deduce the homological identity

$$\Phi_*^{\gamma\alpha}(\{c_k\}) = \{\Phi_\bullet^{\gamma\alpha}(c_k) \cdot c_k\} = \{\big(\Phi_\bullet^{\gamma\beta} \circ \Phi_\bullet^{\beta\alpha}\big)(c_k)\} = \big(\Phi_*^{\gamma\beta} \circ \Phi_*^{\beta\alpha}\big)(\{c_k\})$$

from Proposition 4.6 for each class

$$\{c_k\} \in H_k^\sigma(f^\alpha), \ k \in \mathbb{Z} \ .$$

Thus, without loss of generality, we may start from a fixed $c_k = x_\alpha \in \mathrm{Crit}_k f^\alpha$, so that there is only a finite number of pairs $(x_\beta, x_\gamma) \in \mathrm{Crit}_k f^\beta \times \mathrm{Crit}_k f^\gamma$ with

$$\mathcal{M}_{x_\alpha, x_\beta}^{h^{\alpha\beta}} \times \mathcal{M}_{x_\beta, x_\gamma}^{h^{\beta\gamma}} \neq \emptyset \ .$$

We set the constant

$$R = \max R(x_\alpha, x_\beta, x_\gamma)$$

according to Corollary 2.61. Thus, the identity

$$\begin{aligned}
&\sum_{x_\gamma}\sum_{x_\beta} \sum_{\substack{u_{\alpha\beta}\in\mathcal{M}_{x_\alpha,x_\beta}\\ u_{\beta\gamma}\in\mathcal{M}_{x_\beta,x_\gamma}}} \tau_\sigma(u_{\alpha\beta})\cdot\tau_\sigma(u_{\beta\gamma})\, x_\gamma \\
(4.17)\qquad = &\sum_{x_\gamma}\sum_{x_\beta} \sum_{\substack{u_{\alpha\beta}\in\mathcal{M}_{x_\alpha,x_\beta}\\ u_{\beta\gamma}\in\mathcal{M}_{x_\beta,x_\gamma}}} \tau_\sigma(u_{\alpha\beta}\#_R u_{\beta\gamma})
\end{aligned}$$

together with the bijection

$$\#_R : \mathcal{M}^{h^{\alpha\beta}}_{x_\alpha,x_\beta} \times \mathcal{M}^{h^{\alpha\beta}}_{x_\beta,x_\gamma} \xrightarrow{\cong} \mathcal{M}^{h^{\alpha\beta}}_{x_\alpha,x_\gamma}$$

from Corollary 2.61, proves the assertion. Here, the identity

$$\tau_\sigma(u_{\alpha\beta})\cdot\tau_\sigma(u_{\beta\gamma}) = \tau_\sigma(u_{\alpha\beta}\#_R u_{\beta\gamma}) = \tau_\sigma(u_{\alpha\gamma})$$

follows from the relation between gluing and induced orientation as already analysed above. □

Let us now sum up the results from the last three propositions in the second main theorem of Morse homology:

Theorem 8 *Given a fixed coherent orientation σ on the manifold M and two arbitrary Morse functions f^α and f^β, there is a canonical isomorphism between the associated families of homology groups*

$$\Phi^{\beta\alpha}_* : H^\sigma_*(f^\alpha) \xrightarrow{\cong} H^\sigma_*(f^\beta) \ ,$$

and the following functorial relations are fulfilled for the family $\{\Phi^{\beta\alpha}_* \mid f^\alpha,$ f^β Morse functions$\}$:

- $\Phi^{\gamma\beta}_* \circ \Phi^{\beta\alpha}_* = \Phi^{\gamma\alpha}_*$
- $\Phi^{\alpha\alpha}_* = \mathrm{id}_{H^\sigma_*(f^\alpha)}$.

Proof. First, the isomorphism property is a consequence of these functorial relations as we see from

$$\Phi^{\alpha\beta}_* \circ \Phi^{\beta\alpha}_* = \mathrm{id}_{H^\sigma_*(f^\alpha)}$$

and vice versa. The fact that $\Phi^{\alpha\alpha}_*$ actually reproduces the identity is concluded immediately from the constant Morse homotopy

$$f^\alpha \overset{f^\alpha}{\simeq} f^\alpha \ .$$

□

Definition 4.8 *Finally, this functorial concept leads us to the independence of our homology theory from the choice of a concrete Morse function. Considering the product groups*

$$\widetilde{H}_k^\sigma = \prod_{f^\alpha \text{ Morse fct.}} H_k^\sigma(f^\alpha), \quad k \in \mathbb{Z} \ ,$$

we define H_k^σ *as the subgroup*

$$H_k^\sigma = \left\{ (\ldots, \{c_k^\alpha\}, \ldots, \{d_k^\beta\}, \ldots) \in \widetilde{H}_k^\sigma \,\middle|\, \{d_k^\beta\} = \Phi_k^{\beta\alpha} \cdot \{c_k^\alpha\}, \ldots \right\} ,$$

which is well-defined due to Theorem 8. This homology group, independent of the Morse functions, actually describes the same as the inverse limit $\varprojlim H_k^\sigma$ (f^α) *with respect to the isomorphisms* $\Phi_*^{\beta\alpha}$. *This concept will be called the identification process.*

4.1.4 Topology and Coherent Orientation

Throughout the rest of this chapter we assume without loss of generality that M is a connected manifold. Since we intend to relieve our Morse homology concept from as many noncanonical assumptions and inputs as possible, the last item we have to get rid of is the coherent orientation σ. Therefore, we have to analyse how the homology groups $H_*^{\sigma_i}$ change if we replace σ_1 by σ_2. In a way comparable to the case of Morse homotopies, we will look for canonical isomorphisms, which enable us to identify uniquely the elements of the groups sequence

$$\{ H_*^\sigma \}_{\{\sigma \text{ coh. orientation}\}}$$

as in the inverse limit process. Actually, the key for this problem lies within the group action described in Proposition 3.16,

$$\Gamma \times \mathcal{C}_\Lambda \to \mathcal{C}_\Lambda \ .$$

Our aim is to extract homology isomorphisms from these transformations $\varphi \in \Gamma$ of coherent orientations.

For the sake of simplicity we first illustrate the procedure with a fixed Morse function f and consider the associated homology groups $H_*^\sigma(f)$. Obviously, we may put all coherent orientations $\sigma \in \mathcal{C}_\Lambda$ coinciding on the isolated trajectories of the negative gradient flow of f together in one equivalence class, that is

$$\{\sigma_0\} = \left\{ \sigma \in \mathcal{C}_\Lambda \,\middle|\, \begin{array}{c} \sigma([u, D_u]) = \sigma_0([u, D_u]) \\ \text{for all } u \in \mathcal{M}_{x,y}^f, \ x, y \in \text{Crit} f \end{array} \right\} . \tag{4.18}$$

Hence we define the set of equivalence classes as $\widetilde{\mathcal{C}_\Lambda} = \mathcal{C}_\Lambda/_\sim$. Let us now consider the group of transformations

$$G(f) = \{\pm 1\}^{\mathrm{Crit}\, f}$$

together with the action

$$\begin{aligned}
&G(f) \times \widetilde{\mathcal{C}_\Lambda} \to \widetilde{\mathcal{C}_\Lambda}, \\
&(\varphi, \sigma) \mapsto \varphi\,\sigma, \\
&(\varphi\,\sigma)[u_{x,y}] = \varphi(x)\varphi(y)\sigma[u_{x,y}] \\
&\text{for} \quad [u_{x,y}] = [u_{x,y},\, D_{u_{x,y}}],\ u_{x,y} \in \mathcal{M}^f_{x,y}\ .
\end{aligned} \tag{4.19}$$

In fact, $G(f)/\pm$ forms a subgroup of Γ. The next observation is that, given any such orientation transformation φ, we can indicate uniquely a suitable isomorphism of the associated homology groups $H_*^{\sigma_1}(f)$ and $H_*^{\sigma_2}(f)$: Choosing

$$\begin{aligned}
&\Phi^\varphi : (C_*,\, \partial^\sigma) \to (C_*,\, \partial^{\varphi\sigma}) \\
&\Phi^\varphi \cdot x = \varphi(x) \cdot x,\ x \in \mathrm{Crit}\, f\ ,
\end{aligned} \tag{4.20}$$

we compute

$$\begin{aligned}
\tau_\sigma^{\varphi\sigma}(u_{x,y}) &= \varphi(x)\,\varphi(y)\,\tau_\sigma^\sigma(u_{x,y}), \\
\partial^\sigma x = \sum_y \langle x,\, y\rangle^\sigma y &= \varphi(x) \sum_y \langle x,\, y\rangle^{\varphi\sigma} \varphi(y) \cdot y\ .
\end{aligned}$$

This leads to the identity

$$\begin{aligned}
\Phi^\varphi \partial^\sigma x &= \varphi^2(x) \sum_y \langle x,\, y\rangle^{\varphi\sigma} \varphi(y) \cdot y \\
&= \sum_y \langle x,\, y\rangle^{\varphi\sigma} \Phi^\varphi \cdot y \\
&= \partial^{\varphi\sigma} \Phi^\varphi y\ .
\end{aligned} \tag{4.21}$$

Thus, Φ^φ represents a chain homomorphism, which is one-to-one due to the involutivity $\varphi^2 = \mathrm{id}$. Consequently, each transformation $\varphi \in G(f)$ gives rise to a homology isomorphism. However, there is the problem that the endpoints of the $[u,\, K] \in \mathcal{C}_\Lambda$ merely determine the transformations φ up to the sign $\pm\varphi$. This amounts to stating that the associated isomorphisms of homology are only determined up to the sign of their degree of modulus 1. Fortunately, by introducing an additional parameter, we are able to remedy this flaw within our orientation concept.

At this stage we shall discuss in a more particular way the coherent orientation of equivalent operators, which coincide at their endpoints, but whose supporting curves may be connected to an n-dimensional 'Möbius band' $(u \cdot v^{-1})^*TM$. We remember that we might put the original definition of the equivalence of such Fredholm operators along curves in a more general form, such that operators fulfilling the condition $w_1\big((u \cdot v^{-1})^*TM\big) = 0$ are admitted for simultaneous orientation. Already this weaker condition, which admits more possible coherent orientations, implies the fundamental identity $\partial^\sigma \circ \partial^\sigma = 0$ of a boundary operator.

Example Let us consider the non-orientable manifold $\mathbb{P}^2(\mathbb{R})$ endowed with a suitable Morse function as indicated in Figure 4.8. Then, it is due to the chain

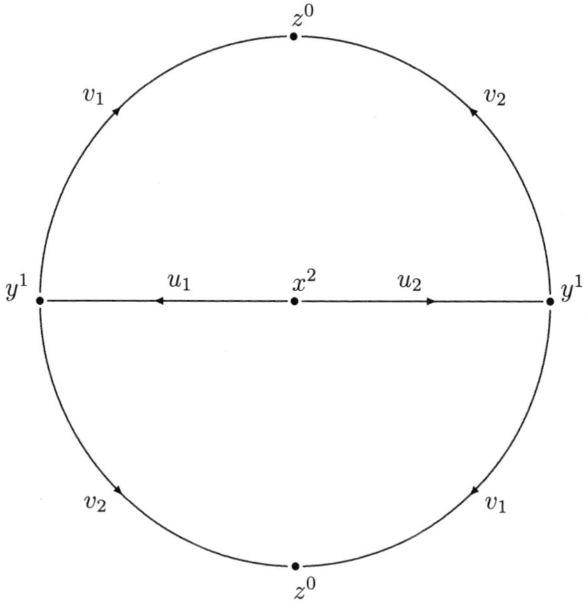

Figure 4.8: ∂-trajectories on $\mathbb{P}^2(\mathbb{R})$

complex property $\partial^2 = 0$ for any arbitrary coherent orientation with respect to the weaker equivalence relation just proposed above that we compute the following relations for the characteristic signs of isolated trajectories:

$$\begin{aligned} \tau_\sigma(u_1) \cdot \tau_\sigma(v_2) &= -\tau_\sigma(u_2) \cdot \tau_\sigma(v_1) \\ \tau_\sigma(u_1) \cdot \tau_\sigma(v_1) &= -\tau_\sigma(u_2) \cdot \tau_\sigma(v_2) \ . \end{aligned} \tag{4.22}$$

These equations do not, however, determine the characteristic signs $\tau_\sigma(u_i)$, $\tau_\sigma(v_i)$ uniquely. Since $u_1 \cdot u_2^{-1}$ and $v_1 \cdot v_2^{-1}$ represent Möbius bands, the operators

$$\begin{aligned}(u_1, D_{u_1}) &\not\sim (u_2, D_{u_2})\\ (v_1, D_{v_1}) &\not\sim (v_2, D_{v_2})\end{aligned}$$

cannot be equivalent according to the weaker version despite the coincidence of the endpoints. In fact, we may suggest exactly two different coherent orientations up to some transformation from $G(f)$, which yield non-isomorphic homologies, namely:

(a) $$\left(H_0^{\sigma_1},\, H_1^{\sigma_1},\, H_2^{\sigma_1} \right) \cong \left(\mathbb{Z},\, \mathbb{Z}_2,\, 0 \right) \quad \text{and}$$

(b) $$\left(H_0^{\sigma_2},\, H_1^{\sigma_2},\, H_2^{\sigma_2} \right) \cong \left(\mathbb{Z}_2,\, 0,\, \mathbb{Z} \right) .$$

If we make use of the original equivalence relation, however, which orients the 'Möbius bands' in a unique way, exactly one of these two possibilities is selected. Referring to this example, we observe that possibility (b) violates the known identity for the homology of connected manifolds, $H_0(M) \cong \mathbb{Z}$, which is principally related to the dimension axiom. Actually, case (b) yields nothing more than the classical cohomology of $\mathbb{P}^2(\mathbb{R})$ turned around by reflection, that is

$$(\mathbb{Z}_2, 0, \mathbb{Z}) \cong (H^2, H^1, H^0) .$$

This means that, already at this stage, we find a hint of Poincaré duality, which is immanent to Morse homology. It is closely connected to the choice between positive or negative gradient flow for the definition of boundary operator. As to the example, in order to comply with the dimension axiom, we have to decide on case (a). As we shall see in Lemma 4.13, this is related to the condition of the equivalence relation for operators in Definition 3.7. That is the condition that we simply demand the coincidence of the trivializations at the negative end, provided that we start from the Morse homology associated to negative gradient flow. Otherwise, choosing the positive end for the coincidence condition would lead us to the cohomology turned around by reflection.

Definition 4.9 *At first we define an appropriate generalization of $G(f)$ with respect to arbitrary Morse functions. Hence, we have to consider the set of Hessians*

$$\mathcal{E} = \left\{ A_m \in \mathrm{End}(TM) \;\middle|\; \begin{array}{c} A_m : T_mM \to T_mM \\ \textit{non-deg. and conj. self-adjoint} \end{array} \right\}$$

instead of $\mathrm{Crit}\, f$. *Let*

$$\Delta = \{\pm 1\}^{\mathcal{E}}$$

be a group with respect to pointwise multiplication together with the action

$$\begin{aligned}&\Delta \times \mathcal{C}_\Lambda \to \mathcal{C}_\Lambda\\ (4.23)\qquad &(\varphi,\, \sigma) \mapsto \varphi\sigma\\ &(\varphi\sigma)\,[u_{x,y},\, K_x^-,\, K_y^+] = \varphi(K_x^-)\varphi(K_y^+)\sigma[u_{x,y},\, K_x^-,\, K_y^+] \;.\end{aligned}$$

This yields a generalization of $\{\pm 1\}^{\mathrm{Crit}\, f}$ in the sense of

$$\begin{aligned} \mathrm{Crit}\, f &\hookrightarrow \mathcal{E} \\ x &\mapsto H^2 f(x) \ . \end{aligned}$$

Once again by analogy with (4.20) and (4.21) the $\varphi \in \Delta$ give rise to isomorphisms

$$\Phi^{\varphi} : \left(C_*(f), \partial^{\sigma}\right) \xrightarrow{\cong} \left(C_*(f), \partial^{\varphi\sigma}\right) \ , \tag{4.24}$$

which additionally are compatible with the isomorphisms

$$\Phi_*^{\beta\alpha} : H_*^{\sigma}(f^{\alpha}) \xrightarrow{\cong} H_*^{\sigma}(f^{\beta})$$

on the level of the homology groups. Hence, the problem indicated above may be expressed as follows:

$$\Delta_{/\pm} = \Gamma \ ;$$

that is, there are exactly two transformations $\varphi \in \Delta$ associated to a given pair (σ_1, σ_2) of coherent orientations, such that the identity $\sigma_2 = \varphi\sigma_1$ is satisfied, namely $\pm\varphi$. In order to attain a unique identification of the homology groups by means of the isomorphisms $\varphi_* = \Phi^{\varphi}$ associated to the $\varphi \in \Delta$, we crucially require the functorial behaviour

$$(\varphi' \cdot \varphi)_* = \varphi'_* \circ \varphi_* \ . \tag{4.25}$$

This condition prevents us from choosing erratically between $+\varphi$ and $-\varphi$. Actually, the problem of a natural selection process with respect to functorial behaviour lies in the relative constitution of a coherent orientation and its transformations. These transformations are given by $\varphi(x)\varphi(y)$ with respect to the pairs (x,y) of endpoints. Thus, the relative character of these transformations is in contrast to the absolute character of a mapping $\varphi : \mathcal{E} \to \{\pm 1\}$. In order to fix such a mapping φ uniquely, we need a characterization of φ by Fredholm operators, which are endowed with only one end within $\mathcal{E}$. This means that, instead of merely orienting the relative, connecting trajectories from $\mathcal{M}_{x,y} = W^u(x) \pitchfork W^s(y)$, we have to find an absolute fixing, for example by means of an orientation of some stable manifold $W^s(y)$. Thus we shall develop the following concept of the so-called 'one-sided' operators to find a solution for this problem.

Let us first regard the trivial case, $M = \mathbb{R}^n$. Let $A^+ \in \mathcal{S}$ be a conjugated self-adjoint operator on $\mathbb{R}^n$ and $A \in C^{\infty}\big([0,\infty], \mathrm{End}\,(n,\mathbb{R})\big)$ endowed with $A(+\infty) = A^+$. In this situation, we consider $[0,\infty] \subset \overline{\mathbb{R}}$ to bear the differentiable structure as a submanifold with boundary within $\overline{\mathbb{R}}$. We consequently define the one-sided operator

$$\begin{aligned} S_A : X = H^{1,2}\big([0,\infty), \mathbb{R}^n\big) &\to Y = L^2\big([0,\infty), \mathbb{R}^n\big) \\ u &\mapsto \dot{u} + A \cdot u \ . \end{aligned} \tag{4.26}$$

Referring to the methods which were developed throughout the Fredholm Section 2.2.1, it is straightforward to show that S_A represents a Fredholm operator with index $n - \mu(A^+)$. We may conclude similarly that

$$\Theta(A^+) = \{ S_A \in \mathcal{F}(X;Y) \mid A(+\infty) = A^+ \} \tag{4.27}$$

is a contractible space such that we are able to equip the determinant bundle Det on $\Theta(A^+)$ with a well-defined orientation. This concept of a one-sided operator in the trivial framework can be transferred to the manifold M as follows.

Let $s \in C^\infty\big([0,\infty), M\big)$ have the fixed end $s(\infty) = x$ and let $A_x \in \mathcal{E}$ be fixed. Then we consider the pair (s, S_A), where

$$\begin{aligned} S_A : H^{1,2}(s^*TM_{|\,[0,\infty)}) &\rightarrow L^2(s^*TM_{|\,[0,\infty)}) \\ u &\mapsto \nabla_t u + A \cdot u \end{aligned} \tag{4.28}$$

is a Fredholm operator of type (4.26), endowed with the fixed end $A(+\infty) = A_x$. This set $\{(s, S_A)\}$ of one-sided operators with fixed end A_x may be assembled in a uniformly orientable equivalence class $[S_{A_x}]$. The uniform orientability is due to the fact that, by analogy to Definition 3.10, the choice of orientation of an arbitrary representative (s, S_A) together with a trivialization of s^*TM induces an orientation of $\Theta(A^+_{x,\mathrm{triv}})$. Hence in a reverse and unique manner, this induces orientations for all the other representatives, so that they do not depend on the actual choice of trivializations. This may be concluded by means of a lemma which is strictly analogous to Lemma 3.8. Moreover, it proves to be consistent with the fact that the space of all curves s with a fixed end $s(\infty) = x$ is contractible to the constant curve $s \equiv x$.

Additionally, we are able to analyse a gluing operation for such one-sided operators together with the 'two-sided' operators $(u, K) \in \Sigma_{u^*TM}$ of the former type:

$$[S_{A_x}]\#[u,K] = [S_{K_+}] \text{ for } A_x = K_- \ .$$

Starting from a given coherent orientation with respect to the classes of operators $[u, K] \in \Lambda$, we derive a unique orientation of $[S_{K_+}]$ from a given orientation of $[S_{A_x}]$ by means of this gluing operation. In order to obtain uniqueness guaranteed, we obviously need independence from the chosen curve u associated to the fixed endpoints. This is the very point where we remark the necessity of a simultaneous orientability of all operators (u, K) belonging to given ends K_x^-, K_y^+. Let us state the consequence of this uniqueness as

Proposition 4.10 *A coherent orientation of the relative classes of operators*[5] $[K_x^-, K_y^+]$ *together with an orientation of a given fixed* $[S_{A_x}]$, $A_x \in \mathcal{E}$ *induce*

5with respect to the equivalence of orientations by means of trivializations with coincidence at the negative end *unique orientations of all positively one-sided classes of operators* $[S_{K_y^+}]$, $K_y^+ \in \mathcal{E}$.

On the other hand, however, let us regard what we have to change, if we wish to endow the classes $[S_{K_x^-}]$, $K_x^- \in \mathcal{E}$ of negative half-sided operators with a uniform orientation, that is in principle an orientation of the unstable manifolds. In the case of a non-orientable manifold M we would also have to change the definition of equivalence of the relative operators $(u, K) \sim (v, L)$. Namely, we would have to choose the coincidence of the trivializations of u^*TM and v^*TM at the positive end $u(+\infty) = v(+\infty)$ as far as 'n-dimensional Möbius bands' $w_1\big((u \cdot v^{-1})^*TM\big) \neq 0$ are concerned. Actually, the question of whether we have to endow either the 'stable' or the 'unstable' manifolds with a uniform orientation corresponds to the decision between negative and positive gradient flow. It is exactly one of the two alternatives, namely the choice of the stable manifolds in the case of the negative gradient flow, which reproduces the necessary identity for the classical homology of connected manifolds (see below), $H_0(M) \cong \mathbb{Z}$, also in the situation of a non-orientable manifold.

Proof of proposition 4.10. Without loss of generality, let $s \in C^\infty\big([0,\infty], M\big)$ and $u_1, u_2 \in C^\infty(\overline{\mathbb{R}}, M)$ be asymptotically constant and satisfy the identities

$$\begin{aligned} s(+\infty) = u_1(-\infty) = u_2(-\infty) = x \\ \text{and } u_1(+\infty) = u_2(+\infty) = y \ . \end{aligned}$$

Moreover, let us assume a positive half-sided operator S_x on s together with relative operators K_1, K_2 on u_1 and u_2, respectively, once again asymptotically constant with

$$S_x^+ = K_1^- = K_2^-, \ \ K_1^+ = K_2^+ \ .$$

We consequently consider asymptotically constant trivializations

$$\begin{aligned} \phi_x &: s^*TM \xrightarrow{\cong} [0,\infty] \times \mathbb{R}^n, \\ \phi_1 &: u_1^*TM \xrightarrow{\cong} \overline{\mathbb{R}} \times \mathbb{R}^n, \\ \phi_2 &: u_2^*TM \xrightarrow{\cong} \overline{\mathbb{R}} \times \mathbb{R}^n \end{aligned}$$

satisfying $\phi_x(+\infty) = \phi_1(-\infty) = \phi_2(-\infty)$, such that the pair (ϕ_1, ϕ_2) is admissible and the non-degenerate, conjugated self-adjoint endpoint operators[6]

$$A_x = \phi_x(S_x)(+\infty) = \phi_{1,2}(K_{1,2})(-\infty), \ \ A_y = \phi_{1,2}(K_{1,2})(+\infty) \in \mathrm{GL}(n, \mathbb{R})$$

are diagonal matrices. If the identity $\phi_1(\pm\infty) = \phi_2(\pm\infty)$ holds at both ends, the proof is straightforward. Thus, let us assume the actually interesting case

---

6$\phi_x(S_x) = \phi_x \circ S_x \circ \phi_x^{-1} \in \mathcal{L}(X;Y)$, see 4.26

$\phi_1(+\infty) = S \circ \phi_2(+\infty)$, where S denotes the reflection matrix

$$S = \begin{pmatrix} -1 & & & \\ & +1 & & \\ & & \ddots & \\ & & & +1 \end{pmatrix}$$

from Definition 3.7. Let us henceforth use the following notations:

$$\begin{aligned} \phi_x^+ &= \phi_x, & \phi_x^- &= S \circ \phi_x, \\ \phi_1^{++} &= \phi_1, & \phi_1^- &= S \circ \phi_1, \\ \phi_2^{+-} &= \phi_2, & \phi_2^{-+} &= S \circ \phi_2 \ . \end{aligned}$$

We compute $(\phi_x^- \#^o \phi_2^{-+})(+\infty) = (\phi_x^+ \#^o \phi_1^{++})(+\infty)$. Now let $[S_x]$ be oriented firmly by $o(S_x)$ as well as $K_1 \sim K_2$ by $o(K_1)$ and $o(K_2)$. Then, the relation between the orientations

$$o(S_x \#^o K_1) \equiv o(S_x) \# o(K_1) \simeq o(S_x) \# o(K_2) \equiv o(S_x \#^o K_2) \tag{4.29}$$

amounts to relating

$$(\phi_x^+ \#^o \phi_1^{++})\big(o(S_x \#^o K_1)\big) \simeq (\phi_x^- \#^o \phi_2^{-+})\big(o(S_x \#^o K_2)\big) \tag{4.30}$$

on the contractible class $\Theta(A_y)$ in trivialized form. Thus, we have to verify that this relation is equivalent to the compatibility of the orientations $o(K_1) \simeq o(K_2)$ according to definition 3.10, that is

$$\phi_1^{++}\big(o(K_1)\big) \simeq \phi_2^{+-}\big(o(K_2)\big) \quad \text{in} \quad \Theta(A_x, A_y) \subset \Sigma_{\text{triv}} \ , \tag{4.31}$$

since $(\phi_1^{++}, \phi_2^{+-})$ is admissible.[7] We notice that we may state relation (4.30) in the trivialized case alternatively as

$$\phi_x^+\big(o(S_x)\big) \# \phi_1^{++}\big(o(K_1)\big) \simeq \phi_x^-\big(o(S_x)\big) \# \phi_2^{-+}\big(o(K_2)\big) \ . \tag{4.32}$$

Now let L_t be any homotopy between $\phi_1^{++}(K_1)$ and $\phi_2^{-+}(K_2)$ in the contractible space $\Theta(A_x, A_y)$. Given a large enough gluing parameter, it induces a homotopy $\phi_x^+(S_x) \# L_t$ between $\phi_x^+(S_x) \# \phi_1^{++}(K_1)$ und $\phi_x^+(S_x) \# \phi_2^{-+}(K_2)$. Then the identity

$$\phi_x^+\big(o(S_x)\big) = (S \circ \phi_x^-)\big(o(S_x)\big) = -\phi_x^-\big(o(S_x)\big)$$

together with the equivalence of orientations (4.32) imply the relation

$$\phi_1^{++}\big(o(K_1)\big) \simeq -\phi_2^{-+}\big(o(K_2)\big) = \phi_2^{+-}\big(o(K_2)\big) \ . \tag{4.33}$$

7See Definition 3.7.

□

Before we are able to relieve the homology groups $H_k^\sigma(M)$ from their parameter σ, we first have to introduce even an additional parameter. Choosing any class $[S_A]$, $A \in \mathcal{E}$, of half-sided operators with fixed positive end and specifying any orientation $o(S_A)$ of this class, we expand the parameter dependence,

$$H_*^\sigma \rightsquigarrow H_*^{(\sigma, o(S_A))} \ .$$

For instance, if we choose the Hessian $A = H^2 f(x)$, $x \in \operatorname{Crit} f$ for a given Morse function f, specifying the orientation $o(S_A)$ amounts in principle to fixing an orientation of the stable manifold $W^s(x)$. Hence, the coherent orientation σ gives rise particularly to an orientation of connecting orbits $u_{x,y} \in \widehat{\mathcal{M}}_{x,y}^f = W^u(x) \pitchfork W^s(y)$ thus leading to orientations of all the other stable manifolds.

Now, let $(\sigma, o(S_A))$ and $(\sigma', o'(S'_A))$ be pairs of parameters, such that there is exactly one transformation pair $\pm\varphi$ (see above). It is due to the coherence of the given orientations and to Proposition 4.10 that we may start without loss of generality from identical classes $[S_A] = [S'_A]$. Let us choose the sign of $\pm\varphi$ according to the case distinction

$$\varphi(A) = \begin{cases} +1, & \text{if } o(S_A) = o'(S_A) \\ -1, & \text{if } o(S_A) = -o'(S_A) \ . \end{cases} \tag{4.34}$$

To put it in other terms, we extend the action

$$\{\pm 1\}^{\mathcal{E}} \times \mathcal{C}_\Lambda \to \mathcal{C}_\Lambda$$

to

$$\varphi \bullet (\sigma, o(S_A)) = (\varphi\,\sigma, \varphi(A) \cdot o(S_A)) \ ,$$

so that we deduce the existence of a now unique transformation map $\varphi \in \{\pm 1\}^{\mathcal{E}}$ between $(\sigma, o(S_A))$ and $(\sigma', o'(S'_A))$ from the coherence condition for σ and σ'. At last, due to the construction, this mapping

$$\Big((\sigma, o(S_A)), (\sigma', o'(S'_A))\Big) \mapsto \varphi^{\sigma'\sigma}$$

satisfies the functorial property

$$\varphi^{\sigma_3\sigma_2} \circ \varphi^{\sigma_2\sigma_1} = \varphi^{\sigma_3\sigma_1}, \ \varphi^{\sigma\sigma} = \mathrm{id} \ , \tag{4.35}$$

which was demanded in (4.25).

Thus, coherent orientations give rise to isomorphical homologies, which may be identified with each other by means of the isomorphisms $\Phi^{\varphi^{\sigma'\sigma}}$ in a unique way due to (4.35). As a consequence, this identification process provides us with homology groups $H_k(M)$ which are finally independent of any other parameters than the order k and the smooth manifold M.

Definition 4.11 *By analogy with definition 4.8 we define the Morse homology groups as the inverse limit*

$$H_k^{\text{Morse}}(M) = \lim_{\leftarrow} H_k^\sigma(M)$$

with respect to the isomorphisms $\varphi_*^{\sigma'\sigma}$.

Referring back to definition 3.7, our argument for the construction of the non-trivialized Fredholm-classes on the one hand was founded on the relation for the halfsided operators as described in proposition 4.10 and on the other hand on the algebro-topological relation $H_0(M) \cong \mathbb{Z}$ for connected manifolds. The discussion of the latter, which is equivalent to the dimension axiom within the axiomatic framework is now anticipated to the next section. We verify the consistence of our orientation construction with this crucial relation as follows.

Proposition 4.12 *Let M be a connected, smooth manifold equipped with a Morse function f, which possesses exactly one local miminum y_0. Then, the Morse homology group of order 0 with respect to f satisfies the relation*

$$H_0(f) \cong \mathbb{Z} \ .$$

The proof is based on the following fact:

Lemma 4.13 *Given two different, isolated trajectories u_1 and u_2 from $\mathcal{M}_{x,y}^f$ with $(x, y) \in \text{Crit}_1 f \times \text{Crit}_0 f$, that is*

$$(u_1, D_{u_1}) \sim (u_2, D_{u_2}) \ ,$$

any coherent orientation σ leads to the identity

$$\tau_\sigma(u_1) = -\tau_\sigma(u_2) \ .$$

Proof. We consider the connection of the curves at the point x referring to a suitable reparametrization which yields

$$\gamma = u_1^{-1} \cdot u_2 \in C^\infty([-1, 1], M) \ .$$

This is nothing else than a suitable parametrization of the 1-dimensional unstable manifold $W^u(x)$ as a 1-cell. We now choose a trivialization of γ^*TM, such that on the one hand the Hessians $H^2 f(x) = D_{u_1}^- = D_{u_2}^-$ and $H^2 f(y) = D_{u_1}^+ = D_{u_2}^+$ become diagonal matrices with respect to this trivialization, and such that on the other hand the induced isomorphisms $T_y M = T_{\gamma(\pm 1)} M \xrightarrow{\cong} \mathbb{R}^n$ differ by the reflection S defined above if they differ at all. Hence, this trivialization gives rise to an admissible pair with respect to $(u_1, D_{u_1}) \sim (u_2, D_{u_2})$. Moreover, we may construct this trivialization in accordance with the splitting-off of a one-dimensional subbundle η of $\xi = \gamma^* TM$,

$$\xi = \eta \oplus \zeta^{n-1} .$$

Here, η denotes the one-dimensional bundle on $[-1, 1]$ induced by $TW^u(x)$ with the fibre $\eta_0 = T_x W^u(x)$ which is identical with the eigenspace of $H^2 f(x)$ associated to the unique negative eigenvalue. Up to an appropriate parametrization, the generating sections $\dot{u}_1, \dot{u}_2$ for the kernels of D_{u_1} and D_{u_2}, respectively, lie exactly in this subbundle η. Thus, starting from a fixed orientation on ζ^{n-1}, we can derive orientations of the bundle $\xi = (u_1^{-1} \cdot u_2)^* TM$ from the canonical orientations $[\dot{u}_1]$ and $[\dot{u}_2]$, which must be opposite each other. This proves the lemma. □

Proof of the proposition. The unique local minimum y_0 of the Morse function f provides us with a generator of the cycle group $Z_0(f)$ for the Morse complex of f. This generator is unique up to sign. We deduce from the lemma that, given any coherent orientation σ, the boundary $\partial^\sigma x_1$ of each generator $x_1 \in C_1(f)$ vanishes. Thus, the assertion follows from $B_0(f) = \{0\}$. □

It seems natural to expect that the proposition remains true if we do not impose the special condition on f to possess only one local minimum. The proof in this general setting requires some more steps and will be given in the next section in relation to the dimension axiom.

4.2 The Eilenberg-Steenrod Axioms

In this section, we shall carry out the final step of the development of Morse homology theory. We shall verify the functorial properties and the accordance with the axioms of a homology theory as they were set up by Eilenberg and Steenrod in [E-S], namely,

- the existence of a long exact homology sequence,
- the homotopy axiom,
- the excision axiom and
- the dimension axiom.

For the sake of simplicity, we shall first derive the functorial construction and homotopy invariance for the concept of the absolute homology groups $H_*^{\text{Morse}}(M)$, before we treat the somewhat more involved relative groups H_*^{Morse} (M, A) associated to admissible pairs of manifolds (M, A). In this framework a natural induction of chain maps for Morse complexes by closed embeddings of manifolds proves essential for functorial construction. In order to discuss this construction, however, we first need some technical preparation, namely suitable extension lemmata for Morse functions.[8]

4.2.1 Extension of Morse Functions and Induced Morse Functions on Vector Bundles

The problem of the extension of Morse functions from submanifolds of M to the whole manifold M may be separated into a technical one on the one hand, the extension of the coercivity property and of the regularity of Morse functions on open submanifolds, and on the other hand, a problem related to dimension with respect to closed submanifolds.

The first lemma yields the extension of the coercivity property. We notice that the coercivity of a continuous function $f : M \to \mathbb{R}$, i.e.

$$M^a = \{ x \in M \mid f(x) \leqslant a \} \quad \text{compact for all } a \in \mathbb{R} \ ,$$

is equivalent to the following property:

> There is a constant $c > 0$ such that the continuous function $f_c = f + c : M \to [0, \infty)$ is non-negative and proper, i.e. $f_c^{-1}(K)$ is compact in M for all compact $K \subset [0, \infty)$.

Lemma 4.14 *Let M be a smooth manifold, $A \subset M$ be a closed subset and $f : A \to [0, \infty)$ be continuous and proper. Then there is a continuous and proper extension*

$$g : M \to [0, \infty) \;\; g_{|A} = f \ .$$

Proof. The proof is basically founded on an application of Tietze's extension theorem for continuous functions:

> Let X be a normal topological space, $A \subset X$ be a closed subset and $f : A \to [0, 1]$ be a continuous function. Then there is a continuous extension
>
> $$f^* : X \to [0, 1], \;\; f^*{}_{|A} = f \ .$$

Since we are able to embed M as a closed submanifold in an $\mathbb{R}^N$ with N chosen suitably large, we may start from the assumption that $M = \mathbb{R}^N$ without loss of generality. Denoting the closed ball with radius r by

$$B_r = \{x \in \mathbb{R}^N \mid \|x\| \leqslant r \}, \ r > 0$$

8for coercive Morse functions!

we define

(4.36) $$f_n = f_{|\,A \cap B_n} \ .$$

Consequently, there is an increasing sequence of positive numbers $(r_n)_{n\in\mathbb{N}} \subset \mathbb{R}_+$, such that the condition

$$f_n(A \cap B_n) \subset [0,\, r_n]$$

is true for all $n \in \mathbb{N}$. This enables us to apply Tietze's theorem inductively for n: For each $n \in \mathbb{N}$ there is a

(4.37) $$\begin{aligned} &g_n^o : B_n \to [0,\, r_n] \text{ satisfying} \\ &g^o_{n|\,A\cap B_n} = f_n \quad \text{and} \quad g^o_{n|\,B_{n-1}} = g^o_{n-1} \ . \end{aligned}$$

Defining

$$h_n = \arctan g_n^o : B_n \to [0,\, \frac{\pi}{2}),\ n \in \mathbb{N}\ ,$$

we obtain a sequence of continuous functions once again according to Tietze's theorem,

$$k_n : \mathbb{R}^N \to [0,\, \frac{\pi}{2}] \quad \text{with} \quad k_{n|\,B_n} = h_n \ \text{and} \lim_{\|x\|\to\infty} k_n = \frac{\pi}{2} \ .$$

It is due to the construction in (4.36) and (4.37) that we may state the uniform convergence on compact sets,

(4.38) $$k_n \overset{C^0_{\mathrm{loc}}}{\longrightarrow} k \in C^0\big(\mathbb{R}^N,\, [0,\, \frac{\pi}{2})\big) \quad \text{with} \lim_{\|x\|\to\infty} k(x) = \frac{\pi}{2} \ .$$

This implies a continuous and proper extension of f by

$$g = \tan k \,\in C^0\big(\mathbb{R}^N,\, [0,\infty)\big),\ g_{|\,A} = f \ .$$

□

We now have to transfer the extension result to the dense subset

$$C^\infty(M,\, \mathbb{R}) \subset C^0(M,\, \mathbb{R})$$

of the smooth functions. Furthermore, we have to add the regularity condition

$$df \pitchfork M \subset T^*M \ .$$

We refer essentially to the results of differential topology as they can be found in [Hi]. Hence, we take over the notations $C^k_S(M,\mathbb{R})$, $\mathcal{N}(f,\,\epsilon)$, etc., for the strong or, equivalently, the Whitney topology and the basis sets on $C^0_S(M,\mathbb{R})$

$$\mathcal{N}(f,\,\epsilon) = \{\, g \mid |g(x) - f(x)| < \epsilon(x) \text{ for all } x \in M \,\},\ \ \epsilon \in C^0(M,\, \mathbb{R}_+) \ .$$

An analogous definition holds for $C^k_S(M,\mathbb{R})$. Then the extension result for smooth Morse functions can be stated as

Lemma 4.15 *Let M be a smooth manifold, let A be a closed and W an open subset and let $f \in C^\infty(W, \mathbb{R})$ be a Morse function on the submanifold W, such that*

$$\operatorname{Crit} f \subset \mathring{A} \subset A \subset W \subset M \ .$$

Then there is a smooth Morse function g on M which extends f, that is,

$$g_{|A} = f_{|A} \ .$$

Proof. Since M is a normal topological space, we find an open subset $U \subset M$ satisfying

$$A \subset U \subset \overline{U} \subset W \ .$$

Due to Lemma 4.14 there is a continuous coercive extension

$$g_0 : M \to \mathbb{R}, \quad g_{0|\overline{U}} = f_{|\overline{U}} \ . \tag{4.39}$$

The set of continuous and proper functions represents an open subset with respect to the Whitney topology (see [Hi]),

$$\operatorname{Prop}^0_S(M, \mathbb{R}) \subset C^0_S(M, \mathbb{R}), \text{ open subset } .$$

Therefore, we find a convex neighbourhood

$$\mathcal{N}_\epsilon(g_0) \subset \operatorname{Prop}^0_S(M, \mathbb{R}) \subset C^0_S(M, \mathbb{R}) \ , \tag{4.40}$$

which contains merely coercive functions. Furthermore, referring to the C^2-Whitney topology, we can find a convex, open neighbourhood

$$\begin{aligned} &\widetilde{\mathcal{N}}_{\tilde\epsilon}(g_0) \subset C^2_S(M,\mathbb{R}) \quad \text{with} \\ &\operatorname{Crit} \tilde h \cap U \backslash A = \emptyset \quad \text{for all } \tilde h \in \widetilde{\mathcal{N}}_{\tilde\epsilon} \ . \end{aligned} \tag{4.41}$$

Thus, the intersection $\mathcal{N} = \widetilde{\mathcal{N}}_{\tilde\epsilon} \cap \mathcal{N}_\epsilon$ is an open and convex subset of $C^2_S(M, \mathbb{R})$, too.

It is clear from the well-known transversality theorems in differential topology (see [Hi]), that

$$X = \{ f \in C^2(M, \mathbb{R}) \mid df \pitchfork M \subset T^*M \}$$

forms an open and dense subset of the Baire space $C^2_S(M, \mathbb{R})$. Since $C^\infty(M, \mathbb{R})$ lies dense in $C^2_S(M, \mathbb{R})$, too, there is a

$$g' \in C^\infty(M, \mathbb{R}) \cap \mathcal{N} \cap X \quad \text{arbitrarily close to } g_0 \text{ in } C^2_S(U\backslash A, \mathbb{R}) \ .$$

Now let $\alpha : M \to [0, 1]$ be a smooth function satisfying

$$\alpha_{|A} = 0 \quad \text{and} \quad \alpha_{|M\backslash U} = 1 \ .$$

We consequently define

$$g = (1 - \alpha) \cdot g_0 + \alpha \cdot g' \ . \tag{4.42}$$

Due to the concrete definition of the convex neighbourhoods $\mathcal{N}_{\dots}$ and the identity of g_0 and f when restricted to U, g satisfies the properties

$$g \in C^\infty(M, \mathbb{R}) \cap \mathcal{N} \cap X \quad \text{and} \quad g_{|A} = f_{|A} \ ,$$

provided that $g'_{|U \setminus A}$ has been chosen close enough to $g_{0|U \setminus A}$. □

As pointed out above, the next step is to develop an extension result for closed submanifolds. Here, we require an additional construction method for Morse functions, which appears very naturally. A feature which proves crucial for the whole Morse homology is the canonical tensorial behaviour of Morse complexes with respect to product operations: Given two manifolds equipped with Morse functions (M, f) and (N, g), the operation

$$\big(f \oplus g\big)\,(m,\, n) = f(m) + g(n) \tag{4.43}$$

naturally provides a Morse function

$$f \oplus g \ \in C^\infty(M \times N,\, \mathbb{R})$$

together with the canonical identification

$$\mathrm{Crit}_k(f \oplus g) = \bigcup_{i+j=k} \mathrm{Crit}_i f \times \mathrm{Crit}_j g \ \in M \times N \ , \tag{4.44}$$

because the Morse index behaves additively with respect to the operation $\oplus$.

At this stage, regarding our current concern, we need a generalization toward smooth vector bundles

$$\varphi : E \to M \ .$$

Let $\langle \cdot, \cdot \rangle_E$ be a Riemannian metric on E with the associated quadratic form

$$q(v_m) = \langle\, v_m,\, v_m \,\rangle_m$$

and let f be an arbitrary Morse function on M. Then we obtain the Morse function

$$\begin{aligned} &f_E : E \to \mathbb{R} \\ &f_E(v_m) = f(m) + q(v_m) \end{aligned} \tag{4.45}$$

on E with the property

$$f_{E|\,M} = f \quad \text{and} \quad \mathrm{Crit}_* f_E \cong \mathrm{Crit}_* f \ . \tag{4.46}$$

On the one hand, the latter is due to the local product structure of E, so that the results (4.43) and (4.44) may be transferred to this special bundle situation. On the other hand, we observe that q restricted to the fibers is a trivial Morse function, that is

$$\mathrm{Crit}_* q_m = \mathrm{Crit}_0 q_m = \{0_m\} \ .$$

This feature enables us to derive the following extension result:

Proposition 4.16 *Given a closed submanifold $A \subset M$ and a Morse function $f_A \in C^\infty(A, \mathbb{R})$ on A, there is a Morse function $f \in C^\infty(M, \mathbb{R})$ extending f_A,*

$$f_{|A} = f_A \ .$$

Proof. Having in mind the above preparations for naturally induced Morse functions on smooth vector bundles, we consider a tubular neighbourhood W of A, which always exists for any closed submanifold (without boundary):

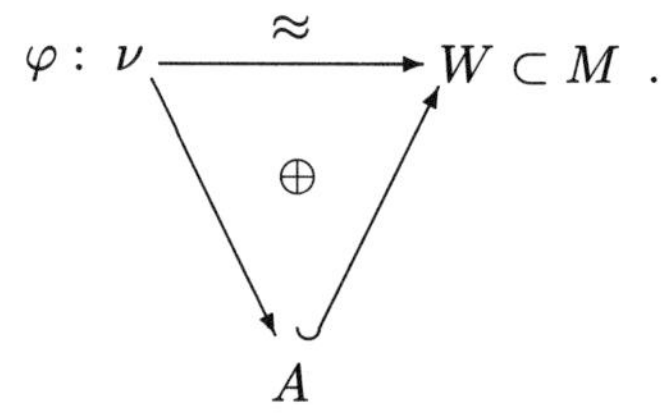

Here, ν is the normal bundle of A and W is an open submanifold. Thus, the diffeomorphism φ induces a Morse function on W,

$$f_W = f_\nu \circ \varphi^{-1} \ , \tag{4.47}$$

which extends $f_A : f_{W|A} = f_A$. The preconditions of Lemma 4.15 are fulfilled if we choose a suitable restriction: Let $R > 0$ and $A_R = \varphi\left(q_\nu^{-1}([0, R])\right)$, that is,

$$A_R \subset A_{R+\epsilon} \subset W \subset M, \ \ A_R \text{ closed} \ .$$

Then, Lemma 4.15 gives rise to a Morse function f on M satisfying

$$f_{|A_{R+\epsilon}} = f_{W|A_{R+\epsilon}} \ ,$$

that is, in particular, $f_{|A} = f_A$. □

Actually, we may gain a further important result from this special extension construction on closed submanifolds:

Corollary 4.17 *The extension f of f_A in the above proposition can be chosen in such a way that there are no trajectories for the negative gradient flow of f leaving A.*

Proof. It is due to construction by means of quadratic extension on the normal bundle ν that the negative gradient field $-\nabla f_\nu$ represents an inner normal vector field on the boundary $\partial q_\nu^{-1}([0,\epsilon])$ of the manifold with boundary $q_\nu^{-1}([0,\epsilon])$ for any $\epsilon > 0$. This means that $-\nabla f_W$ is an inner normal vector field for ∂A_ϵ in A_ϵ. From this we obtain the corollary. □

4.2.2 The Homology Functor and Homotopy Invariance

The Diffeomorphical Functor

First, we wish to discuss the homomorphisms between the respective homology groups induced by diffeomorphisms. We assume

$$\varphi : M \xrightarrow{\approx} N$$

to be a smooth diffeomorphism between two manifolds and f to be a Morse function on M. Obviously, φ gives rise to the Morse function

$$\varphi_* f = f \circ \varphi^{-1} \tag{4.48}$$

on N, so that we obtain the one-to-one correspondence

$$\varphi : \mathrm{Crit}_* f \xrightarrow{\cong} \mathrm{Crit}_*(\varphi_* f) \ . \tag{4.49}$$

However, before we are led to a chain map between the associated Morse complexes

$$\varphi_\bullet : C_*(f) \to C_*(\varphi_* f)$$

we have to verify the condition

$$\partial \varphi_\bullet = \varphi_\bullet \partial \ . \tag{4.50}$$

Provided that the generic Riemannian metrics on M and N have already been fixed, it is generally merely an isometry φ which maps isolated trajectories for f onto isolated trajectories for $\varphi_* f$. But we are able to choose the generic metric g on M in a way that also provides a generic metric on N via induction by φ. This means that we may choose the metrics generically such that φ becomes isometrical. Remember that the transformations of the Morse complexes associated to fixed Morse functions by means of suitable Morse homotopies. Analogously, we may regard the homology groups as independent of the Riemannian metric and we therefore assume without loss of generality that the given φ is isometrical. The only item left to verify with respect to (4.50) is the consistency of the coherent orientations which have been chosen for M and N independent of φ. This can be achieved by means of an appropriate transformation by an $\alpha \in \{\pm 1\}^{\mathrm{Crit}_* f}$ as we described in the last section, that is

$$\begin{aligned} &\varphi_\bullet : C_*(f) \to C_*(\varphi_* f) \\ &\varphi_\bullet x_k = \alpha(x_k) \cdot \varphi(x_k), \ x_k \in \mathrm{Crit}_k f \ . \end{aligned} \tag{4.51}$$

To sum up, up to a respectively appropriate homotopy of the Riemannian metric each diffeomorphism $\varphi : M \xrightarrow{\cong} N$ induces a homology isomorphism

$$\varphi_* : H_*^{\sigma_M}(M,\, f) \xrightarrow{\cong} H_*^{\sigma_N}(N,\, \varphi_* f) \ .$$

Now, regarding a homotopy of Morse functions

$$f^\alpha \xrightarrow{h_t} f^\beta \ ,$$

we can confirm functorial consistency. φ induces a homotopy

$$\varphi_* f^\alpha \xrightarrow{\varphi_* h_t} \varphi_* f^\beta \ ,$$

such that in the case of an isometry the h-trajectories associated to

$$\dot\gamma = -\frac{\nabla h_t}{\sqrt{1 + |\dot h_t|^2 |\nabla h_t|^2}} \circ \gamma$$

are mapped by φ onto the corresponding h-trajectories belonging to $\varphi_* h_t = h_t \circ \varphi^{-1}$. In a nutshell, the equality

$$\Phi_\bullet(\varphi_* f^\beta, \varphi_* f^\alpha) \circ \varphi_\bullet = \varphi_\bullet \circ \Phi_\bullet(f^\beta, f^\alpha) \tag{4.52}$$

implies that φ in fact induces a homomorphism

$$\varphi_* : H_*(M) \to H_*(N)$$

according to the definition of $H_*(M) = \varprojlim H_*(M,\, f^\alpha)$. Obviously, the construction guarantees consistency with the compositions of the morphisms

$$(\varphi \circ \psi)_* = \varphi_* \circ \psi_* \ , \tag{4.53}$$

so that we obtain a functor as far as smooth manifolds together with diffeomorphisms are concerned.

The Embedding Functor and the Homotopy Lemma

By means of the extension results and this diffeomorphism functor we are now able to accomplish the extension to closed embeddings. Let $\varphi : M \hookrightarrow N$ be an embedding such that $\varphi(M)$ represents a closed submanifold within

N. We henceforth denote by $\varphi_*^o f$ the Morse function on $\varphi(M)$ which is induced by f and φ and which can be extended to a Morse function $\varphi_* f$ on the whole manifold N without any negative gradient flow trajectory leaving the submanifold $\varphi(M)$ according to the corollary of Proposition 4.16. Actually, this condition allows us to identify the Morse complex $C_*\left(\varphi_*^o f\right)$ on $\varphi(M)$ with a subcomplex of $C_*\left(\varphi_* f\right)$ on N, that is

$$\partial^{\varphi_* f}{}_{| C_*(\varphi_*^o f)} = \partial^{\varphi_*^o f} \ . \tag{4.54}$$

Here we assume a suitable extension of the coherent orientation of $\varphi(M)$ to the whole N. Thus, the identification with the subcomplex together with the diffeomorphical functor with respect to the diffeomorphism

$$\varphi : M \xrightarrow{\approx} \varphi(M)$$

gives rise to the chain map

$$\varphi_\bullet : C_*(M,\, f) \to C_*(N,\, \varphi_* f)$$

and consequently to the homomorphisms

$$\varphi_* : H_*(M,\, f) \to H_*(N,\, \varphi_* f)$$

and

$$\varphi_* : H_*(M) \to H_*(N)$$

with respect to the identification process. Once again, we verify the composition rule, so that we obtain the sequence of functors

$$H_k : \big(\text{manifolds, closed embeddeddings}\big) \to \mathcal{AB},^{9} \ \ k \in \mathbb{N}_0 \ . \tag{4.55}$$

Before the next step of the generalization aiming at arbitrary smooth mappings, we first prove the crucial

Lemma 4.18 (homotopy lemma) *A smooth 1-parameter family of closed embeddings* $(\varphi_t : M \hookrightarrow N)_{t\in[0,1]}$ *implies the identity*

$$\varphi_{0*} = \varphi_{1*} : H_*(M) \to H_*(N) \ .$$

The proof of this lemma can be reduced to the following special case:

Auxiliary Proposition 4.19 *Let* f *be a Morse function on* M *and* $i^0,\, i^1 :$ $M \hookrightarrow \mathbb{R} \times M$ *be the embeddings given by*

$$i^\nu(m) = (\nu,\, m) \ \ \textit{for}\ \nu = 0,1 \ .$$

9category of Abelian groups

Additionally, we consider the quadratic functions

$$\begin{aligned} &q_\nu : \mathbb{R} \to \mathbb{R} \\ &q_\nu(x) = (x-\nu)^2, \ \nu = 0, 1 \ . \end{aligned}$$

Then the isomorphisms

$$i_*^\nu : H_*(f) \xrightarrow{\cong} H_*(q_\nu \oplus f)$$

induced by the canonical identification $\operatorname{Crit}_* f \cong \operatorname{Crit}_*(q_\nu \oplus f)$ *and the canonical homology isomorphism*

$$\Phi_*^{10} : H_*(q_0 \oplus f) \xrightarrow{\cong} H_*(q_1 \oplus f)$$

fit in with the commutative diagram

$$\begin{array}{ccc} & & H_*(q_0 + f) \\ & \overset{i_*^0}{\nearrow} & \\ H_*(f) & \oplus & \Big\downarrow \Phi_*^{10} \\ & \underset{i_*^1}{\searrow} & \\ & & H_*(q_1 + f) \ . \end{array}$$

Conclusion: If we carry out the identification of the homology groups $H_*(\tilde{f})$ for Morse functions $\tilde{f}$ on $\mathbb{R} \times M$, we obtain the identity

$$i_*^0 = i_*^1 : H_*(M) \to H_*(\mathbb{R} \times M) \tag{4.56}$$

from this commutative diagram.

Proof of the auxiliary proposition 4.19: We have to verify the existence of a Morse homotopy

$$q_0 \oplus f \overset{h_t}{\simeq} q_1 \oplus f \ ,$$

such that the associated homomorphisms $\Phi_k^{h_t} : C_k(q_0 \oplus f) \to C_k(q_1 \oplus f)$ map the generating critical points to each other according to the formula

$$\Phi_k^{h_t}(0, y_k) = (1, y_k) \quad \text{for all } y_k \in \operatorname{Crit}_k f, \ k \in \mathbb{N} \ . \tag{4.57}$$

We consider a smooth, monotone function $\alpha \in C^\infty(\mathbb{R}, [0,1])$ satisfying

$$\alpha(t) = \begin{cases} 0, & t \leqslant 0 \\ 1, & t \geqslant 0 \end{cases}$$

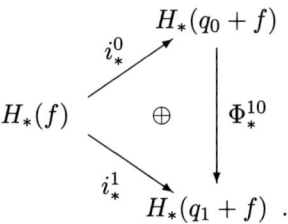

and we define the homotopy

$$h_t : \mathbb{R} \times M \to \mathbb{R}, \ t \in [0,1]$$
$$h_t(x, y) = \big(x - \alpha(t)\big)^2 + f(y) \ . \tag{4.58}$$

It is clear that this 'translation' represents a Morse homotopy between $q_0 \oplus f$ and $q_1 \oplus f$. For the sake of simplicity we content ourselves with a consideration of the homomorphism $\Phi_k^{h_t}$ with respect to the isolated h-trajectories for the time-dependent negative gradient flow

$$\big(\dot\beta(t), \dot\gamma(t)\big) = -\nabla h_t \circ \big(\beta(t), \gamma(t)\big) \ . \tag{4.59}$$

We leave out the norming by the scaling factor $\left(1 + |\dot h_t|^2 |\nabla h_t|^2\right)^{-\frac{1}{2}}$. This is admissible for the following reasons: First, we will derive explicitly the necessary compactness results for this special Morse homotopy. This feature was the only reason in general for the scaling. Second, by means of a homotopy using an additional parameter λ, we may verify that the homomorphism Φ_k on the level of homology is independent of the special form of this norming.

The isolated h-trajectories

$$(\beta, \gamma) \in \mathcal{M}^{h_t}_{z_0, z_1}, \ \ z_\nu = (\nu, y_\nu^k) \in \operatorname{Crit}_k(q_\nu \oplus f), \ \nu = 0, 1$$

are determined uniquely by (4.59) as solutions of the differential equation

$$\big(\dot\beta(t), \dot\gamma(t)\big) = \Big(2\big(\alpha(t) - \beta(t)\big), -\nabla f \circ \gamma(t)\Big) \ . \tag{4.60}$$

Now let Ψ be the isomorphism

$$\Psi = \frac{\partial}{\partial t} + 2 : H^{1,2}(\mathbb{R}, \mathbb{R}) \xrightarrow{\cong} L^2(\mathbb{R}, \mathbb{R}), \ \ \Psi \in \Sigma_{\text{triv}} \ .$$

Since $\dot\alpha$ is a smooth function with compact support according to the construction, we observe that

$$\dot\alpha \in C_0^\infty(\mathbb{R}, \mathbb{R}) \subset L^2(\mathbb{R}, \mathbb{R}) \ .$$

As a consequence, $\Psi^{-1}(\dot\alpha)$ is determined uniquely. Thus, $\beta_0 = \alpha - \Psi^{-1}(\dot\alpha) \in C^\infty(\mathbb{R}, \mathbb{R})$ solves the equation

$$\left(\frac{\partial}{\partial t} + 2\right) \cdot \beta_0 = \Psi(\alpha) - \dot\alpha = 2\alpha \ . \tag{4.61}$$

To put it in simpler terms, β_0 is the unique solution of

$$\dot\beta_0(t) = 2\left(\alpha(t) - \beta_0(t)\right), \ t \in \mathbb{R}; \ \ \beta_0 \in \mathcal{P}^{1,2}_{0,1}(\mathbb{R}, \mathbb{R}) \ .$$

This proves the statement that each isolated trajectory $(\beta_0, \gamma) \in \mathcal{M}^{h_t}_{z_0,z_1}$ corresponds to exactly one trajectory of the time-independent negative gradient flow for f. This amounts to the one-to-one correspondence

$$\begin{array}{ccc} \mathcal{M}^f_{x,y} & \xrightarrow{\cong} & \mathcal{M}^{h_t}_{z_0,z_1}, \\ \gamma & \mapsto & (\beta_0, \gamma) \end{array}$$

with $z_0 = (0, x)$ und $z_1 = (1, y)$. Due to the assumption we are considering the relative Morse index $\mu(x) = \mu(z_0) = \mu(z_1) = \mu(y)$. Thus, the time-independent f-trajectories in question have to be constant, such that $\mathcal{M}^{h_t}_{z_0,z_1}$ consists exactly of the trajectories

$$\mathcal{M}^{h_t}_{z_0,z_1} = \left\{ (\beta_0, y^k) \in \mathcal{P}^{1,2}_{z_0,z_1}(\mathbb{R} \times M) \;\middle|\; y^k \in \operatorname{Crit}_k f \right\} .$$

From this we easily conclude the assertion. □

Proof of the homotopy lemma. Let us consider the commutative diagram of closed embeddings for any $t \in [0, 1]$:

$$\begin{array}{ccc} M & \xhookrightarrow{i^t_M} & \mathbb{R} \times M \\ \varphi_t \Big\downarrow & \oplus & \Big\downarrow \phi \\ N & \xhookrightarrow{i^t_N} & \mathbb{R} \times N \end{array}$$

where

$$\begin{aligned} \phi(t) &= \big(t, \varphi_t(x)\big) , \\ i^t_M(x) &= (t, x), \; i^t_N(y) = (t, y) . \end{aligned}$$

Since we are already allowed to deal with the covariant homological functor for closed smooth embeddings, we deduce the identity

$$\phi_* \circ i^t_{M*} = i^t_{N*} \circ \varphi_{t*} : H_*(M) \to H_*(\mathbb{R} \times N) . \tag{4.62}$$

The auxiliary proposition states that isomorphisms i^t_{M*} and i^t_{N*} do not depend on the parameter t. Hence we obtain the identity

$$\varphi_{t*} = {i_N}_*^{-1} \circ \phi_* \circ i_{M*}$$

independent of $t \in [0, 1]$, also. □

The Homology Functor for Arbitrary Smooth Mappings

In the next passages we concentrate on projection mappings of the type

$$p : M \times N \to N \ ,$$

which play an essential role second to that of closed embeddings. The idea is to extract the functorial behaviour of the homology morphisms step by step from the projections $p^k : \mathbb{R}^k \times M \to M$. By analogy with the 1-dimensional case in the auxiliary Proposition 4.19 we consider the canonical identification

$$\mathrm{Crit}_j(f) \cong \mathrm{Crit}_j(q_k \oplus f), \ j = 0, \ldots, n \, ,$$

with respect to the canonical quadratic form

$$q_k(x) = \langle\, x, \, x \,\rangle, \ x \in \mathbb{R}^k \ .$$

First, we again obtain the isomorphisms $i^k_* : H_*(f) \xrightarrow{\cong} H_*(q_k \oplus f)$ yielding

$$i^k_* : H_*(M) \xrightarrow{\cong} H_*(\mathbb{R}^k \times M)$$

independent of the Morse function f. This is due to compatibility with Morse homotopies. Thus, the inverse mapping by means of the identification of critical points represents the homomorphism p^k_* induced by the projection map p^k,

$$p^k_* = \left(i^k_*\right)^{-1} : H_*(\mathbb{R}^k \times M) \xrightarrow{\cong} H_*(M) \ . \tag{4.63}$$

Using the notation

$$\begin{array}{rcl} i^{k+l,k} : \mathbb{R}^k \times M & \hookrightarrow & \mathbb{R}^{k+l} \times M \\ \big((x_1, \ldots, x_k), \, m\big) & \mapsto & \big((x_1, \ldots, x_k, 0, \ldots, 0), \, m\big) \end{array}$$

we may immediately set up the commutative diagram

(4.64)

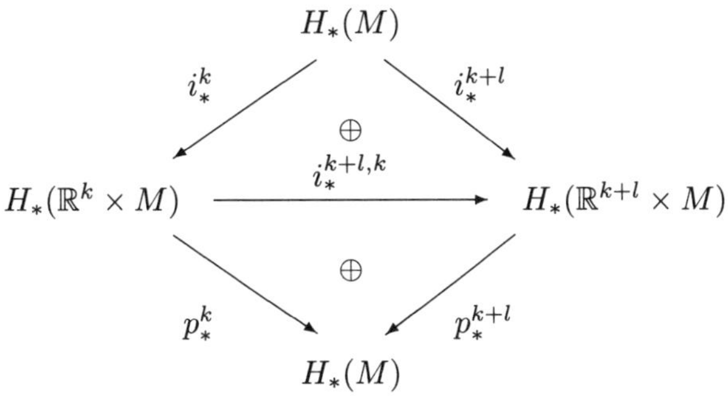

which describes the relations between the inclusions and the projections as far as the trivial extensions by $\mathbb{R}^\nu$ are concerned.

Definition 4.20 *Resorting to a suitable closed embedding* $\varphi^k : M \hookrightarrow \mathbb{R}^k$ *we are able to factor the projection map* $p : M \times N \to N$ *into*

$$p = p^k \circ \left(\varphi^k \times \mathrm{id}_N\right) .$$

According to this factoring we define

$$\begin{aligned} p_* &: H_*(M \times N) \to H_*(N) \\ p_* &= p^k_* \circ \left(\varphi^k \times \mathrm{id}_N\right)_* . \end{aligned}$$

Obviously, we have to verify that this homomorphism p_* does not depend on the noncanonical choice of the embedding φ^k. For this item we rely on the homotopy lemma.

Let $\psi^l : M \hookrightarrow \mathbb{R}^l$ be another closed embedding. Extending both given embeddings by means of the inclusions $i^{k+l,k}$ and $i^{k+l,l}$, i.e.

$$i^{k+l,k} \circ \varphi^k \text{ and } i^{k+l,l} \circ \psi^l : M \hookrightarrow \mathbb{R}^{k+l} ,$$

and regarding the functorial behaviour of the homological morphisms with respect to closed embeddings together with relation (4.64), we may start from two closed embeddings $\varphi, \psi : M \hookrightarrow \mathbb{R}^n$ without loss of generality. Due to the homotopy lemma and the uniform product structure it is sufficient to construct a homotopy $\varphi \simeq \psi$. Since we are already allowed to exploit the functorial behaviour with respect to the composition $p^{n,2n} \circ i^{2n,n} = \mathrm{id}_{\mathbb{R}^n}$, we are free to compose embeddings φ and ψ with the inclusion

$$i^{2n,n} : \mathbb{R}^n \hookrightarrow \mathbb{R}^{2n} .$$

Summing up, it is sufficient to find a homotopy

$$\varphi \overset{\phi_t}{\simeq} \psi, \ \phi_t : M \hookrightarrow \mathbb{R}^{2n} \text{ closed embedding for all } t \in [0,1] . \tag{4.65}$$

Let $\tau : [0,1] \to [0,1]$ be a smooth monotone mapping, locally constant at 0 and 1 satisfying $\tau(0) = 0$ und $\tau(1) = 1$. Then we define the homotopies

$$\begin{aligned} &\phi^1 : M \times [0,1] \to \mathbb{R}^n \oplus \mathbb{R}^n, \\ &\phi^1(\cdot, t) = \Big(\big(1 - \tau(t)\big) \cdot \varphi, \ \tau(t) \cdot \varphi\Big) \end{aligned}$$

and

$$\begin{aligned} &\phi^2 : M \times [0,1] \to \mathbb{R}^n \oplus \mathbb{R}^n, \\ &\phi^2(\cdot, t) = \Big(\tau(t) \cdot \psi, \ \big(1 - \tau(t)\big) \cdot \varphi\Big) . \end{aligned}$$

Thus we obtain smooth homotopies of closed embeddings for each fixed parameter t satisfying

$$(\varphi, 0) \overset{\phi^1_t}{\simeq} (0, \varphi) \overset{\phi^2_t}{\simeq} (\psi, 0) .$$

We finally reparametrize this composition of homotopies obtaining

$$\phi_t = \begin{cases} \phi^1_{2t}, & 0 \leqslant t \leqslant \frac{1}{2} \\ \phi^2_{2t-1}, & \frac{1}{2} \leqslant t \leqslant 1 \end{cases}$$

as a smooth homotopy due to the definition of τ. Obviously, ϕ_t fulfills the required properties.

By means of this method of factoring through a closed embedding $\varphi : M \hookrightarrow \mathbb{R}^k$ we may derive the identity

$$p_* \circ i_{c*} = \mathrm{id}_{H_*(N)} \tag{4.66}$$

for arbitrary inclusions

$$\begin{aligned} i_c : N &\hookrightarrow M \times N, \\ y &\mapsto (c,\, y), \;\; c \in M \text{ fixed}, \end{aligned}$$

from the homotopy lemma.

Now we are able to associate to any arbitrary smooth mapping between manifolds a homomorphism between the respective Morse homology groups.

Definition 4.21 *Let $f \in C^\infty(M,\, N)$ be a smooth mapping. Then the graph-embedding*

$$(\mathrm{id}, f) : M \hookrightarrow M \times N$$

represents a closed embedding, so that

$$(\mathrm{id}, f)_* : H_*(M) \to H_*(M \times N)$$

is well-defined. Resorting to the projection homomorphism constructed above, $p_* : H_*(M \times N) \to H_*(N)$, we now define

$$f_* = p_* \circ (\mathrm{id}, f)_* : H_*(M) \to H_*(N)\ . \tag{4.67}$$

To sum up the functorial construction, we choose indirect way toward the general homological functor via the factorization $f = p \circ (\mathrm{id},\, f)$ into mappings which can be treated more easily. We now have to verify that this definition is compatible with the already existing functor for closed embeddings and projection maps and that the cofunctor property of a covariant functor

$$(g \circ f)_* = g_* \circ f_*$$

is satisfied.

Auxiliary Proposition 4.22 *Given a closed embedding $f : M \hookrightarrow N$, the directly induced homomorphism $(\varphi_f)_*$ from (4.55) is identical to f_*:*

$$f_* = (\varphi_f)_* : H_*(M) \to H_*(N) \ .$$

Proof. Let $\varphi^M : M \hookrightarrow \mathbb{R}^m$ be a closed embedding. We consequently obtain the commutative diagram

$$\begin{array}{ccccc} M & \overset{f}{\hookrightarrow} & N & & \\ {\scriptstyle (\mathrm{id}, f)} \downarrow & \oplus \quad \begin{array}{c}(0, \mathrm{id}) \\ = i\end{array} & \downarrow & \oplus & \searrow {\scriptstyle \mathrm{id}} \\ M \times N & \overset{\varphi^M \times \mathrm{id}}{\hookrightarrow} & \mathbb{R}^m \times N & \overset{p}{\longrightarrow} & N \ . \end{array} \tag{4.68}$$

The two closed embeddings

$$\phi_0 = i \circ f \quad \text{and} \quad \phi_1 = (\varphi \times \mathrm{id}) \circ (\mathrm{id},\, f)$$

can be related by means of a homotopy through closed embeddings

$$\phi_t = (t \cdot \varphi,\, f) : M \to \mathbb{R}^m \times N,\ t \in [0,1] \ .$$

Then the homotopy lemma implies the identity $\phi_{0*} = \phi_{1*}$ from which the proof follows immediately due to the equality $p_* \circ i_* = \mathrm{id}_*$. □

We are able to prove analogously

Auxiliary Proposition 4.23 *The projection map $p : M \times N \to N$ directly gives rise to the homomorphism $(\varphi_p)_*$ from Definition 4.20, and this homomorphism is identical to p_*:*

$$p_* = (\varphi_p)_* : H_*(M \times N) \to H_*(N) \ .$$

The last step toward the covariant homological functor follows from

Auxiliary Proposition 4.24 *Any two smooth mappings $f \in C^\infty(M, N)$ and $g \in C^\infty(N, P)$ satisfy the composition rule*

$$(g \circ f)_* = g_* \circ f_* \ .$$

Proof. At first, let us consider the commutative diagram

$$
\begin{array}{ccccc}
N & \xrightarrow[= i]{(0,\mathrm{id})} & \mathbb{R}^m \times N & \xrightarrow{\mathrm{id}\times(\mathrm{id},g)} & \mathbb{R}^m \times N \times P \\
\Big\downarrow (\mathrm{id}, g) & & \oplus & & \Big\downarrow p_3 \\
N \times P & & \xrightarrow{\qquad p_2 \qquad} & & P \; .
\end{array}
\tag{4.69}
$$

Due to the definition of i_*, p_*, etc., the following identity is true:

$$p_{2*} \circ (\mathrm{id},\, g)_* = p_{3*} \circ \big(\mathrm{id} \times (\mathrm{id},\, g)\big)_* \circ i_* \; . \tag{4.70}$$

As to the projection map $p_2' : \mathbb{R}^m \times N \to N$, we obtain the isomorphism $i_* = \left(p'_{2*}\right)^{-1} : H_*(N) \xrightarrow{\cong} H_*(\mathbb{R}^m \times N)$ and hence the identity

$$g_* \circ p'_{2*} = p_{3*} \circ \big(\mathrm{id} \times (\mathrm{id},\, g)\big)_* : H_*(\mathbb{R}^m \times N) \to H_*(P) \tag{4.71}$$

from (4.70) according to the definition $g_* = p_{2*} \circ (\mathrm{id},\, g)_*$. Setting $p_3 = p \circ (\mathrm{id} \times \varphi^N \times \mathrm{id})$ and choosing suitable closed embeddings $\varphi^M : M \hookrightarrow \mathbb{R}^m$ and $\varphi^N : N \hookrightarrow \mathbb{R}^n$ we are led to the commutative diagram

$$
\begin{array}{ccccc}
M \times N & \xrightarrow{\varphi^M \times \mathrm{id}} & \mathbb{R}^m \times N & \xrightarrow{\mathrm{id}\times(\mathrm{id},g)} & \mathbb{R}^m \times N \times P \\
(\mathrm{id}, f) \Big\uparrow & \oplus & p_2' \Big\downarrow \Big\uparrow i & \oplus & \Big\downarrow p_3 \\
M & \xrightarrow{\quad f \quad} & N & \xrightarrow{\quad g \quad} & P
\end{array}
$$

Thus, we deduce the identity

$$g_* \circ f_* = p_{3*} \circ \big(\mathrm{id} \times (\mathrm{id},\, g)\big)_* \circ \left(\varphi^M \times \mathrm{id}\right)_* \circ (\mathrm{id},\, f)_* \tag{4.72}$$

from (4.71) together with this diagram. Altogether, we come to the factorization

$$g_* \circ f_* = p_* \circ \phi_{0*} \; , \tag{4.73}$$

where $\phi_0 = \left(\varphi^M,\, \varphi^N \circ f,\, g \circ f\right)$ describes a closed embedding. From the other direction, we consider the expression

$$(g \circ f)_* = p_* \circ \phi_{1*} \tag{4.74}$$

using the closed embedding

$$\phi_1 = \left(\varphi^M \times \mathrm{id}\right) \circ (\mathrm{id},\, g \circ f) : M \hookrightarrow \mathbb{R}^m \times \mathbb{R}^n \times P \; .$$

Relying on (4.65) we may once again set up a homotopy $\phi_0 \overset{\phi_t}{\simeq} \phi_1$ through closed embeddings $\phi_t : M \hookrightarrow \left(\mathbb{R}^m \times \mathbb{R}^n\right)^2$, so that the assertion follows from the homotopy lemma. □

Proposition 4.25 *The Morse homology groups*

$$H_k^{\text{Morse}}(M) = \varprojlim H_k(M,\, f), \quad k \in \mathbb{N}_0$$

can be organized as a family of covariant functors from the category of smooth manifolds without boundary together with smooth maps as morphisms into the category of Abelian groups. Moreover, homotopical maps induce identical group homomorphisms, i.e. H_k *is a covariant functor for each* $k \in \mathbb{N}_0$

$$\begin{array}{rcll} H_k : & \left(C^\infty\text{-}mfds., \begin{array}{c} homotopy\ classes \\ of\ C^\infty\text{-}maps \end{array} \right) & \to & \mathcal{AB} \\ & \left(M \xrightarrow{[f]} N \right) & \mapsto & \left(f_* : H_k(M) \to H_k(N)\right) . \end{array}$$

Proof. We have developed all the details except the explicit homotopy invariance for arbitrary smooth mappings. But it is due to the definition of f_* by means of the factorization that this follows immediately from the homotopy between the closed embeddings $(\mathrm{id},\, f) \simeq (\mathrm{id},\, g)$ induced by $f \simeq g$ and from the homotopy lemma. □

4.2.3 Relative Morse Homology

In this section we finally analyse the feature which is still lacking with respect to an axiomatic homology theory. That is the concept of relative homology groups associated to admissible pairs (M, A) of manifolds. The underlying idea is to deduce such homology groups from a chain complex, which is now generated by critical points of a Morse function f on M outside of A. The appropriate algebraic object is a quotient complex analogous to the singular theory,

$$C_*(f,\, f_A) = C_*(f) \big/ C_*(f_A) \ .$$

In this situation, the Morse complex $C_*(f_A)$ of a Morse function on A has to admit a canonical identification with a subcomplex of $C_*(f)$. Altogether, we once again require a functorial concept which guarantees at each stage compatibility with the axiomatically required long exact homology sequence. As to this concept of a relative homology theory, we first have to compile some technical preparations.

Relative Morse Functions and the Relative Morse Complex

Definition 4.26 *We call a pair of smooth manifolds without boundary (M, A), where A is a submanifold of M, admissible, if either A is already a closed submanifold or the topological boundary ∂A is a 1-codimensional, orientable, closed submanifold of M. If (M, A) is admissible and of the latter type, we call a smooth function $f \in C^\infty(M, \mathbb{R})$ steep w.r.t. ∂A, if $-\nabla f$ represents an inner normal field with respect to ∂A, that is:*

- $\nabla f \pitchfork \partial A \subset M$ *and*
- *the negative gradient flow of f on ∂A flows into A.*

Since we have already treated sufficiently the aspect of suitable Morse functions for closed submanifolds, we shall henceforth consider the situation of admissible pairs of manifolds where A is an open submanifold whenever we do not give an explicit specification.

Remarks

(a) The just defined steepness of a function with respect to ∂A is obviously a convex property, that is: If $f, g \in C^\infty(M, \mathbb{R})$ are steep w.r.t. ∂A, the same is true for all $(1-t)f + tg$, $t \in [0, 1]$.

(b) If f is steep w.r.t. ∂A, it holds that $\operatorname{Crit} f \cap \partial A = \emptyset$. Moreover, since M is a normal topological space, there is a neighbourhood $U(A)$, which separates A from $\operatorname{Crit} f$, $U(A) \cap \operatorname{Crit} f = \emptyset$.

Definition 4.27 *We now call $f \in C^\infty(M, \mathbb{R})$ a Morse function on (M, A), if f is a Morse function on M and additionally steep w.r.t. ∂A. In the situation of a closed submanifold $A \subset M$, instead of steepness, f has to yield a subcomplex $C_*(f_{|A})$ of $C_*(f)$ in the sense of Corollary 4.17, i.e. there are no f-trajectories leaving A.*

In principle this definition establishes the coercivity property for the Morse function f on M relative to A. Since due to the above assumption A is an open submanifold of M, $f_{|A}$ cannot be a Morse function on A if $\partial A \neq \emptyset$ holds. In this case the coercivity property is no longer satisfied, as not all sublevel-sets are compact.[10] However, since we are able to isolate the critical

10Referring to Definition 2.40 concerning the compactness analysis we restate that we have to consider manifolds with complete Riemannian metrics. Since the submanifold A with non-void boundary cannot be complete with respect to the induced Riemannian metric, we have to alter this metric near the boundary. Similar to the following argument, this can be accomplished without changing the Morse-homological information, i.e. $\operatorname{Crit}_*(f_A)$ and the connecting isolated trajectories.

points of f from the boundary ∂A, it is possible to transform $f_{|A}$ on a suitable neighbourhood of ∂A to a Morse function f_A such that the respective Morse complex, that is, the critical points and the associated trajectory spaces within A, remains unchanged.

Given the closed submanifold $\partial A \subset M$ we choose a tubular neighbourhood

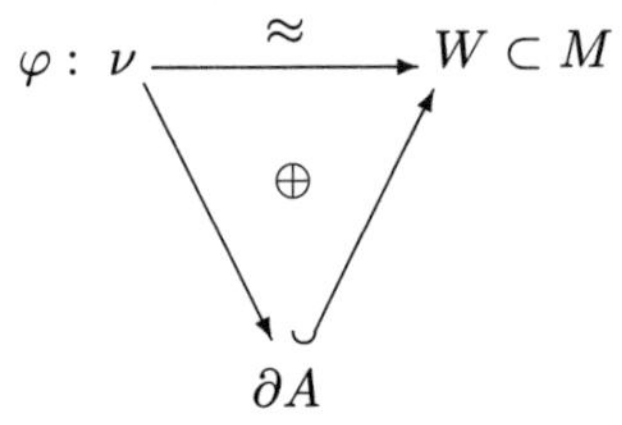

and a Riemmanian metric on the normal bundle ν. Then starting from

$$N_\epsilon^0 = \{\, v_x \in \nu \mid \|v_x\| < \epsilon(x) \ \text{ for all } x \in \partial A \,\}, \quad \epsilon \in C^\infty\big(\partial A,\, (0,1)\big)$$

φ induces a topological basis of neighbourhoods $\{\, N_\epsilon = \varphi(N_\epsilon^0) \,\}_{\epsilon \leqslant \epsilon_0}$ of ∂A for $\epsilon_0 = \epsilon_0(\partial A) \in C^\infty\big(\partial A, (0,1)\big)$ small enough.

Lemma 4.28 *Given a Morse function f on (M, A), there is an $\epsilon \in C^\infty\big(\partial A,$ $(0,1)\big)$, $\epsilon \leqslant \epsilon_0$ and a Morse function f_A on A, such that*

$$f_{A|\,A\setminus N_\epsilon} = f_{|\,A\setminus N_\epsilon} \quad \textit{and} \quad \operatorname{Crit} f_A \cap N_\epsilon = \emptyset \;.$$

Proof. Due to the assumption f is steep w.r.t. ∂A and $\{N_\epsilon\}_{\epsilon \leqslant \epsilon_0}$ represents a neighbourhood basis of ∂A. Hence, referring to remark (b) above we may start from an $\epsilon_0 \in C^\infty\big(\partial A,\, (0,1)\big)$ such that $\operatorname{Crit} f \cap N_\epsilon = \emptyset$ for all $\epsilon \leqslant \epsilon_0$. Setting $N_\epsilon^+ = N_\epsilon \cap A$, we define the smooth function

$$\begin{aligned} g_\epsilon^0 :\ & \varphi^{-1}(N_\epsilon^+) \to \mathbb{R}, \\ & v_x \mapsto \mathrm{e}^{\frac{1}{\|v_x\|_x} - \frac{1}{\epsilon(x) - \|v_x\|_x}} \end{aligned}$$

on $\varphi^{-1}(N_\epsilon^+) = \{\, v_x \in \nu^+ \mid \|v_x\|_x < \epsilon \}$. Thus, composing $g_\epsilon = g_\epsilon^0 \circ \varphi^{-1}$, we obtain a function $g_\epsilon \in C^\infty\big(N_\epsilon^+,\, (0,1)\big)$, such that

$$\begin{aligned} & f_\epsilon : A \to \mathbb{R}, \\ & f_\epsilon(x) = \begin{cases} f(x) + g_\epsilon(x), & x \in N_\epsilon^+ \\ f(x), & x \in A \setminus N_\epsilon \end{cases} \end{aligned}$$

is well-defined. Due to the construction of g_ϵ, f_ϵ is a smooth and coercive function on A, which is identical to f if restricted to $A \setminus N_\epsilon$. Since $-\nabla f$ represents an inner normal field of ∂A, we may choose ϵ small enough, such that ∇f_ϵ does not vanish within N_ϵ^+, too. Thus, $f_A = f_\epsilon$ is a suitable Morse function on A according to the assertion. □

Definition 4.29 *Let (M, A) be an admissible pair of manifolds. Then we call two Morse functions f and f' on (M, A) equivalent relative to A, if they are identical on $\overline{M\backslash A}$,*

$$f \sim_A f' \overset{\text{def}}{\Longleftrightarrow} f_{|\overline{M\backslash A}} = f'_{|\overline{M\backslash A}} \ .$$

The equivalence class $[f]_A$ is called a relative Morse function on (M, A). Note that this definition remains reasonable in the case when A is a closed submanifold of M.

Lemma 4.30 *We can identify the set of relative Morse functions*

$$\{\, [f]_A \,|\, f \text{ a Morse function on } (M, A)\}$$

with the set of smooth functions $f \in C^\infty(\overline{M\backslash A}, \mathbb{R})$,[11] which – in the situation of an open submanifold A – are steep w.r.t. ∂A, coercive on $M\backslash A$ and whose critical points within $M\backslash\overline{A}$ are regular.

Proof. The proof of this lemma is straightforward if one uses the extension results from the first section of this chapter. □

Example In the final chapter we shall briefly touch on the question of the representation of characteristic classes of vector bundles within the framework of this Morse homology theory. Let $\pi : E \to M$ be a smooth, n-dimensional vector bundle. Then, as we already know, any Morse function f on M together with the quadratic form q_n with respect to a Riemannian metric on E induces the Morse function $f \oplus q_n$ on E together with the canonical identification

$$\mathrm{Crit}_* f \cong \mathrm{Crit}_*(f \oplus q_n) \ .$$

In contrast let us now consider the negative quadratic form $-q_n$. It yields a fibrewise relative Morse function $-q_{n,x}$ on $\big(E_x, E_x\backslash U(0_x)\big)$ for each suitably small, connected and closed neighbourhood of zero $U(0_x)$ such that this pair is admissible. By analogy with above, we obtain a relative Morse function for the admissible pair $\big(E, E\backslash U(M)\big)$, where $U(M) \subset E$ is a suitably regular but arbitrarily small closed neighbourhood of the zero section. Moreover, we once again obtain a canonical identification of critical sets, but this time with a shifting in the grading,

$$\mathrm{Crit}_* f \cong \mathrm{Crit}_{*+n}\big(f \oplus (-q_n)\big) \ .$$

We guess that this is the inital step toward the Thom isomorphism.

11i.e. smooth on a neighbourhood of $\overline{M\backslash A}$ within M

Definition 4.31 *Due to Lemma 4.28, we can find such a pair of Morse functions (f, f_A) on (M, A) for each relative Morse function $[f]_A$, that the Morse complex $C_*(f_A)$ represents a subcomplex of $C_*(f)$. This means that we obtain a short exact sequence of chain complexes:*

$$0 \longrightarrow C_*(f_A) \xrightarrow{i_\bullet} C_*(f) \xrightarrow{j_\bullet} C_*(f)/_{C_*(f_A)} \longrightarrow 0 \ .$$

Consequently, we can define the quotient complex

$$C_k([f]_A) = C_k(f, f_A) = C_k(f)/_{C_k(f_A)}, \ k \in \mathbb{N}_0 \ ,$$

where $C_k(f, f_A)$ is generated by the critical points $\mathrm{Crit}_k f \backslash \mathrm{Crit}_k f_A$ lying within $M\backslash\overline{A}$. Since these critical points are already uniquely determined by the relative Morse function $[f]_A$, the relative Morse complex $C_*([f]_A)$ is well-defined, i.e. independent of the concrete choice of the pair (f, f_A) representing the fixed relative Morse function. We notice that due to the assumption of steepness w.r.t. ∂A, the ∂-operator on $C_*(f, f_A)$ is already uniquely determined by $f_{|\overline{M\backslash A}}$. It is now this relative chain complex from which we gain the relative homology groups associated to $[f]_A$,

$$H_k(f, f_A) = H_k([f]_A) = H_k(C_*(f, f_A)), \ k \in \mathbb{N}_0 \ ,$$

together with a long exact homology sequence

$$\ldots \longrightarrow H_k(f_A) \xrightarrow{(i_\bullet)_*} H_k(f) \xrightarrow{(j_\bullet)_*} H_k(f, f_A) \xrightarrow{\partial_*} H_{k-1}(f_A) \longrightarrow \ldots \ .$$

We finish this introduction of a relative Morse complex with the following lemma:

Lemma 4.32 *There is a relative Morse function $[f]_A$ for each admissible pair of manifolds (M, A).*

Proof. In the situation of a closed submanifold $A \subset M$ we may refer immediately to the extension result in Proposition 4.16. This is due to classical Morse theory [M1], which states that we can find a Morse function on any given smooth manifold. Let us now regard the case where A is an open submanifold. We use the condition that the normal bundle of ∂A within M is orientable, so that we may construct a smooth function on the tubular neighbourhood of ∂A, which is coercive and steep with respect to ∂A. Again due to the extension lemmata we can extend this function to a Morse function on (M, A). □

The Relative Homology Functor

The next step is to deduce independence from the concrete relative Morse functions as in the situation of the absolute Morse homology groups. Actually, the possibility of change between suitably chosen Morse functions step by step allows us to build up a functorial concept. Now, the analogous limit process of the identification of the special homology groups depending on the functions becomes a bit more involved. Namely, we additionally require compatibility with the long exact homology sequence specified above.

Definition 4.33 *Let $[f]_A$ and $[g]_A$ be relative Morse functions on (M, A). We call a homotopy $[h_t]_A$ satisfying*

$$h_t : \overline{M\backslash A} \to \mathbb{R},\ t \in [0,1],$$
$$f_{|\overline{M\backslash A}} \overset{h_t}{\simeq} g_{|\overline{M\backslash A}}$$

a relative Morse homotopy w.r.t. A, if there are representatives f, g and h_t on M such that h_t describes a regular Morse homotopy $f \overset{h_t}{\simeq} g$ and h_t is steep w.r.t. ∂A for all $t \in \mathbb{R}$.

Lemma 4.34 *Every two relative Morse functions $[f]_A$ and $[g]_A$ admit a relative Morse homotopy $[h_t]_A$.*

Proof. Considering the regularity condition for Morse homotopies from Definition 2.29, we begin construction with the trivial homotopy

$$h_t^o = \beta(t) \cdot f + \big(1 - \beta(t)\big) \cdot g\ .$$

Here, $\beta : \mathbb{R} \to [0,1]$ denotes a smooth function satisfying

$$\beta(t) = \begin{cases} 1, & t \leqslant -1\,, \\ 0, & t \geqslant 1\ . \end{cases}$$

Since steepness is a convex property as we already remarked above, the requirement that h_t^o be steep w.r.t. ∂A for all $t \in \mathbb{R}$ is satisfied. Thus, it particularly holds that

$$\operatorname{Crit} h_t^o \cap U(\partial A) = \emptyset \quad \text{for all } t \in \mathbb{R}$$

for an appropriate neighbourhood of ∂A. Then, given a suitably small $N_\epsilon \subset U$ from the neighbourhood basis specified above, we define the following function:

$$h : \big(\mathbb{R} \times N_\epsilon\big) \cup \big(\mathbb{R}\backslash(-1,1) \times M\big) \to \mathbb{R}$$
$$h(t,x) = \begin{cases} f(x), & t \leqslant -1 \\ g(x), & t \geqslant 1 \\ h_t^o(x), & x \in N_\epsilon\ . \end{cases}$$

This can finally be extended yielding a regular Morse homotopy such that the steepness condition is in fact fulfilled. □

Lemma 4.35 *Each representative $f \overset{h_t}{\simeq} g$ of a relative Morse homotopy $[h_t]_A$ admits Morse functions f_A, g_A and a Morse homotopy $h_{A,t}$ on A such that the identity*

$$h_{A,t}|_{A \backslash N_\epsilon} = h_t|_{A \backslash N_\epsilon}$$

holds for some $\epsilon \in C^\infty\big(\partial A,\, (0,1)\big)$.

Proof. The proof of this lemma may be carried out by the same means as in Lemmata 4.28 and 4.34. □

Thus, the induced homotopy morphisms

$$\begin{aligned} \Phi^{h_t} :&\quad C_*(f) \to C_*(g) \quad \text{and} \\ \Phi^{h_{A,t}} :&\quad C_*(f_A) \to C_*(g_A) \end{aligned}$$

satisfy the consistency condition

$$\Phi^{h_t}|_{C_*(f_A)} = \Phi^{h_{A,t}} . \tag{4.75}$$

Hence, we obtain a morphism between the respective short exact sequences of chain complexes,

$$\begin{array}{ccccccccc} 0 & \longrightarrow & C_*(f_A) & \xrightarrow{i_\bullet} & C_*(f) & \xrightarrow{j_\bullet} & C_*([f]_A) & \longrightarrow & 0 \\ & & \downarrow \Phi^{h_{A,t}} & \oplus & \downarrow \Phi^{h_t} & \oplus & \downarrow \Phi^{[h_t]_A} & & \\ 0 & \longrightarrow & C_*(g_A) & \xrightarrow{i_\bullet} & C_*(g) & \xrightarrow{j_\bullet} & C_*([g]_A) & \longrightarrow & 0 \end{array} \tag{4.76}$$

Due to the steepness condition of h_t, $\Phi^{[h_t]_A}$ is well-defined, independently of the actual extension $(h_t,\, h_{A,t})$. This morphism of short exact sequences of chain complexes gives rise to a homotopy morphism between the associated long exact homology sequences

$$\begin{array}{cccccccccc} \cdots & \longrightarrow & H_k(f_A) & \longrightarrow & H_k(f) & \longrightarrow & H_k(f, f_A) & \longrightarrow & H_{k-1}(f_A) & \longrightarrow \cdots \\ & & \downarrow \Phi_*^{h_A} & \oplus & \downarrow \Phi_*^{h} & \oplus & \downarrow \Phi_*^{[h]_A} & \oplus & \downarrow \Phi_*^{h_A} & \\ \cdots & \longrightarrow & H_k(g_A) & \longrightarrow & H_k(g) & \longrightarrow & H_k(g, g_A) & \longrightarrow & H_{k-1}(g_A) & \longrightarrow \cdots . \end{array} \tag{4.77}$$

However, one problem remains. On the one hand, $\Phi_*^{h_{A,t}}$ and $\Phi_*^{h_t}$ fit in with the functorial concept, by means of which we establish the identification process. We consequently obtain independency from the concrete Morse functions, and we conclude from the algebraic five lemma that the $\Phi_*^{[h_t]_A}$ describe isomorphisms, too. On the other hand, we cannot necessarily conclude the crucial functorial condition

$$\begin{aligned} \Phi_*^{[h^{\gamma\beta}]_A} \circ \Phi_*^{[h^{\beta\alpha}]_A} &= \Phi_*^{[h^{\gamma\alpha}]_A} \\ \Phi_*^{[h^{\alpha\alpha}]_A} &= \mathrm{id}_{H_*(f,f_A)} \end{aligned} \tag{4.78}$$

for these isomorphisms. This has to lead back once more to the deduction from the λ-homotopies and the associated chain homotopy operators.

Resorting to the methods of extension in the previous lemmata, we can show that, given any two relative Morse homotopies $[h_0^{\alpha\beta}]_A$ and $[h_1^{\alpha\beta}]_A$, there are pairs of representatives $(h_0,\, h_{0,A})$ and $(h_1,\, h_{1,A})$ as well as a pair of λ-homotopies $(H_\lambda^{\alpha\beta},\, H_\lambda^{\alpha\beta})$ between these pairs of Morse homotopies, such that the condition of compatibility

$$\begin{array}{ccc} C_*(f_A^\alpha) & \xrightarrow{i_\bullet} & C_*(f^\alpha) \\ \Big\downarrow \Psi_A^{\beta\alpha} & \oplus & \Big\downarrow \Psi^{\beta\alpha} \\ C_{*+1}(f_A^\beta) & \xrightarrow{i_\bullet} & C_{*+1}(f^\beta) \end{array} \tag{4.79}$$

is satisfied by the associated chain homotopy operators $\Psi^{\beta\alpha}$ and $\Psi_A^{\beta\alpha}$.

Definition 4.36 *As a consequence, relation (4.79) induces a chain homotopy operator also for the relative Morse complexes*

$$\Psi_{\mathrm{rel}}^{\beta\alpha} : C_*(f^\alpha, f_A^\alpha) \to C_{*+1}(f^\beta, f_A^\beta)$$

with the same properties as in the absolute case. Hence, the isomorphism $\Phi_*^{[h]_A}$ of the homology groups is independent of the actual relative Morse homotopy, and relation (4.78) is true. Thus, the family

$$\big\{ H_*(f, f_A) \,\big|\, [f]_A \text{ rel. Morse fct.} \big\}$$

meets the identification requirement by means of the isomorphisms $\Phi_*^{[h^{\beta\alpha}]_A}$. Moreover, this process is compatible with the respective long exact sequence,

$$H_k(M, A) = \varprojlim H_k\big([f]_A\big),\ k \in \mathbb{N}_0 \ ,$$

so that we obtain the long exact sequence of Morse homology groups

$$\ldots \to H_k(A) \xrightarrow{(i_\bullet)_*} H_k(M) \xrightarrow{(j_\bullet)_*} H_k(M, A) \xrightarrow{\partial_*} H_{k-1}(A) \to \ldots \ .$$

The only feature which still has to be verified is the identity of the homomorphism $(i_\bullet)_*$ induced by the pairs (f, f_A) and the homomorphism i_*, which was directly deduced above by means of the factorization techniques. This is not obvious, as we generally have to start from open submanifolds $A \subset M$ with regard to the relative Morse homology.

Lemma 4.37 *Both homomorphisms $(i_\bullet)_*$ and i_* deduced from the inclusion $A \hookrightarrow M$ are identical.*

Proof. In the case of a closed submanifold, the proof was carried out above in the auxiliary Proposition 4.22. Hence, let A be given as an open submanifold. Recalling the definition, we obtain

$$i_* = p_* \circ (\varphi \times \mathrm{id}_M)_* \circ (\mathrm{id}_A \times i)_* : H_*(A) \to H_*(M) \ , \tag{4.80}$$

where φ represents a closed embedding $\varphi : A \hookrightarrow \mathbb{R}^k$. Let (f, f_A) be a representing pair belonging to a relative Morse function on (M, A) and let $g_A = (\varphi \times \mathrm{id}_A)_* f_A$ be a Morse function on $\mathbb{R}^k \times A$ induced by the closed embedding $\varphi \times \mathrm{id}_A$. We may construct a Morse function g on $\mathbb{R}^k \times M$ in an appropriate way, which is steep w.r.t. $\mathbb{R}^k \times \partial A$ and yields a relative Morse function together with g_A on $(\mathbb{R}^k \times M, \mathbb{R}^k \times A)$ in the sense of Lemma 4.28. Since A is embedded into $\mathbb{R}^k$ as a closed submanifold, we can likewise construct a relative Morse homotopy $[h_t]_A$ represented by a pair

$$(h_t, h_{A,t}) \ : \ g \overset{h_t}{\simeq} q_k \oplus f, \ g_A \overset{h_{A,t}}{\simeq} q_k \oplus f_A \ .$$

The first step of the proof now consists of the following commutative diagram

$$\begin{array}{ccccc} & & H_*(f_A) & & \\ & \swarrow^{(\{0\} \times \mathrm{id}_A)_*} & \oplus & \searrow^{(\varphi \times \mathrm{id}_A)_*} & \\ H_*(q^k \oplus f_A) & & \xleftarrow{\Phi_*^{h_A}} & & H_*(g_A) \ . \end{array} \tag{4.81}$$

This follows from the homotopy lemma because $\{0\} \times \mathrm{id}_A$ and $\varphi \times \mathrm{id}_A$ are homotopical closed embeddings of A into $\mathbb{R}^k \times A$. Fortunately, it does not matter that the associated homotopy of embeddings does not directly induce the previously fixed Morse homotopy h_A, because the homomorphism $\Phi_*^{h_A}$ on the level of the homology groups is independent of the concrete homotopy. The next diagram is an immediate consequence from the morphism (4.77) of the long exact sequences:

(4.82)
$$\begin{array}{ccc} H_*(q^k \oplus f_A) & \xleftarrow{\Phi_*^{h_A}} & H_*(g_A) \\ \Big\downarrow {\scriptstyle ((\mathrm{id} \times i)_\bullet)_*} & \oplus & \Big\downarrow {\scriptstyle ((\mathrm{id} \times i)_\bullet)_*} \\ H_*(q^k \oplus f) & \xleftarrow{\Phi_*^{h}} & H_*(g) \end{array} .$$

According to the definition, the homomorphisms $(\varphi \times i)_*$ from the closed embedding and $\big((\mathrm{id} \times i)_\bullet\big)_* \circ (\varphi \times \mathrm{id}_A)_*$ are identical. Therefore, the identification process by means of the Φ_*^h together with the combination of (4.81) and (4.82) imply the identity

(4.83) $(\varphi \times i)_* = \big((\mathrm{id} \times i)_\bullet\big)_* \circ \big(\{0\} \times \mathrm{id}_A\big)_* : H_*(A) \to H_*(\mathbb{R}^k \times M)$.

Since it holds that

$$(i_\bullet)_* = p_* \circ \big((\mathrm{id} \times i)_\bullet\big)_* \circ \big(\{0\} \times \mathrm{id}_A\big)_* : H_*(A) \to H_*(M)$$

due to the definition, the assertion follows immediately from the composition of (4.83) with the projection map $p : \mathbb{R}^k \times M \to M$. □

In order to complete the deduction, we still have to go through the steps where we develop the relative homological functor by analogy with the absolute case. Moreover, each step must be taken in accordance with the long exact homology sequence. This means that we first have to consider closed embeddings

$$\begin{aligned} &\varphi : (M, A) \hookrightarrow (N, B) \\ &\varphi : M \hookrightarrow N \text{ closed} \quad \text{and } \varphi(A) \subset B \ , \end{aligned}$$

and then projection maps $p : (M \times N, M \times B) \to (N, B)$. Finally, we must go through the factorizations

$$\begin{array}{ccccc} & & (M, A) & & \\ & \swarrow {\scriptstyle ((\mathrm{id}, f), (i, f))} & \oplus & {\scriptstyle (f, f_{|A})} \searrow & \\ (M \times N, M \times B) & & \xrightarrow{\quad p \quad} & & (N, B) \end{array} .$$

The crucial point is the possibility of changing the relative Morse function via an appropriate homotopy, so that the associated isomorphism $\Phi_*^{[h]_A}$ becomes the identity after the identification process. However, by analogy with Lemma 4.37, we have to verify again that homomorphism $(j_\bullet)_*$ is identical to homomorphism j_*, which is induced by the inclusion of pairs $j : (M, \emptyset) \hookrightarrow (M, A)$.

Actually, all proofs within this framework for relative Morse functions can be carried out analogously by means of the extension and homotopy techniques. Summing up, we obtain

Proposition 4.38 *The relative homology groups* $\big(H_k(M,A)\big)_{k\in\mathbb{N}_0}$ *for admissible pairs of manifolds and the long exact homology sequence fit in with the concept of a family of covariant functors*

$$H_k : \begin{pmatrix} \textit{admissible pairs of mfds.,} \\ \textit{smooth mappings of pairs} \end{pmatrix} \to \mathcal{AB}, \;\; k \in \mathbb{N}_0$$
$$\big((M,A) \xrightarrow{f} (N,B)\big) \mapsto \big(f_* : H_k(M,A) \to H_k(N,B)\big) \ .$$

By analogy with the absolute case and relying on the methods of relative Morse homology, we are able to prove the homotopy invariance:

> Homotopical mappings of pairs $f \simeq g : (M,A) \to (N,B)$ induce the same morphism $f_* = g_*$ for the long exact homology sequences.

With regard to the intended axiomatic homology theory, we finally have to verify the excision axiom and the dimension axiom.

The Excision Axiom

Proposition 4.39 *Let* (M,A) *be an admissible pair of manifolds and let* B *be a closed subset* $B \subset \overset{o}{A}$ *such that* $(M\backslash B, A\backslash B)$ *again represents an admissible pair. Then the inclusion* $i : (M\backslash B, A\backslash B) \hookrightarrow (M,A)$ *induces an isomorphism*

$$i_* : H_*(M\backslash B, A\backslash B) \xrightarrow{\cong} H_*(M,A) \ .$$

Proof. Obviously, it is sufficient to treat the case of an open submanifold A. According to the concept of relative Morse homology, we can choose a concrete, appropriate Morse function in order to compute the inclusion homomorphism and the actual homology groups. Considering a Morse function $f \in C^\infty_{\text{Morse}}(\overline{M\backslash A}, \mathbb{R})$ such that the underlying relative Morse complexes of the respective homology groups are generated by the same set of critical points of f within $M\backslash A$, we deduce the assertion immediately from the canonical isomorphism

$$C_*(f_M, f_A) \cong C_*(f_{M\backslash B}, f_{A\backslash B}) \ .$$

□

The Dimension Axiom

The precise dimension axiom in the framework of Morse homology is a trivial fact: Each Morse function on a space consisting of one point $\{x\}$ has exactly one critical point, and this point has index 0. Hence we obtain the identity

$$H_*\big(\{x\}\big) \cong (\mathbb{Z}, 0, \ldots) \ .$$

The argument for a finite-dimensional vector space, for instance $\mathbb{R}^n$, is similar. Let q_n be the quadratic form with respect to a scalar product. Then the only critical point is a minimum and it generates the homology

$$H_*(\mathbb{R}^n) \cong (\mathbb{Z}, 0, \ldots) \ .$$

We shall now verify a relation which in fact is not trivial, but which is characteristic for any standard homology theory on manifolds.

Proposition 4.40 *Any smooth and connected manifold M satisfies the relation*

$$H_0^{\mathrm{Morse}}(M) \cong \mathbb{Z} \ .$$

The main part of the proof, in which the property of connectedness plays a crucial role, is based upon the following

Lemma 4.41 *Let $y, y' \in \mathrm{Crit}_0 f$ be local minima and let $\gamma \in C^0\big([-1,1], M\big)$ be a connecting path, $\gamma(-1) = y$, $\gamma(1) = y'$. Then there is a division homotopical to γ consisting of isolated trajectories*

$$u_i = u_{x_i,y_i} \in \mathcal{M}^f_{x_i,y_i}, \ (x_i, y_i) \in \mathrm{Crit}_1 f \times \mathrm{Crit}_0 f \ .$$

This means that there are isolated trajectories $u_0, \ldots, u_k$ satisfying

$$u_0(-\infty) = y, \ u_i(+\infty) = u_{i+1}(-\infty), \ i = 0, \ldots, k-1 \ \text{and} \ u_k(+\infty) = y' \ ,$$

such that the composition of suitably reparametrized curves u_i is homotopical to γ,

$$\big(\ldots(\tilde{u}_0 \cdot \tilde{u}_1) \cdot \ldots \cdot \tilde{u}_k\big) \simeq \gamma \ .$$

Proof. Since f is coercive, i.e. M^a is compact for all $a \in \mathbb{R}$, the number

$$\nu(\gamma) = \#\big\{ x \in \mathrm{Crit}_1 f \ \big| \ f(x) \leqslant \max_{[-1,1]} f \circ \gamma \big\}$$

is well-defined and finite. The proof of this lemma is carried out by induction on $\nu(\gamma)$.

Within the following steps of the proof we shall make use on the one hand of homotopies, which may be specified concretely in local coordinates, and on the other hand of the global, negative gradient semi-flow on the compact subset M^a

$$\Psi : \mathbb{R} \times M^a \to M^a \quad \text{with} \quad a > \max_{[-1,1]} f \circ \gamma \ .$$

The beginning of the induction follows immediately from the application of a well-known lemma from classical variational calculus (see [Cou]), from the so-called mountain-pass lemma:

> Given two different local minima y and y' of a coercive, smooth function f, there is a further critical point $x \neq y, y'$ satisfying the minimax-property
>
> $$f(x) = \inf_{p \in \mathcal{P}} \max_{z \in p} f(z),$$
> $$\mathcal{P} = \{ p \subset M \mid y, y' \in p, \, p \text{ compact and connected} \} \ .$$

In our situation we can additionally compute the Morse index, that is $x \in \mathrm{Crit}_1 f$. Thus, the beginning of the induction by $\nu(\gamma) = 0$ implies the identity of the endpoints $y = y'$.

Regarding the induction conclusion, we may assume without loss of generality that the function $f \circ \gamma : [-1,1] \to \mathbb{R}$ takes its maximum in exactly one 'moment' $t_0 \in [-1,1]$ and that

$$\gamma(t_0) \in \mathrm{Crit}_1 f \ . \tag{4.84}$$

Otherwise we consider a homotopy of γ by means of the flow Ψ_τ, so that $\max f \circ \gamma$ decreases strictly until the maximal point is close enough to a critical point such that we are able to use the local coordinates provided by the well-known classical Morse lemma (see [M1]). If the respective critical point should have a Morse index larger than 1, we find a homotopy in these coordinates, which 'pushes' γ past this critical point, so that the new curve γ no longer intersect the stable manifold of this critical point. Note that this is due to the codimension of this stable manifold which is exactly the Morse index. This process can be iterated finitely many times, until condition (4.84) is fulfilled at last. If there are more than one $t_0 \in [-1,1]$ with $f\big(\gamma(t_0)\big) = \max f \circ \gamma$, we will go through the following process sufficiently often.

Given critical point $x = \gamma(t_0) \in \mathrm{Crit}_1 f$, we split γ into parts $\gamma = \gamma_{yx} \cdot \gamma_{xy'}$. The Morse index $\mu(x) = 1$ implies that there are exactly two isolated trajectories u_{xy_1}, u_{xy_2} of the type specified in the assertion, which leave x. Regarding the homotopy class,

$$\gamma \simeq \gamma_{yx} \cdot u_{xy_1} \cdot u_{xy_1}^{-1} \cdot u_{xy_2} \cdot u_{xy_2}^{-1} \cdot \gamma_{xy'} \ , \tag{4.85}$$

because $u_{xy_1} \cdot \ldots \cdot u_{xy_2}^{-1} \simeq \mathrm{const} = x$ is contractible. Analysing the situation at x by means of the local coordinates provided by the Morse lemma, we observe that the order of the parts $u_{xy_i}^{\pm 1}$ can be chosen in such a way that the negative gradient flow Ψ_τ pushes the curves $\gamma_1 = \gamma_{yx} \cdot u_{xy_1}$ and $\gamma_2 = u_{xy_2}^{-1} \cdot \gamma_{xy'}$ below the level $f(x)$, that is,

$$\max_{[-1,1]} f(\Psi_\tau \circ \gamma_{1,2}) < f(x) \quad \text{for some } \tau > 0 \ . \tag{4.86}$$

After finitely many division steps of this type for each t_0, the parts γ_i, which still do not necessarily consist of isolated trajectories, satisfy the condition

$$\nu(\gamma_i) < \nu(\gamma) \ ,$$

upon which we have founded the induction process. □

Corollary 4.42 *Given any two local minima y and y' for the Morse function f from the same connected component of M, there is a connecting curve consisting of isolated trajectories of the type*

$$u_{x_i,y_i} \in \mathcal{M}^f_{x_i,y_i}, \ (x_i, y_i) \in \mathrm{Crit}_1 f \times \mathrm{Crit}_0 f \ .$$

Proof of proposition 4.40. This corollary yields the fact, that each two local minima of a fixed Morse function y and y' generate the same cyclic subgroup of H_0,

$$\{y\} = \pm\{y'\} \ .$$

It thus follows that

$$H_0(M) \cong \mathbb{Z}_p \text{ or } \mathbb{Z} \ . \tag{4.87}$$

This is concluded completely from the condition of connectedness on M. On the other hand, we have already concluded the fact that $H_0(M)$ is infinitely cyclic from special Morse functions in Proposition 4.12. This led back to the problem of orientation. Resorting to the functorial concept, we are now able to prove this in general as a consequence of the dimension axiom.

Let $c : M \to \{x\}$ be a constant mapping. We consequently obtain a non-vanishing group homomorphism

$$c_* : H_0(M) \to H_0(\{x\}) \cong \mathbb{Z} \ .$$

Hence, $c_* \neq 0$ together with result (4.87) from connectedness yields the assertion $H_0(M) \cong \mathbb{Z}$. □

4.2.4 Summary

Summarizing the above results, we have analysed all elements of a homology theory which satisfies the axioms set up by Eilenberg and Steenrod. We have obtained a family $(H_k)_{k\in\mathbb{N}_0}$ of covariant functors from the category of admissible pairs of smooth manifolds (M, A) together with the smooth mappings of pairs

$$\varphi : (M, A) \to (N, B),\ \varphi(A) \subset B$$

as morphisms into the category of Abelian groups, so that the following axioms are satisfied:

1. The existence of a natural, long exact homology sequence,
2. the homotopy invariance,
3. the invariance under excision and
4. the normalization of the point space:

$$H_k\big(\{x\}\big) \cong \begin{cases} \mathbb{Z}, & k = 0 \\ 0, & \text{sonst} \end{cases} .$$

4.3 The Uniqueness Result

In this section we will investigate the isomorphism result, which relates the Morse homology constructed in this chapter to any other axiomatic homology theory such as, for instance, singular theory. This equivalence between homology theories will be accomplished by means of the homology axioms verified above. Following Eilenberg and Steenrod [E-S] we will draw this final conclusion from the axiomatic approach in terms of a uniqueness result, which expresses the existence of a natural isomorphism to any other axiomatic homology theory provided that both theories in question be considered for appropriate categories of topological spaces. This so-called uniqueness category[12] in the case treated within this work will be the category of pairs of finite CW-complexes. We state the main conclusion as follows:

Theorem 9 *Let $\tilde{H}_*$ be any axiomatic homology theory defined on the category of finite CW-pairs. Then given a fixed isomorphism*

$$h_o : H_0^{\text{Morse}}\big(\{pt\}\big) \xrightarrow{\cong} \tilde{H}_0\big(\{pt\}\big)$$

for a fixed point manifold $M = \{pt\}$, there exists a unique, natural equivalence of homology theories

$$\Phi_* : H_*^{\text{Morse}} \xrightarrow{\cong} \tilde{H}_* \qquad \textit{with} \qquad \Phi_0\big(\{pt\}\big) = h_o ,$$

where H_*^{Morse} and $\tilde{H}_*$ are restricted to admissible pairs of manifolds which are finite CW-pairs.

12The reader is referred to [E-S].

For one who is not familiar with this axiomatic approach set up by Eilenberg and Steenrod it might be surprising that no natural homomorphism defined explicitly in terms of the two homology theories in question is needed. This arises from a constructive existence result for just such a natural homomorphism which automatically implies uniqueness. The strategy for proving the theorem is to exploit this uniqueness property of the category of CW-spaces. But in order to make use of this feature we at least must extend the Morse homology deduced for admissible pairs of manifolds to a CW-subcategory which gives rise to the same uniqueness result. Therefore, we will restrict ourselves to the following pairs of CW-complexes (X, Y).

Definition 4.43 *Let us denote by* $\mathrm{CW}_{\mathrm{reg}}$ *the subcategory of CW-pairs which are embedded smoothly into finite-dimensional manifolds* M, *i.e. the characteristic maps associated to the cells are relative diffeomorphisms. Additionally, these CW-pairs must have open neighbourhoods* (U, R) *in* M *which are admissible and of which they are strong deformation retracts.*

Remarks

- *Without loss of generality we may assume that any two of such neighbourhood pairs are homotopically equivalent by smooth homotopies.*
- *If we focus on manifolds* M *endowed with such a smooth CW-structure,* $M = \{X_k\}_{k=0,\dots,n}$, *we see that the pairs of skeletons* (X_k, X_l) *have this regularity property because of the well-known theorem that subcomplexes of CW-complexes are strong neighbourhood deformation retracts.*

Finally, this class of objects in $\mathrm{CW}_{\mathrm{reg}}$ will suffice for providing the proof of the main theorem, and this subcategory of sufficiently regular CW-pairs lends itself to the following extension of the Morse homology as an axiomatic homology functor.

Definition 4.44 *Given a pair* $(X, Y) \subset M$ *from the category* $\mathrm{CW}_{\mathrm{reg}}$ *we define* $\mathfrak{U}_{(X,Y)}$ *to be the family of neighbourhoods* (U, R) *of* (X, A), *of which* (X, A) *is a deformation retract in the above sense and which are itself admissible pairs of manifolds. Observing that* $\mathfrak{U}_{(X,Y)}$ *is a directed, quasi-ordered set by inclusions, which by assumption are homotopy equivalences, we define*

$$\check{H}_*(X, Y) = \lim_{\longleftarrow} \{ H_*^{\mathrm{Morse}}(U, R) \,|\, (U, R) \in \mathfrak{U}_{(X,Y)} \}$$

as the inverse limit with respect to the inclusion isomorphisms

$$i_{*\,(V,S)}^{\;(U,R)} : H_*(U, R) \xrightarrow{\cong} H_*(V, S), \ (U, R) \subset (V, S) \ .$$

Remark *There is some minor technical work left to show that the property of these neighbourhood pairs to be admissible does not disturb the directedness of* $\mathfrak{U}$.

By a restriction to neighbourhoods which are homotopically equivalent to (X, Y) by inclusion, we are able to deduce the natural isomorphisms

$$I_*^{(U,R)} : \check{H}_*(X,Y) \xrightarrow{\cong} H_*(U,R),\ (U,R) \in \mathfrak{U}_{(X,Y)}$$

immediately. This feature guarantees that exactness of sequences is preserved under this special inverse limit process. In general, this is not true and usually has to be provided in terms of a Mittag-Leffler condition. Obviously, by tubular neighbourhood constructions, we have a natural isomorphism

$$\check{H}_*(M,A) \cong H_*^{\mathrm{Morse}}(M,A)$$

for pairs (M, A) which are equally admissible and regular CW-complexes due to the above definition. It is straightforward from the properties of the inverse system defined above to derive a functor $\check{H}_*$ for the pairs of the category $\mathrm{CW}_{\mathrm{reg}}$. Actually, by noting that the restriction to the homotopically equivalent neighbourhoods yields the exactness of the long exact homology sequence

$$\ldots \xrightarrow{j} \check{H}_{k+1}(X,Y) \xrightarrow{\partial} \check{H}_k(Y) \xrightarrow{i} \check{H}_k(X) \xrightarrow{j} \check{H}_k(X,Y) \xrightarrow{\partial} \ldots ,$$

one is able to verify easily the axioms for this extended homology theory $\check{H}_*$. In the future, we will drop the indices $\check{}$ and ^{Morse} from the notation and use H_* with a sense depending on the context.

We now move toward the construction of the natural homomorphism based upon the category $\mathrm{CW}_{\mathrm{reg}}$.

Proposition 4.45 *Let H_* and H'_* be two axiomatic homology theories defined on $\mathrm{CW}_{\mathrm{reg}}$ and let*

$$h_o : H_0(\{pt\}) \xrightarrow{\cong} H'_0(\{pt\})$$

be an isomorphism for a fixed point manifold $\{pt\}$. Then there is a unique extension of h_o to a natural isomorphism $h_* : H_* \xrightarrow{\cong} H'_*$.

Proof. The main step of the proof relies on the construction of CW-homology. As this is standard knowledge from algebraic topology the reader is referred to any fundamental book on this topic.

At first, we will deduce unique isomorphisms h_* for n-cells e_n and n-spheres S^n from the given homomorphism h_o. This will be done by means of the homology axioms and particularly by induction based on the Mayer-Vietoris sequence, for which exactness holds because all pairs appearing in the following argument give rise to proper triads. Let us recall this essential and constructive tool from homology theory. Given a proper triad $(X; X_1, X_2)$, i.e. the inclusions

$$i_1 : (X_1, X_1 \cap X_2) \hookrightarrow (X_1 \cup X_2, X_2) \quad \text{and}$$
$$i_2 : (X_2, X_1 \cap X_2) \hookrightarrow (X_1 \cup X_2, X_1)$$

inducing isomorphisms of the homology groups in all dimensions, together with relations $X = X_1 \cup X_2$, $A = X_1 \cap X_2$, the associated Mayer-Vietoris sequence[13]

$$\ldots \longrightarrow H_q(A) \xrightarrow{\Psi} H_q(X_1) \oplus H_q(X_2) \xrightarrow{\Phi} H_q(X) \xrightarrow{\Delta} H_{q-1}(A) \longrightarrow \ldots$$

is exact.

Now, by naturality with respect to the homology axioms, it is straightforward to construct isomorphisms

$$h_* : H_*(X) \xrightarrow{\cong} H'_*(X)$$

for all spaces X which are finite disjoint unions of open or closed n-cells. Let X be a single point space $X = \{x\}$ and let $\varphi : \{pt\} \xrightarrow{\approx} \{x\}$ be the unique diffeomorphism. This gives rise to

$$h_0(\{x\}) = \varphi_* \circ h_o \circ \varphi_*^{-1} : H_0(\{x\}) \xrightarrow{\cong} H'_0(\{x\})$$

in a unique and natural way. From the Mayer-Vietoris sequence we deduce h_* for all finite disjoint unions of points and the result for n-cells follows from the homotopy axiom.

The next step is to construct the extension of these h_* on arbitrary cells to arbitrary n-spheres. This can be done inductively by means of the Mayer-Vietoris sequence with respect to the standard decomposition of the n-sphere into the two closed hemispheres D^+ and D^- with the equator $S^{n-1} = D^+ \cap D^-$ as intersection. Again by naturality and induction we find unique isomorphisms $h_* : H_*(X) \xrightarrow{\cong} H'_*(X)$, where X is allowed to be an arbitrary n-cell or n-sphere from $\mathrm{CW}_{\mathrm{reg}}$. Further, making use of the axiomatic long exact homology sequence we infer the extension of h_* to all pairs of the diffeomorphism type of (D^n, S^{n-1}) denoting the closed n-disk with its boundary sphere. Finally, the Mayer-Vietoris sequence again implies the extension to one-point unions of n-spheres $\bigvee_{i=1}^k S_i^n$, which are objects in $\mathrm{CW}_{\mathrm{reg}}$, also.

We now have developed the natural homomorphism h_* for a family of cellular spaces already large enough to carry out the crucial step toward CW-homology. We now denote by X a cellular complex from $\mathrm{CW}_{\mathrm{reg}}$. As is well-known from standard algebraic topology, the groups

$$C_q(X) = H_q(X^q, X^{q-1}),\ q \geqslant 0$$

13For the definition of Ψ, Φ and Δ and more details, see [E-S].

are free Abelian of finite rank, where $(X^q)_{q\geqslant 0}$ denotes the family of q-skeletons of X for $q \geqslant 0$. If $F^e : (D^q, S^{q-1}) \to (X^q, X^{q-1})$ describes the characteristic map associated to the q-cell $e \subset X$, we can take the family

$$\{\, F^e_*(\{D^q\}_{S^{q-1}}) \in H_q(X^q, X^{q-1}) \,|\, e \;\; q\text{-cell in } X \,\}$$

as a basis for $C_q(X)$. By means of this basis, the homomorphisms F^e_* and the above investigations about the extension of h_o to finite unions of pairs (D^q, S^{q-1}) and one-point unions of spheres, we immediately deduce natural isomorphisms

$$h_q : H_q(X^q, X^{q-1}) \xrightarrow{\cong} H'_q(X^q, X^{q-1}), \;\; q \geqslant 0 \;.$$

Thus we have already established a natural equivalence between the CW-homology theories

$$h_q : H_q\big(C(X)\big) \xrightarrow{\cong} H_q\big(C'(X)\big), \;\; q \geqslant 0 \tag{4.88}$$

associated to H_* and H'_*. Moreover, we see from the above constructions that this equivalence is unique with respect to $\big(h_o, \{pt\}\big)$. Now, using the canonical and natural isomorphisms associated to CW-homology

$$\begin{aligned} T : &\quad H_q\big(C(X)\big) \xrightarrow{\cong} H_q(X),\; q \geqslant 0 \quad \text{and} \\ T' : &\quad H_q\big(C'(X)\big) \xrightarrow{\cong} H'_q(X),\; q \geqslant 0 \;, \end{aligned}$$

we can accomplish the proof by deriving the natural isomorphism h_* between H_* and H'_* on $\mathrm{CW}_{\mathrm{reg}}$ as the composition

$$h_* = T' \circ h_*^{\mathrm{CW}} \circ T^{-1} \;. \tag{4.89}$$

Here, h_*^{CW} means the cellular equivalence given in (4.88). The naturality of h_* in general follows from the naturality with respect to cellular maps and cellular approximation together with homotopy invariance. Thus we have proven that $\mathrm{CW}_{\mathrm{reg}}$ represents a uniqueness category in the sense of Eilenberg and Steenrod. □

Let us now consider the above extension of Morse homology to an axiomatic homology theory on $\mathrm{CW}_{\mathrm{reg}}$ together with the uniqueness property of this category. Since this uniqueness category contains the admissible pairs of manifolds equipped with a finite CW-structure, we obtain the proof of Theorem 9. Of course, it should be possible to derive this uniqueness of Morse homology for a far larger class of spaces. But within the intended frame of this work we shall be content with this result.

Chapter 5

Extensions

In this last chapter we present a brief outline of further elements of algebraic topology within the framework of Morse homology. In this context we emphasize the Poincaré duality arising rather immediately and the feature of product operations appearing naturally in the Morse theoretic approach. At first, we briefly introduce the dual concept of Morse homology.

5.1 Morse Cohomology

By strict analogy to other known homology theories which are deduced from chain complexes, we obtain a cohomological version of Morse homology by means of a dualization concept.

Definition 5.1 *Given a chain complex* $\big(C_k(f), \partial_k\big)_{k\in\mathbb{N}_0}$ *as considered in the context of Morse homology, we define a cochain complex* $\big(C^k(f), \delta^k\big)_{k\in\mathbb{N}_0}$ *as follows:*

$$\begin{aligned} &C^k(f) = \operatorname{Hom}\big(C_k(f), \mathbb{Z}\big), \\ &\langle\, \delta^k \alpha^k,\, x_{k+1}\,\rangle = \langle\, \alpha^k,\, \partial_{k+1} x_{k+1}\,\rangle \\ &\text{for all } \alpha^k \in C^k(f),\ x_{k+1} \in \operatorname{Crit}_{k+1} f,\ k \in \mathbb{N}_0\ . \end{aligned}$$

This definition is unique, as $\operatorname{Crit}_{k+1} f$ *comprises exactly the generators of* $C_{k+1}(f)$.

Remark Discussing this dualization process we must be aware of the fact that $C^k(f)$ need not necessarily be finitely generated any more, as we assumed for $C_k(f)$. Nevertheless, if the set $\mathrm{Crit}_k f$ is finite, we are supplied with the canonical isomorphism

$$\begin{array}{ccc} C_k(f) & \xrightarrow{\cong} & C^k(f) \\ x_k & \mapsto & \left(\langle x_k, \cdot \rangle : y_k \mapsto \begin{cases} 1, & y_k = x_k \\ 0, & \text{otherwise} \end{cases} \right) . \end{array} \tag{5.1}$$

Moreover, we compute the coboundary operator

$$\delta^k x_k = \sum_{\mu(y)=k+1} \langle \partial y, x_k \rangle y \ ,$$

if in addition $\mathrm{Crit}_{k+1} f$ is finite.

In principle, we resort to the same isolated trajectories together with their characteristic signs determined by a coherent orientation as in the definition of the boundary operator. As indicated by the increasing Morse index, this amounts to replacing negative gradient flow by positive gradient flow. But two problematic items still remain. First, $-f$ need not be coercive. We recall that it was coercivity by which we established the compactness result for the trajectory spaces. In fact, if the underlying manifold is not compact, changing from f to $-f$ necessarily abolishes coercivity. The other problem arises in the case of non-orientable manifolds as remarked in Section 4.1.4. Actually, both flaws within our idea of symmetry concern exactly the requirements of the Poincaré duality. This will be analysed further in the following section.

Summing up the present discussion we observe that the dualization concept known as the Hom-functor allows us to construct an axiomatic Morse cohomology. Actually, the application of this functor together with the known universal coefficient theorem gives rise to a transfer of the identification process (yielding cohomology groups independent of concrete Morse functions) as well as of all functorial and axiomatic properties to the contravariant functor H^*. The reader not familiar with this dualization process may refer to the analogous treatment of singular or simplicial cohomology.

5.2 Poincaré Duality

As we noted above this Morse cohomology theory has the disadvantage that the cohomology classes cannot necessarily be represented by the critical points of a Morse function. This is merely possible when the cochain complexes are finitely generated. We observed that the compactness assumption for the manifold M within classical Poincaré duality provides a sufficient condition for this finiteness of the set of generators. Then, coercivity is trivial. On compact manifolds we are naturally provided with a symmetry which is expressed by the lack of a canonical distinction between negative and positive gradient flow. If we additionally assume the orientability of M, both requirements for Poincaré duality are complete. On the other side, the problem of the noncanonical construction method of a coherent orientation vanishes, with respect to the condition of admissible pairs of trivializations in Definition 3.7. Under these circumstances, Morse homology cannot be set apart from Morse cohomology with regard to the construction.

We are provided with a natural isomorphism in the case of an n-dimensional, closed and orientable manifold M

$$PD : H_k(M) \xrightarrow{\cong} H^{n-k}(M),\ k = 0, \dots, n$$

as follows:

Let f be any Morse function on M. Then, $-f$ is also a Morse function and we can easily find a regular Morse homotopy

$$f \overset{h}{\simeq} -f \ .$$

This induces the canonical isomorphism

$$\Phi_*^h : H_k(f) \xrightarrow{\cong} H_k(-f),\ k = 0, \dots, n \ . \tag{5.2}$$

Noting now that we may use finitely many critical points from $\mathrm{Crit}_k f$ as generators of $C^k(f) = \mathrm{Hom}\big(C_k(f), \mathbb{Z}\big)$, we are led to the identity

$$\begin{array}{rcl} \mathrm{Crit}_k(-f) & \to & \mathrm{Crit}_{n-k} f, \\ x & \mapsto & x \ . \end{array} \tag{5.3}$$

This identity gives rise to the likewise canonical isomorphism

$$\Gamma : C_k(-f) \xrightarrow{\cong} C^{n-k}(f) \ ,$$

which is compatible with the ∂- and δ-operator, equally. This means that it is well-defined on the level of homology and cohomology. Thus we canonically obtain the isomorphisms

$$\begin{array}{rrcl} \Phi_* : & H_k(f) & \xrightarrow{\cong} & H_k(-f) \text{ and} \\ \Gamma_* : & H_k(-f) & \xrightarrow{\cong} & H^{n-k}(f) \ , \end{array}$$

and hence by composition

$$PD = \Gamma_* \circ \Phi_* : H_k(f) \xrightarrow{\cong} H^{n-k}(f),\ k = 0, \dots, n \ . \tag{5.4}$$

Actually, due to the construction, this proves compatible with the isomorphisms induced by Morse homotopies leading to the identification processes

$$H_*(M) = \varprojlim H_*(M, f), \text{ etc.}$$

Altogether we naturally obtain the Poincaré duality for closed and orientable manifolds. The same is true for the case of $\mathbb{Z}_2$ coefficients where non-orientable closed manifolds are admitted. Then, the problem of a coherent orientation becomes insignificant.

5.3 Products

Here we recall the naturally induced construction of the Morse function $f \oplus g$ on the product manifold $M \times N$, provided that we are given the manifolds with Morse functions (M, f) and (N, g). In the last chapter, we have already benefited from the canonical identification of the critical points of this induced Morse function with respective pairs of critical points of single Morse functions. If we now use the tensor product for chain complexes in the form

$$(5.5) \qquad \big(C_*(f) \otimes C_*(g)\big)_k = \bigoplus_{i=0}^{k} C_i(f) \otimes C_{k-i}(g)$$

together with the boundary operator

$$(5.6) \qquad \begin{aligned} &\partial_k(\sigma_i \otimes \tau_{k-i}) = \partial_i^f \sigma_i \otimes \tau_{k-i} + (-1)^i \sigma_i \otimes \partial_{k-i}^g \tau_{k-i}, \\ &\qquad \text{for} \quad \sigma_i \otimes \tau_{k-i} \in C_i(f) \otimes C_{k-i}(g) \ , \end{aligned}$$

we are able to state

Lemma 5.2 *Let f and g be any Morse functions on manifolds M and N, respectively. Then, with regard to the induced Morse function $f \oplus g$ on $M \times N$, there is an isomorphism between the respective chain complexes for any arbitrary coherent orientations,*

$$P_\bullet : C_*(f) \otimes C_*(g) \xrightarrow{\cong} C_*(f \oplus g) \ .$$

Moreover, this isomorphism appears to be canonical up to transformations of the coherent orientations.

Proof. As mentioned above we start with the identification

$$\mathrm{Crit}_k(f \oplus g) = \bigcup_{i+j=k} \mathrm{Crit}_i f \times \mathrm{Crit}_j g \ .$$

Thus we define a chain group isomorphism

$$P_k^\alpha : \big(C_*(f) \otimes C_*(g)\big)_k \xrightarrow{\cong} C_k(f \oplus g),\ k \in \mathbb{N}_0$$

by means of the tensorial extension of the mapping

$$x_i \otimes y_j \mapsto \alpha(x_i, y_j) \cdot (x_i, y_j)$$

defined on the generating elements $(x_i, y_j) \in \operatorname{Crit}_i f \times \operatorname{Crit}_j g$. Here we denote by $\alpha \in \{\pm 1\}^{\operatorname{Crit}_*(f \oplus g)}$ an equivalence transformation of the coherent orientation on $M \times N$.

We now have to verify that $P_\bullet^\alpha$ in fact yields a chain homomorphism, provided that we have chosen a suitable transformation α. This choice of transformation is related to the coherent orientations on M, N and $M \times N$ which are supplied in an noncanonical manner. As to the ∂-operator on $C_*(f \oplus g)$, let us consider the gluing situation sketched in Figure 5.1. Note that we

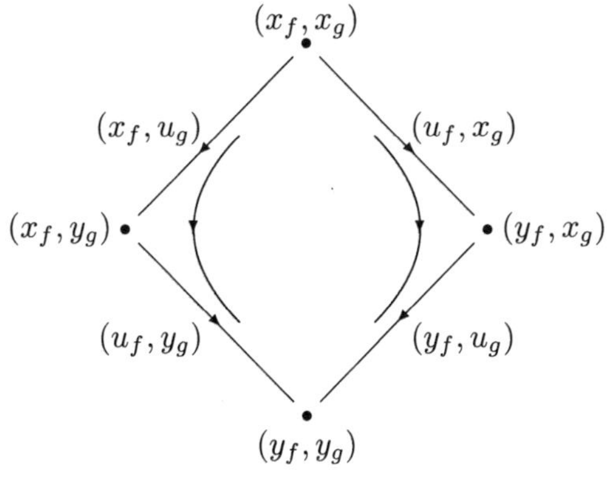

Figure 5.1: Cobordism equivalence in the product case

deal with isolated trajectories in $M \times N$, that is

$$\mu(x_{\dots}) - \mu(y_{\dots}) = 1 \ .$$

Due to our analysis of the gluing process we derive the equation

$$\tau_\alpha(u_f, x_g) \cdot \tau_\alpha(y_f, u_g) = -\tau_\alpha(x_f, u_g) \cdot \tau_\alpha(u_f, y_g) \tag{5.7}$$

in a way which is entirely independent of the coherent orientation on $M \times N$. This means that (5.7) holds for all transformations α. The actual problem consists of relating the characteristic signs $\tau_\alpha(u_f, x_g)$, etc., of the isolated trajectories on $M \times N$ to the signs $\tau(u_f)$, etc., of the associated trajectories on M and respectively N, that is,

$$\begin{aligned} &\tau_\alpha(u_f, x_g) = \varphi(u_f, x_g)\tau(u_f) \quad \text{and} \\ &\tau_\alpha(x_f, u_g) = \varphi(x_f, u_g)\tau(u_g) \\ &\text{with} \quad \varphi(\cdot, \cdot) \in \{\pm 1\} \ . \end{aligned}$$

Since we are free choose a suitable transformation $\alpha \in \{\pm 1\}^{\mathrm{Crit}_*(f \oplus g)}$, the relating signs $\varphi(\cdot, \cdot)$ are independent of the concrete trajectories and critical points. They can depend merely on the Morse indices and - this is the crucial point - on the order of the pair of the isolated non-trivial trajectory and the constant trajectory. Actually, already in the definition of the tensorial chain complex $C_*(f) \otimes C_*(g)$ we have different equivalent possibilities for the boundary operator. We have selected the option

$$\partial(x_f \otimes x_g) = \partial x_f \otimes x_g + (-1)^{\mu(x_f)} x_f \otimes \partial x_g \ .$$

Thus, choosing an appropriate transformation α, we obtain the relations

$$(5.8) \qquad \begin{aligned} \tau(u_f, x_g) &= \tau(u_f) \quad \text{and} \\ \tau(x_f, u_g) &= (-1)^{\mu(x_f)} \tau(u_g) \ . \end{aligned}$$

Hence, the identity

$$P_\bullet\, x_f \otimes x_g = (x_f, x_g)$$

holds up to equivalence transformations of the noncanonical coherent orientations. □

Definition 5.3 *Since Morse homology groups have been constructed in such a way that they are invariant under equivalence transformations of the coherent orientation, the above isomorphism $P_\bullet$ induces an isomorphism P_* at the homology level. This isomorphism, which is therefore canonical, appears to be the cardinal feature, upon which we found the definition of the homology cross product. It is accomplished by means of the natural homomorphism*

$$\lambda : H_p(f) \otimes H_q(g) \to H_{p+q}(C_*(f) \otimes C_*(g))$$

provided for any tensorial chain complexes. λ usually appears in relation with the well-known Künneth formula. We define the homology cross product as

$$\times = P_* \circ \lambda : H_p(f) \otimes H_q(g) \to H_{p+q}(f \oplus g) \ .$$

Due to the naturality of the homomorphism λ we are able to transfer the entire identification concept based on the isomorphisms Φ_*^h induced by Morse homotopies. Hence, we obtain the cross product for Morse homology independent of *the Morse functions:*

$$H_p(M) \otimes H_q(N) \xrightarrow{\times} H_{p+q}(M \times N) \ .$$

Since $P_\bullet$ is a chain isomorphism, resorting to the duality functor Hom allows us to construct likewise a cross product for the cohomology groups:

$$\times : H^p(f) \otimes H^q(g) \to H^{p+q}\big(C_*(f) \otimes C_*(g)\big) \overset{\left(P^{-1}\right)^*}{\longrightarrow} H^{p+q}(f \oplus g) \ .$$

Once again we obtain the associated product for the cohomology independent of the concrete Morse functions,

$$H^p(M) \otimes H^q(N) \xrightarrow{\times} H^{p+q}(M \times N) \ .$$

Definition 5.4 *Having defined the cohomology cross product by means of a relation which is naturally immanent in Morse theory, we now are able to define the cup product. Here, we use a classical method based on the fact that cohomology is a contravariant functor. Namely, we consider the diagonal embedding*

$$\triangle : M \to M \times M \ ,$$

which is a closed embedding. Hence, it immediately gives rise to the homomorphism

$$\triangle^* : H^k(M \times M) \to H^k(M) \ .$$

To give more details, we obtain this homomorphism from a Morse homotopy

$$\triangle_* f \overset{h_t}{\simeq} f \oplus f \ ,$$

where $\triangle_* f$ denotes the extension of the Morse function $f \oplus q_n$ from the tubular neighbourhood of the diagonal M to $M \times M$. For $\{\sigma\} \in H^k(M \times M)$,

$$\triangle^*\{\sigma\} = \big\{\triangle^\bullet \sigma\big\} \in H^k(M) \ ,$$

where $\triangle^\bullet \sigma \in \mathrm{Hom}\big(C_k(f), \mathbb{Z}\big)$ is determined by

$$\big\langle \triangle^\bullet \sigma, \, x_k \big\rangle = \big\langle \sigma, \, \Phi^{h_t} \cdot \triangle(x_k) \big\rangle \ \textit{for} \ x_k \in \mathrm{Crit}_k f \ .$$

Thus, we define the cup product as the composition

$$H^p(M) \otimes H^q(M) \xrightarrow{\times} H^{p+q}(M \times M) \xrightarrow{\triangle^*} H^{p+q}(M),$$
$$\alpha \cup \beta = \triangle^*(\alpha \times \beta) \ .$$

Finally, we intend to present a brief outlook into the possibilities of extending this homology and cohomology theory gained from classical Morse theory even further, for instance toward the topic of vector bundles.

Proposition 5.5 *Let M be a smooth manifold and let $\pi : E \to M$ be a smooth n-dimensional vector bundle on M. Furthermore, let $E_0 = E\backslash U(M)$ be the complement of a closed neighbourhood of the zero section M within E, so that it is contractible to the submanifold M and (E, E_0) is an admissible pair of manifolds. Then there is a natural isomorphism*

$$T : H^i(M) \xrightarrow{\cong} H^{i+n}(E, E_0)$$

for all $i \in \mathbb{N}_0$.

Proof. Let f be an arbitrary Morse function on M and let q_n be the quadratic form associated to a Riemannian metric on E. Referring to the example which we have already analysed with respect to relative Morse functions, we observe that $f \oplus (-q_n)$ represents a relative Morse function on (E, E_0) with its critical points identical to those within $\mathrm{Crit}_* f$ lying on the submanifold M. In the same way we conclude that all isolated trajectories counting for the relative boundary operator of the relative Morse complex associated to $f \oplus (-q_n)$ on (E, E_0) are located on this submanifold. Actually, we immediately obtain a chain isomorphism between the absolute Morse complex of f on M and this relative complex. Since all stages of the identification process via Morse homotopies can be accomplished in a manner that is compatible with this construction, the assertion follows from this identification isomorphism T between the absolute complex of M and the relative complex of (E, E_0). Here, the grading of the relative complex is shifted by the constant n given as the dimension of the vector bundle. □

Referring to the classical framework of characteristic classes one would now have to verify that this isomorphism T describes the Thom isomorphism. Moreover, the following diagram should be given:

$$\begin{array}{ccc} H^i(E) & \xrightarrow{\cdot \cup u} & H^{i+n}(E, E_0) \\ {\scriptstyle \pi^*}\uparrow & \oplus \nearrow {\scriptstyle T} & \\ H^i(M) & & \end{array}$$

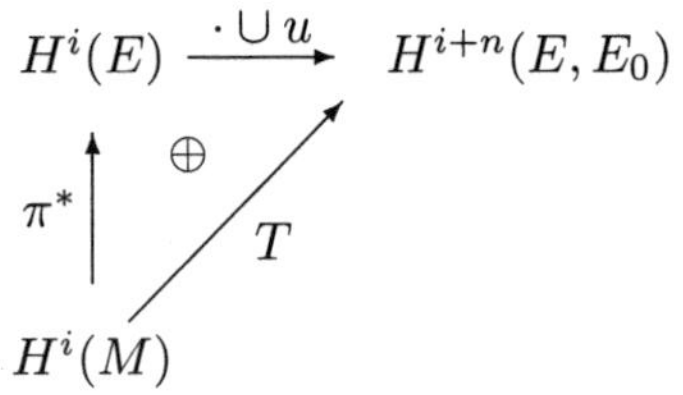

Here, $u \in H^n(E, E_0)$ denotes the Thom class of the bundle E.

Let us finish the discussion with the guess that one should be able to obtain a representative for this Thom class within the framework of Morse cohomology theory by means of the critical points $(x_0, 0) \in \mathrm{Crit}_n\bigl(f \oplus (-q_n)\bigr)$ for $x_0 \in \mathrm{Crit}_0 f$.

Appendix A

Curve Spaces and Banach Bundles

In this chapter we present a brief introduction of Banach manifolds and associated Banach bundles which form the foundation of the entire analytic approach to Morse homology chosen in this monograph. We discuss explicitly the structure of an infinite-dimensional Banach manifold defined on curve spaces as was stated in Proposition 2.7. The first section of this chapter deals with these manifolds themselves. The second section comprises an analysis of certain Banach bundles on these manifolds. We shall partly use notations and schemes of proofs taken from [Eli]. However, it is quite important to point out that our more specialized situation differs from that in [Eli] with respect to the non-compactness of the curves' domain $\mathbb{R}$. There are several steps in the discussion where we need a more refined treatment.

A.1 The Manifold of Maps $\mathcal{P}^{1,2}_{x,y}(\mathbb{R}, M)$

Definition A.1 *A covariant functor*

$$\mathfrak{S} : \mathrm{Vec}_{C^\infty}(\overline{\mathbb{R}}) \to \mathrm{Ban}$$

is called a section functor if it associates to each smooth vector bundle ξ on $\overline{\mathbb{R}}$[1] *a vector space $\mathfrak{S}(\xi)$ of sections in ξ together with a Banach space topology, so that $\mathfrak{S}$ maps each smooth bundle homomorphism $\varphi : \xi \to \eta$ to a linear map $\mathfrak{S}_*\varphi \in \mathcal{L}(\mathfrak{S}(\xi); \mathfrak{S}(\eta))$ defined by $(\mathfrak{S}_*\varphi) \cdot s = \varphi \cdot s$, and*

$$\mathfrak{S}_* : C^\infty(\mathrm{Hom}(\xi, \eta)) \to \mathcal{L}(\mathfrak{S}(\xi); \mathfrak{S}(\eta))$$

[1] endowed with the already defined differentiable structure

is continuous.

Auxiliary Proposition A.2 *The mappings* $H^{1,2}_{\overline{\mathbb{R}}}, L^2_{\overline{\mathbb{R}}} : \mathrm{Vec}_{C^\infty}(\overline{\mathbb{R}}) \to \mathrm{Ban}$ *defined in Definition 2.5 are section functors.*

Proof. We have already verified that the Banach space topology on $H^{1,2}_{\overline{\mathbb{R}}}(\xi)$ and $L^2_{\overline{\mathbb{R}}}(\xi)$ is independent of the respectively chosen trivialization ϕ. Each pair of norms $\|\cdot\|^{\phi}_{1,2}$ and $\|\cdot\|^{\phi}_{0,2}$ induced by such trivializations from the canonically normed vector spaces $H^{1,2}(\mathbb{R},\mathbb{R}^n)$ and $L^2(\mathbb{R},\mathbb{R}^n)$, respectively, are equivalent, because ϕ is equipped with asymptotical differentiability. Hence, we may start without loss of generality from the trivial bundle $\xi = \overline{\mathbb{R}} \times \mathbb{R}^n$, i.e.

$$H^{1,2}_{\overline{\mathbb{R}}}(\xi) = H^{1,2}(\mathbb{R},\mathbb{R}^n) \quad \text{and} \quad L^2_{\overline{\mathbb{R}}}(\xi) = L^2(\mathbb{R},\mathbb{R}^n) \ .$$

Now let $A \in C^\infty\big(\mathrm{Hom}(\xi,\eta)\big)$, that is, without loss of generality, $A \in C^\infty\big(\overline{\mathbb{R}}, \mathrm{M}(m\times n,\mathbb{R})\big)$. Then the following estimates hold for $s \in H^{1,2}$ and respectively L^2, where we denote $(As)(t) = A(t)\cdot s(t)$:

$$\text{(A.1)} \qquad \|As\|_{0,2} \leqslant \|A\|_\infty \cdot \|s\|_{0,2}$$

$$\begin{aligned} \|(As)'\|^2_{0,2} &= \|A's + As'\|^2_{0,2} \\ &= \int_{\mathbb{R}} \big(\, |A's|^2 + |As'|^2 + 2\langle\, A's, As'\,\rangle\,\big)\, dt \\ &\leqslant 2\Big(\ \|A's\|^2_{0,2} \ + \ \|As'\|^2_{0,2}\ \Big) \end{aligned}$$

$$\text{(A.2)} \qquad \leqslant 2\Big(\ \|A'\|^2_{0,2}\ \|s\|^2_\infty \ + \ \|A\|^2_\infty\ \|s'\|^2_{0,2}\ \Big) \ .$$

According to Lemma 2.2, $\overline{\mathbb{R}}$-differentiability yields the finite norms $\|A'\|_{0,2}$, $\|A\|_\infty < \infty$. Moreover, the estimate

$$\text{(A.3)} \qquad \|s\|_\infty \leqslant \|s\|_{1,2} \quad \text{for all } s \in H^{1,2}(\mathbb{R},\mathbb{R}^n) \ .$$

follows from the simple calculation

$$\begin{aligned} \Big|\, \big|s(t_1)\big|^2 - \big|s(t_0)\big|^2 \,\Big| &= \Big|\int_{t_0}^{t_1} \frac{d}{dt}\big|s(t)\big|^2\, d\tau \Big| \\ = \Big| \int_{t_0}^{t_1} 2\big\langle s(\tau), \dot s(\tau)\big\rangle\, d\tau \Big| &\leqslant \int_{t_0}^{t_1} \Big(\big|s(\tau)\big|^2 + \big|\dot s(\tau)\big|^2\Big)\, d\tau \\ &\leqslant \|s\|^2_{1,2} \ . \end{aligned}$$

Thus (A.1) and (A.2) give rise to the estimate

$$\|As\|^2_{1,2} \leqslant \mathrm{const}\Big(\ \|A'\|^2_{0,2} \ + \ \|A\|^2_\infty\Big)\ \|s\|^2_{1,2} \ ,$$

that is

$$\text{(A.4)} \qquad \|A\|_{\mathcal{L}(H^{1,2};H^{1,2})} \leqslant \text{const}\sqrt{\|A'\|_{0,2}^2 + \|A\|_\infty^2} \; .$$

Consequently, the mapping $\mathfrak{S}_* : C^\infty(\mathrm{Hom}(\xi,\eta)) \to \mathcal{L}(\mathfrak{S}(\xi);\mathfrak{S}(\eta))$ is continuous. □

Remark It is worth mentioning that the estimate (A.4) with $\|A'\|_{0,2} < \infty$ relies essentially on the special choice of the differentiable structure on $\overline{\mathbb{R}}$!

Corollary A.3 *As to the section functor* $\mathfrak{S} = H^{1,2}_{\mathbb{R}}$, *the estimate* $\|A\|_\infty \leqslant \|A\|_{1,2}$ *for* $A \in H^{1,2}(\mathbb{R}, \mathrm{M}(m \times n, \mathbb{R}))$ *implies the continuity of the map*

$$\mathfrak{S}_* : \mathfrak{S}(\mathrm{Hom}(\xi,\eta)) \to \mathcal{L}(\mathfrak{S}(\xi);\mathfrak{S}(\eta)) \; ,$$

which is stronger than the mere section functor property.

The next step provides us with properties of this special section functor, which correspond to the conditions of a so-called manifold model in [Eli].

Auxiliary Proposition A.4 *The section functor* $\mathfrak{S} = H^{1,2}_{\mathbb{R}}$ *on* $\mathrm{Vec}_{C^\infty}(\overline{\mathbb{R}})$ *has the properties*

(a) $\mathfrak{S}(\xi) \hookrightarrow C^0(\xi)$ *is continuous for each* $\xi \in \mathrm{Vec}_{C^\infty}(\overline{\mathbb{R}})$ *and*

(b) $$\begin{array}{rcl} \mathfrak{S}(\mathrm{Hom}(\xi,\eta)) & \to & \mathcal{L}(\mathfrak{S}(\xi);\mathfrak{S}(\eta)) \\ A & \mapsto & (A_* : s \mapsto As) \end{array} \quad \textit{is continuous}$$
for all $\xi, \eta \in \mathrm{Vec}_{C^\infty}(\overline{\mathbb{R}})$.

(c) Given $\xi, \eta \in \mathrm{Vec}_{C^\infty}(\overline{\mathbb{R}})$ *and an open subset* $\mathcal{O} \subset \xi$ *such that there is a section* $\gamma \in C^0(\xi)$ *with compact support in* $\mathbb{R}$ *and* $\gamma(\overline{\mathbb{R}}) \subset \mathcal{O}$, *each smooth bundle map* $f \in C^\infty(\mathcal{O}, \eta)$ *satisfying* $f(0_{\pm\infty}) = 0_{\pm\infty}$ *induces a well-defined and continuous map on* $\mathfrak{S}(\mathcal{O}) = \{ s \in \mathfrak{S}(\xi) \,|\, s(\overline{\mathbb{R}}) \subset \mathcal{O} \}$ [2]

$$\begin{array}{rcl} f_* : \mathfrak{S}(\mathcal{O}) & \to & \mathfrak{S}(\eta) \\ s & \mapsto & f \circ s \; . \end{array}$$

Proof. (a) has been already proved by (A.3) and (b) follows likewise from (A.3) together with (A.4).
As to (c): Since f is a smooth bundle map, its fiber restrictions f_t are in particular Lipschitz continuous, uniformly in $t \in \mathbb{R}$. Thus, with respect to any trivialization of ξ, we obtain the estimates

$$\text{(A.5)} \qquad |\, f_t(x_t) - f_t(y_t) \,| \leqslant \text{const}\, |\, x_t - y_t \,| \quad \text{for all } t \in \overline{\mathbb{R}}$$

2Note that this subset of sections is open within $\mathfrak{S}(\xi)$ due to item (a).

and

$$\left|f_t\big(s(t)\big)\right| \leqslant \text{const}\left|s(t)\right| + \left|f_t(0)\right| \ .$$

Since $f_t(0) \in C^\infty(\overline{\mathbb{R}}, \eta)$, this estimate together with Corollary 2.4 implies that the map $f_* : H^{1,2}_{\overline{\mathbb{R}}}(\mathcal{O}) \to H^{1,2}_{\overline{\mathbb{R}}}(\eta)$ is well-defined. The continuity of f_* with respect to the section functor $\mathfrak{S} = H^{1,2}_{\overline{\mathbb{R}}}$ is obtained from (A.5) together with an analogous uniform estimate involving first derivatives of the $\overline{\mathbb{R}}$-smooth bundle map f. □

Now these properties enable us to deduce the crucial lemma concerning the construction of the manifold of maps within our framework.

Fundamental Lemma A.5 *Let $\mathfrak{S}, \mathcal{O} \subset \xi$ and $f \in C^\infty(\mathcal{O}, \eta)$ be as in the auxiliary Proposition A.4. Then the map $f_* : \mathfrak{S}(\mathcal{O}) \to \mathfrak{S}(\eta)$ is smooth and the k-th derivative is given by $D^k f_*(s) = \mathfrak{S}_*(F^k f \circ s)$, which is well-defined. Here, $F^k f : \mathcal{O} \to \mathrm{Hom}(\xi \oplus \ldots \oplus \xi\,; \eta)$ denotes the k-th fibre derivative of f.*

Proof. We prove the lemma by induction on k.

$k = 0$: The continuity of f_* has been already verified as item (c) in the auxiliary Proposition A.4.

$k = 1$: Let $\mathcal{O}$ be fibrewise convex without loss of generality and let $s_0 \in \mathfrak{S}(\mathcal{O})$ be fixed. Then, for $x, y \in \mathcal{O}$ from the same fibre, i.e. $\pi(x) = \pi(y)$, we define

$$\text{(A.6)}\qquad \begin{aligned} &\Theta : \mathcal{O} \oplus \mathcal{O} \to \mathrm{Hom}(\xi, \eta) \\ &\Theta(x,y) \cdot z = \left[\int_0^1 Ff\big(x + t(y-x)\big)dt - Ff(x)\right] \cdot z \ . \end{aligned}$$

Since $Ff : \mathcal{O} \to \mathrm{Hom}(\xi, \eta)$ is smooth and fibre respecting, the same is true for Θ. Moreover, we easily verify the equations

$$\text{(A.7)}\quad \Theta(x,y) \cdot (y-x) = f(y) - f(x) - Ff(x) \cdot (y-x), \ \ \Theta(0,0) = 0 \ .$$

The section functor $\mathfrak{S}$ with properties (b) and (c) from the auxiliary Proposition A.4 gives rise to the composition of the continuous maps

$$\mathfrak{S}(\Theta) : \mathfrak{S}(\mathcal{O}) \oplus \mathfrak{S}(\mathcal{O}) \to \mathfrak{S}\big(\mathrm{Hom}(\xi, \eta)\big) \to \mathcal{L}\big(\mathfrak{S}(\xi); \mathfrak{S}(\eta)\big) \ ,$$

so that

$$\text{(A.8)}\qquad \mathfrak{S}(\Theta)\,(s_0, s_0) = 0 \quad \text{and} \quad \lim_{s \to s_0} \mathfrak{S}(\Theta)\,(s_0, s) = 0 \ .$$

Since (A.7) implies the identity

$$f_*(s) - f_*(s_0) - \mathfrak{S}_*(Ff \circ s_0) \cdot (s - s_0) = \mathfrak{S}(\Theta)\,(s_0, s) \cdot (s - s_0) \ ,$$

the map f_* is differentiable at s_0 with the representation of the differential

$$Df_*(s_0) = \mathfrak{S}_*(Ff \circ s_0) \in \mathcal{L}\big(\mathfrak{S}(\xi); \mathfrak{S}(\eta)\big) \ . \tag{A.9}$$

It should be mentioned that, due to (A.4), this is also well-defined if it holds that $Ff(\pm\infty, 0) \neq 0$, i.e. $Ff \circ s_0 \notin \mathfrak{S}\big(\mathrm{Hom}(\xi,\eta)\big)$. The continuity of $Df_* : \mathfrak{S}(\mathcal{O}) \to \mathcal{L}\big(\mathfrak{S}(\xi); \mathfrak{S}(\eta)\big)$ follows from the continuity of the mapping

$$\begin{array}{rcl} \mathfrak{S}(\mathcal{O}) & \to & \mathfrak{S}\big(\mathrm{Hom}(\xi,\eta)\big) \subset \mathcal{L}\big(\mathfrak{S}(\xi); \mathfrak{S}(\eta)\big) \\ s & \mapsto & Ff \circ s - Ff \circ s_0 \ , \end{array}$$

because

$$\begin{array}{rcl} \vartheta : \mathcal{O} & \to & \mathrm{Hom}(\xi,\eta) \\ (t,x) & \mapsto & Ff(t,x) - Ff\big(t, s_0(t)\big) \end{array}$$

is a smooth,[3] fibre respecting map, and it satisfies the condition $\vartheta(\pm\infty, 0) = (\pm\infty, 0)$ as with (c) in the auxiliary Proposition A.4.

$k \rightsquigarrow k+1$: Let us now start from the k-times continuously differentiable map f_* together with

$$D^k f_*(s_0) = \mathfrak{S}_*(F^k f \circ s_0) \in \mathcal{L}\big(\mathfrak{S}(\xi), \ldots, \mathfrak{S}(\xi); \mathfrak{S}(\eta)\big)$$

and let us denote by $f^{(k)}$ the map

$$\begin{array}{l} f^{(k)} : \mathcal{O} \to \mathrm{Hom}(\xi \oplus \ldots \oplus \xi, \eta) \\ f^{(k)}(t,\xi) = F^k f(t,\xi) - F^k f\big(t, s_0(t)\big) \ . \end{array}$$

Then $f^{(k)}$ again satisfies the inital condition on f, namely in particular

$$f^{(k)}(\pm\infty, 0) = 0$$

with $\eta_{\text{new}} = \mathrm{Hom}(\xi \oplus \ldots \oplus \xi, \eta)$. Thus, the step '$k = 1$' yields the continuous differentiability of ${f^{(k)}}_* : \mathfrak{S}(\mathcal{O}) \to \mathcal{L}\big(\mathfrak{S}(\xi), \ldots, \mathfrak{S}(\xi); \mathfrak{S}(\eta)\big)$ together with

$$D{f^{(k)}}_*(\tilde{s}_0) = \mathfrak{S}_*\big(Ff^{(k)} \circ \tilde{s}_0\big) = \mathfrak{S}_*\big(F^{k+1} f \circ \tilde{s}_0\big) \ .$$

□

We now assume (M, g) to be a paracompact, Riemannian C^∞-manifold together with the associated exponential map. Furthermore, let $\mathcal{D}$ denote an open zero section neighbourhood within the tangent bundle $\tau : TM \to M$ such that

$$\begin{array}{rcl} \mathcal{D} & \xrightarrow{\approx} & V(\triangle) \subset M \times M \\ \xi & \mapsto & \big(\tau(\xi), \exp(\xi)\big) \end{array} \tag{A.10}$$

3Here, without loss of generality, $s_0 \in C^\infty(\xi)$.

represents a diffeomorphism onto a diagonal neighbourhood within $M \times M$. We observe that smooth, compact curves $h \in C^\infty(\overline{\mathbb{R}}, M)$ give rise to the pull-back bundles

$$h^*TM = \{ (t,\xi) \in \overline{\mathbb{R}} \times TM \,|\, \tau(\xi) = h(t) \} \in \mathrm{Vec}_{C^\infty}(\overline{\mathbb{R}}) ,$$

on which the above section functor $H^{1,2}_{\mathbb{R}}$ is well-defined. From now on we shall again use the notation $\mathfrak{S}$ for this section functor. We obviously obtain from $\mathcal{D}$ the open zero section neighbourhoods

$$h^*\mathcal{D} = \{ (t,\xi) \in \overline{\mathbb{R}} \times \mathcal{D} \,|\, \tau(\xi) = h(t) \} \overset{\text{open}}{\subset} h^*TM .$$

Definition A.6 *Starting from* $s \in \mathfrak{S}(h^*\mathcal{D}) \subset C^0(h^*\mathcal{D})$, *we define the continuous curve*

$$\exp_h s = \exp \circ s \in C^0(\overline{\mathbb{R}}, M), \; (\exp \circ s)(t) = \exp_{h(t)} s(t) .$$

Thus, provided any fixed endpoints $x, y \in M$, *the set of curves*

$$\mathcal{P}^{1,2}_{x,y}(\mathbb{R}, M) = \{ \exp \circ s \in C^0(\overline{\mathbb{R}}, M) \,|\, s \in \mathfrak{S}(h^*\mathcal{D}), \; h \in C^\infty_{x,y}(\overline{\mathbb{R}}, M) \}$$

is well-defined.

Theorem 10 $\mathcal{P}^{1,2}_{x,y}(\mathbb{R}, M)$ *is equipped with the differentiable structure of an infinite-dimensional Banach manifold. The family*

$$\{ \mathfrak{S}(h^*\mathcal{D}), \mathfrak{S}(\exp_h) \}_{h \in C^\infty_{x,y}(\overline{\mathbb{R}}, M)} ,$$

$$\textit{where} \quad \mathfrak{S}(\exp_h) : \begin{array}{ll} \mathfrak{S}(h^*\mathcal{D}) & \to \mathcal{P}^{1,2}_{x,y} \\ s & \mapsto \exp \circ s \end{array} ,$$

represents an associated atlas of charts. Moreover, the Banach manifold $\mathcal{P}^{1,2}_{x,y}$ *together with this differentiable structure is independent of the Riemannian metric* g *on* M.

Proof. We consider the following smooth mappings with respect to $h \in C^\infty_{x,y}(\overline{\mathbb{R}}, M)$:

$$\begin{aligned} \phi_h : h^*\mathcal{D} &\;\to\; \overline{\mathbb{R}} \times M \\ (t,\xi) &\;\mapsto\; (t, \exp_{h(t)} \xi) . \end{aligned} \tag{A.11}$$

According to the choice of $\mathcal{D}$, this is an embedding onto an open neighbourhood of the graph of h. We consequently define

$$\begin{aligned} \mathcal{U}_h &= \phi_h(h^*\mathcal{D}) \subset \overline{\mathbb{R}} \times M , \\ \mathfrak{S}(\mathcal{U}_h) &= \{ g \in C^0(\overline{\mathbb{R}}, M) \,|\, \mathrm{graph}\, g \subset \mathcal{U}_h, \; \phi_h^{-1} \circ (\mathrm{id}, g) \in \mathfrak{S}(h^*\mathcal{D}) \} \\ &\subset C^0_{x,y}(\overline{\mathbb{R}}, M) \end{aligned}$$

and

$$\text{(A.12)}\qquad \begin{array}{rll} \mathfrak{S}(\phi_h^{-1}): & \mathfrak{S}(\mathcal{U}_h) & \to \mathfrak{S}(h^*\mathcal{D}) \\ & g & \mapsto \phi_h^{-1} \circ (\mathrm{id}, g) \ . \end{array}$$

It holds that

$$\mathcal{P}^{1,2}_{x,y}(\mathbb{R}, M) = \bigcup_{h \in C^\infty_{x,y}(\overline{\mathbb{R}}, M)} \mathfrak{S}(\mathcal{U}_h) \ ,$$

and that $\{\mathfrak{S}(\phi_h^{-1})\}_{h \in C^\infty_{x,y}(\overline{\mathbb{R}}, M)}$ is a family of one-to-one correspondences. Let us now endow this curve space with a topology such that these bijections become homeomorphisms. The fact that this topology is well-defined is particularly due to the following proof of the diffeomorphism property for changes of the charts $\mathfrak{S}(\phi_f^{-1}) \circ \mathfrak{S}(\phi_h^{-1})^{-1}$.

Let us choose any two supporting curves $f, h \in C^\infty_{x,y}(\overline{\mathbb{R}}, M)$ with $\mathcal{U}_h \cap \mathcal{U}_f \neq \emptyset$. Then $\mathcal{O}_h = \phi_h^{-1}(\mathcal{U}_h \cap \mathcal{U}_f) \subset h^*\mathcal{D}$ and $\mathcal{O}_f = \phi_f^{-1}(\mathcal{U}_h \cap \mathcal{U}_f) \subset f^*\mathcal{D}$ are open subsets satisfying the condition in (c) in the auxiliary Proposition A.4. The non-void intersection $\mathfrak{S}(\mathcal{U}_h) \cap \mathfrak{S}(\mathcal{U}_f)$ is open within $\mathfrak{S}(\mathcal{U}_h)$ and $\mathfrak{S}(\mathcal{U}_f)$, respectively. We now consider the mapping

$$\text{(A.13)}\qquad \begin{array}{l} \Phi_{fh} = \phi_f^{-1} \circ \phi_h : h^*TM \supset \mathcal{O}_h \to \mathcal{O}_f \subset f^*TM \\ \Phi_{fh}(t, \xi) = \left(t, \exp_{f(t)}^{-1}(\exp_{h(t)} \cdot \xi)\right) \ , \end{array}$$

which describes the change of the charts. It fulfills the condition $\Phi_{fh}(\pm\infty, 0) = 0$ as we can see from the fixed endpoints $h(\pm\infty) = f(\pm\infty) = x, y$. Thus, the fundamental Lemma A.5 states that

$$\Phi_{fh*} = \mathfrak{S}(\phi_f^{-1}) \circ \mathfrak{S}(\phi_h^{-1})^{-1} : \mathfrak{S}(\mathcal{O}_h) \to \mathfrak{S}(\mathcal{O}_f)$$

is a smooth map with

$$\Phi_{fh*}^{-1} = \Phi_{hf*}$$

as its inverse, hence the diffeomorphism property.

The compatibility of two atlases belonging to different Riemannian metrics follows from the fact that $C^\infty_{x,y}(\overline{\mathbb{R}}, M)$ lies dense within $C^0_{x,y}(\overline{\mathbb{R}}, M)$ and that, for any arbitrary metric g, the exponential $\exp_g$ represents a local diffeomorphism in the sense of (A.10). □

Supplement Moreover, we are able to find a countable subset of $C^\infty_{x,y}(\overline{\mathbb{R}}, M)$ which lies likewise dense within $C^0_{x,y}(\overline{\mathbb{R}}, M)$. This enables us to reduce the atlas given in Theorem 10 to a countable subatlas.

Corollary A.7 *Given Riemannian C^∞-manifolds M and N together with a map $f \in C^\infty(M, N)$,*

$$\begin{array}{rcl} \mathfrak{S}(f) : \mathcal{P}^{1,2}_{x,y}(\mathbb{R}, M) & \to & \mathcal{P}^{1,2}_{f(x),f(y)}(\mathbb{R}, N) \\ \gamma & \mapsto & f \circ \gamma \end{array}$$

is a well-defined, smooth map between Banach manifolds.

Proof. First, any fixed $\gamma \in \mathcal{P}^{1,2}_{x,y}$ is mapped to $f \circ \gamma \in C^0_{f(x),f(y)}(\overline{\mathbb{R}}, N)$. Since $C^\infty_{f(x),f(y)}(\overline{\mathbb{R}}, N)$ lies dense within $C^0_{f(x),f(y)}(\overline{\mathbb{R}}, N)$, we find a representation of the shape

$$f \circ \gamma = \exp_k t,\ k \in C^\infty_{f(x),f(y)}(\overline{\mathbb{R}}, N),\ t \in C^0(k^*\mathcal{D}_N)\ .$$

Secondly, we may express γ as

$$\gamma = \exp_h s,\ h \in C^\infty_{x,y}(\overline{\mathbb{R}}, M),\ s \in \mathfrak{S}(h^*\mathcal{D}_M)\ .$$

Furthermore, we find an open neighbourhood $\widetilde{\mathcal{O}}_h \subset h^*\mathcal{D}_M$, such that the conditions in (c) in the auxiliary Proposition A.4 are satisfied for the smooth fibre respecting map

$$\begin{aligned} \psi^f_{kh} &= \exp_k^{-1} \circ f \circ \exp_h : \widetilde{\mathcal{O}}_h \to k^*\mathcal{D}_N\ , \\ \psi^f_{kh}(\pm\infty, 0) &= (\pm\infty, 0)\ . \end{aligned}$$

Hence it is due to the fundamental Lemma A.5, that $\psi^f_{kh*} : \mathfrak{S}(\widetilde{\mathcal{O}}_h) \to \mathfrak{S}(k^*\mathcal{D}_N)$ is a smooth map. This proves the corollary. □

A.2 Banach Bundles on $\mathcal{P}^{1,2}_{x,y}(\mathbb{R}, M)$

We henceforth use the notation $\mathfrak{S}$ for the section functor $H^{1,2}_{\mathbb{R}}$. Let us fix a Riemannian metric g on the manifold M, as was required in the last section in order to define the atlas on $\mathcal{P}^{1,2}_{x,y}$. Given such a g we are provided canonically with the following features known from Riemannian geometry:

Let

$$K : T(TM) \to TM,\ \ \tau : TM \to M$$

denote the unique Levi-Civita connection and the canonical projection in the tangent bundle of M. As a consequence, K and $D\tau : T(TM) \to TM$ yield a decomposition of $T(TM)$ into a horizontal bundle and a vertical bundle:[4]

$$\text{(A.14)}\qquad \begin{aligned} T_{\xi,v}(TM) &= \ker\big(D\tau(\xi)\big) \\ T_{\xi,h}(TM) &= \ker\big(K(\xi)\big) \end{aligned}\ ,\ \ \xi \in TM\ .$$

4Note that the vertical bundle comes canonically and independent of any specification of a Riemannian metric.

Referring to the exponential map associated to g and K, $\exp : \mathcal{D} \to M$, we obtain isomorphisms at any $\xi \in \mathcal{D}$, where $\mathcal{D}$ once again denotes the injectivity neighbourhood associated to exp:

$$\text{(A.15)} \quad \begin{aligned} \nabla_1 \exp(\xi) &= D\exp(\xi) \circ \left(D\tau_{| T_{\xi,h}(TM)}\right)^{-1} : T_{\tau(\xi)}M \xrightarrow{\cong} T_{\exp(\xi)}M\,, \\ \nabla_2 \exp(\xi) &= D\exp(\xi) \circ \left(K_{| T_{\xi,v}(TM)}\right)^{-1} : \ T_{\tau(\xi)}M \xrightarrow{\cong} T_{\exp(\xi)}M\ . \end{aligned}$$

It holds that $\nabla_1 \exp(0) = \nabla_2 \exp(0) = \mathrm{id}_{T_{\tau(0)}M}$. Furthermore,

$$\text{(A.16)} \quad \begin{aligned} &\Theta : \mathcal{D} \to \mathrm{Hom}(TM, TM) \\ &\Theta(\xi) = \left(\nabla_2 \exp(\xi)\right)^{-1} \circ \nabla_1 \exp(\xi) : T_{\tau(\xi)}M \xrightarrow{\cong} T_{\tau(\xi)}M \end{aligned}$$

is a smooth fibre respecting map which satisfies the identities

- $\Theta(0) = \mathrm{id}$,
- $F\Theta(0) = 0$ and
- $\Theta(\xi)\cdot \xi = \xi$.

Given any $v \in T_pM$ und $X \in C^\infty(TM)$, we obtain the covariant derivative of X in the direction of v from the formula

$$\nabla_v X(p) = K \circ D_p X \cdot v\ .$$

Thus, straightforward computation for $h \in C^\infty_{x,y}$ and $\xi \in C^\infty(h^*TM)$ yields the identity

$$\text{(A.17)} \quad \frac{\partial}{\partial t} \exp_h \xi = \nabla_1 \exp(\xi) \cdot \dot{h} + \nabla_2 \exp(\xi) \cdot \nabla_t \xi\ .$$

Let us consider once again the map from (A.13) representing the change of charts Φ_{fh} for smooth $\overline{\mathbb{R}}$-curves $f, h \in C^\infty_{x,y}(\overline{\mathbb{R}}, M)$. Then the fibre derivative $F\Phi_{fh}(t,\xi) \in \mathrm{Hom}\left(T_{h(t)}M, T_{f(t)}M\right)$ has the representation

$$\text{(A.18)} \quad \begin{aligned} &F\Phi_{fh} : \mathcal{O}_h \to \mathrm{Hom}\left(h^*TM, f^*TM\right) \\ &F\Phi_{fh}(t,\xi) = \nabla_2 \exp\left(\exp^{-1}_{f(t)}(\exp_{h(t)} \xi)\right)^{-1} \circ \nabla_2 \exp(\xi)\ . \end{aligned}$$

Thus, according to the fundamental Lemma A.5,

$$\text{(A.19)} \quad \begin{aligned} \mathfrak{S}(\mathcal{O}_h) &\to \mathcal{L}\big(\mathfrak{S}(h^*TM); \mathfrak{S}(f^*TM)\big) \\ s &\mapsto \mathfrak{S}_*(F\Phi_{fh} \circ s) = D\Phi_{fh*}(s) \end{aligned}$$

is a smooth map.

Remark Similar to the proof of this fundamental lemma, we may verify also that $\Theta : \mathcal{D} \to \mathrm{Hom}\,(TM, TM)$ gives rise to a smooth map

$$\begin{array}{rrcl} \text{(A.20)} & \mathfrak{S}(\Theta) : \mathfrak{S}(\mathcal{O}_h) & \to & \mathcal{L}(\mathfrak{S}(h^*TM); \mathfrak{S}(h^*TM)) \\ & s & \mapsto & \mathfrak{S}_*(\Theta \circ s) \ . \end{array}$$

Now let $\mathfrak{T} : \mathrm{Vec}_{C^\infty}(\overline{\mathbb{R}}) \to \mathrm{Ban}$ be any section functor satisfying the condition that

$$\begin{array}{rcl} \mathfrak{T}_* : \mathfrak{S}\big(\mathrm{Hom}\,(\xi, \eta)\big) & \to & \mathcal{L}\big(\mathfrak{T}(\xi); \mathfrak{T}(\eta)\big) \\ A & \mapsto & (s \mapsto A \cdot s) \end{array}$$

yields a continuous map for each two $\xi, \eta \in \mathrm{Vec}(\overline{\mathbb{R}})$, that is, in short terms,

$$\text{(A.21)} \qquad \mathfrak{S}(\mathrm{Hom}) \subset \mathcal{L}(\mathfrak{T}; \mathfrak{T}) \ .$$

Referring to the estimate (A.1) in the auxiliary proposition A.2, we notice that $\mathfrak{T} = L^2_{\overline{\mathbb{R}}}$ fulfills this condition with respect to $\mathfrak{S} = C^0_{\overline{\mathbb{R}}}$ and thus in particular with respect to $\mathfrak{S} = H^{1,2}_{\overline{\mathbb{R}}}$ due to (A.3).

Given a fixed smooth section $s_0 \in C^\infty_{\overline{\mathbb{R}}}(\mathcal{O}_h)$, we are provided with a smooth map as in the proof of the fundamental lemma,

$$\begin{array}{rrcl} \text{(A.22)} & \vartheta_* : \mathfrak{S}(\mathcal{O}_h) & \to & \mathfrak{S}\big(\mathrm{Hom}\,(h^*TM, f^*TM)\big) \\ & s & \mapsto & F\Phi_{fh} \circ s - F\Phi_{fh} \circ s_0 \ . \end{array}$$

It holds that

$$\begin{array}{rl} & \mathfrak{S}_* \circ \vartheta_*(s) = \mathfrak{S}_*(F\Phi_{fh} \circ s - F\Phi_{fh} \circ s_0) \\ = & \mathfrak{S}_*(F\Phi_{fh} \circ s) - \mathfrak{S}_*(F\Phi_{fh} \circ s_0) \in \mathcal{L}\big(\mathfrak{S}(h^*TM); \mathfrak{S}(f^*TM)\big) \ . \end{array}$$

Thus, property (A.21) of the section functor $\mathfrak{T}$ yields the smooth mapping

$$\begin{array}{rrcl} \text{(A.23)} & \mathfrak{S}(\mathcal{O}_h) & \to & \mathcal{L}\big(\mathfrak{T}(h^*Tm); \mathfrak{T}(f^*TM)\big) \\ & s & \mapsto & \mathfrak{T}_*(F\Phi_{fh} \circ s - F\Phi_{fh} \circ s_0) \ . \end{array}$$

Since $\mathfrak{T}_*(F\Phi_{fh} \circ s_0) \in \mathcal{L}\big(\mathfrak{T}(h^*TM); \mathfrak{T}(f^*TM)\big)$ is already well-defined according to the section functor property, we obtain the smooth map

$$\begin{array}{rrcl} \text{(A.24)} & \mathfrak{T}_* : \mathfrak{S}(\mathcal{O}_h) & \to & \mathcal{L}\big(\mathfrak{T}(h^*TM); \mathfrak{T}(f^*TM)\big) \\ & s & \mapsto & \mathfrak{T}_*(F\Phi_{fh} \circ s) \end{array}$$

in analogy with

$$\mathfrak{S}_* : \mathfrak{S}(\mathcal{O}_h) \to \mathcal{L}\big(\mathfrak{S}(h^*TM); \mathfrak{S}(f^*TM)\big) \ .$$

After these preparations we are able to prove the following

Theorem 11 *The section functor $\mathfrak{T}$ satisfying condition (A.21) admits a unique extension to the continuous vector bundles of the shape*

$$g^*TM,\ \ g \in \mathcal{P}^{1,2}_{x,y}(\mathbb{R}, M) \subset C^0_{x,y}(\overline{\mathbb{R}}, M);\ x, y \in M\ .$$

Moreover,

$$\mathfrak{T}\left(\mathcal{P}^{1,2*}_{x,y}TM\right) = \bigcup_{g\in\mathcal{P}^{1,2}_{x,y}(\mathbb{R},M)} \mathfrak{T}(g^*TM)$$

is a Banach vector bundle on $\mathcal{P}^{1,2}_{x,y}(\mathbb{R}, M)$.

Proof. Let $\alpha, \beta \in C^0_{x,y}(\overline{\mathbb{R}}, M)$ be given, satisfying

$$\beta = \exp\circ\xi,\ \ \xi \in C^0(\alpha^*\mathcal{D}) \subset C^0(\alpha^*TM)\ .$$

We consequently define $J_{\beta\alpha} \in C^0\big(\mathrm{Hom}\,(\alpha^*TM, \beta^*TM)\big)$ by

$$J_{\beta\alpha}(t)\cdot v(t) = \nabla_2 \exp\big(\xi(t)\big)\cdot v(t)\ . \tag{A.25}$$

Thus, $J_{\beta\alpha}(t)$ is a toplinear isomorphism for each $t \in \overline{\mathbb{R}}$. Now let us consider charts based on the supporting curves $h, f \in C^\infty_{x,y}(\overline{\mathbb{R}}, M)$ as in the proof of Theorem 10. Formula (A.18) applied to $s \in \mathfrak{S}(\mathcal{O}_h)$ and $v \in \mathfrak{S}(h^*TM)$ leads to the identity

$$\nabla_2 \exp(\Phi_{fh}\circ s)\circ(F\Phi_{fh}\circ s)\cdot v = \nabla_2 \exp(s)\cdot v\ ,$$

that is

$$\mathfrak{S}_*(F\Phi_{fh}\circ s) = \mathfrak{S}_*\big((J_{gf})^{-1}\circ J_{gh}\big),\quad g = \exp_h s \in \mathcal{P}^{1,2}_{x,y}(\mathbb{R}, M)\ . \tag{A.26}$$

We therefore define the fibre at $g = \exp_h s$ as

$$\mathfrak{T}(g^*TM) = \left\{\, J_{gh}\cdot v \,\middle|\, v \in \mathfrak{T}(h^*TM)\,\right\}\ .$$

This is independent of the choice of h, because (A.26) implies the equation

$$J_{gf}\circ\mathfrak{S}_*(F\Phi_{fh}\circ s)\cdot v = J_{gh}\cdot v\ . \tag{A.27}$$

Thus, referring to (A.24), the Banach space topology on $\mathfrak{T}(g^*TM)$ is defined in a unique way independent of the underlying curve h. The smoothness of the map

$$\begin{array}{rcl} \mathfrak{T}_* : \mathfrak{S}(\mathcal{O}_h) & \to & \mathcal{L}\big(\mathfrak{T}(h^*TM); \mathfrak{T}(f^*TM)\big) \\ s & \mapsto & \mathfrak{T}_*(F\Phi_{fh}\circ s) \end{array}$$

due to (A.24) implies that the maps

$$J_{gh} : \mathfrak{T}(h^*TM) \to \mathfrak{T}(g^*TM)$$

provide us with a well-defined local trivialization of the bundle $\mathfrak{T}\left(\mathcal{P}^{1,2}_{x,y}{}^{*}TM\right)$ together with smooth transition maps. □

Since the transition maps due to (A.27) comply with the identity

$$(J_{gf})^{-1} \circ J_{gh} = \mathfrak{S}_*(F\Phi_{fh} \circ s) = D\Phi_{fh*}(s) \tag{A.28}$$

with respect to the special section functor $\mathfrak{T} = \mathfrak{S}$, the Banach bundle

$$\mathfrak{S}\left(\mathcal{P}^{1,2}_{x,y}{}^{*}TM\right) = \bigcup_{h \in \mathcal{P}^{1,2}_{x,y}(\mathbb{R},M)} \mathfrak{S}(h^*TM)$$

represents exactly the tangent bundle of the Banach manifold $\mathcal{P}^{1,2}_{x,y}(\mathbb{R}, M)$. As a consequence we obtain

Corollary A.8 $T\mathcal{P}^{1,2}_{x,y}(\mathbb{R}, M) = H^{1,2}_{\mathbb{R}}\left(\mathcal{P}^{1,2}_{x,y}{}^{*}TM\right)$.

Remark A more detailed analysis of the exponential map, its derivatives $\nabla_1 \exp$ and $\nabla_2 \exp$ and of the Levi-Civita connection would yield the possibility of discussing the associated exponential map $\exp_T$ on the Riemannian manifold TM. As a consequence, the functorial behaviour of $\mathcal{P}^{1,2}_{\cdot,\cdot}(\mathbb{R}, \cdot)$ as stated in Corollary A.3 applied to the projection map $\tau : TM \to M$ would enable us to find a representation of the tangent bundle $T\mathcal{P}^{1,2}_{x,y}$ in the following form:

$$\begin{aligned} T\mathcal{P}^{1,2}_{x,y}(\mathbb{R}, M) &= \mathcal{P}^{1,2}_{0_x,0_y}(\mathbb{R}, TM), \\ T_\gamma\mathcal{P}^{1,2}_{x,y}(\mathbb{R}, M) &= \left\{ \zeta \in \mathcal{P}^{1,2}_{0_x,0_y}(\mathbb{R}, TM) \,\middle|\, \tau \circ \zeta = \gamma \right\}, \quad \gamma \in \mathcal{P}^{1,2}_{x,y} \ . \end{aligned}$$

According to this representation we may grasp the tangent space $T_\gamma\mathcal{P}^{1,2}_{x,y}$ as the space of the '$H^{1,2}_{\mathbb{R}}$-vector fields along the $H^{1,2}_{\mathbb{R}}$-curve γ' equipped with a Banach space topology.

Theorem 12 *Let $X \in C^\infty(TM)$ be a smooth vector field on M satisfying $X(x) = X(y) = 0$. Then the mapping*

$$C^\infty_{x,y}(\overline{\mathbb{R}}, M) \ni \gamma \mapsto \dot{\gamma} + X \circ \gamma \in C^\infty(\gamma^*TM)$$

can be extended to a smooth section in the Banach bundle $L^2_{\mathbb{R}}(\mathcal{P}^{1,2}_{x,y}{}^{*}TM)$:

$$\begin{aligned} \Gamma : \mathcal{P}^{1,2}_{x,y}(\mathbb{R}, M) &\to L^2_{\mathbb{R}}(\mathcal{P}^{1,2}_{x,y}{}^{*}TM) \\ \gamma &\mapsto \dot{\gamma} + X \circ \gamma \in L^2_{\mathbb{R}}(\gamma^*TM) \ . \end{aligned}$$

Proof. It is sufficient to verify the smoothness of this map with respect to the local representation by a chart together with the overlying local trivialization. We thus consider

$$\begin{aligned} H^{1,2}_{\mathbb{R}}(\mathcal{U}_h) \times L^2_{\mathbb{R}}(h^*TM) &\to L^2_{\mathbb{R}}(\mathcal{P}^{1,2}_{x,y}{}^{*}TM)_{\mid H^{1,2}_{\mathbb{R}}(\mathcal{U}_h)} \\ (\exp_h \xi, \eta) &\mapsto \nabla_2 \exp(\xi) \cdot \eta \ . \end{aligned}$$

The local representation of the section Γ consequently gets the shape

$$\text{(A.29)}\qquad \begin{array}{rcl} \Gamma_{\text{triv},h} : H^{1,2}_{\mathbb{R}}(h^*\mathcal{D}) & \to & L^2_{\mathbb{R}}(h^*\mathcal{D}) \\ \xi & \mapsto & \nabla_2 \exp(\xi)^{-1}\big(\nabla_1 \exp(\xi)\cdot \dot h \\ & & +\nabla_2 \exp(\xi)\cdot \nabla_t\xi + X(\exp_h \xi)\big) \ . \end{array}$$

We may also express this as

$$\text{(A.30)}\qquad \Gamma_{\text{triv},h}(\xi) = \Theta(\xi)\cdot \dot h + \nabla_t\xi + \nabla_2\exp(\xi)^{-1}\cdot (X\circ \exp_h)(\xi) \ .$$

As we already noticed in (A.20), the mapping

$$H^{1,2}_{\mathbb{R}} \ni \xi \mapsto \Theta(\xi) \in \mathcal{L}\big(H^{1,2}_{\mathbb{R}}(h^*TM); H^{1,2}_{\mathbb{R}}(h^*TM)\big)$$

is smooth, therefore in particular the mapping

$$\text{(A.31)}\qquad \xi \mapsto \Theta(\xi)\cdot \dot h \in H^{1,2}_{\mathbb{R}}(h^*TM) \quad \text{for} \quad \dot h \in C^\infty_{\overline{\mathbb{R}}}(h^*TM)$$

is also. We furthermore observe that

$$\text{(A.32)}\qquad H^{1,2}_{\mathbb{R}} \ni \xi \mapsto \nabla_t\xi \in L^2_{\mathbb{R}}$$

is continuously linear and thus also smooth. Considering now the open zero section neighbourhood $\mathcal{O} = h^*\mathcal{D} \subset h^*TM$, we are led to the smooth bundle map

$$\text{(A.33)}\qquad \begin{aligned} &f : \mathcal{O} \to h^*TM \\ &f(v) = \nabla_2\exp(v)^{-1}\cdot X(\exp_h v) \end{aligned}$$

satisfying the condition $f(0_{\pm\infty}) = 0_{\pm\infty}$ due to the assumption $X(x) = X(y) = 0$. Hence, the fundamental Lemma A.5 guarantees the smoothness of the map

$$f_* : H^{1,2}_{\mathbb{R}}(h^*\mathcal{D}) \to H^{1,2}_{\mathbb{R}}(h^*TM) \ .$$

Piecing all this together proves the assertion. □

Corollary A.9 *Given a smooth Morse function f on M, the mapping*

$$\begin{array}{rcl} F : \mathcal{P}^{1,2}_{x,y} & \to & L^2_{\mathbb{R}}(\mathcal{P}^{1,2\,*}_{x,y}TM), \quad x,y \in \operatorname{Crit} f \\ s & \mapsto & \dot s + \nabla f\circ s \end{array}$$

describes a smooth section in the specified L^2-Banach bundle. The local representation at $\gamma \in C^\infty_{x,y}$ is of the type

$$\begin{aligned} &F_{\text{loc},\gamma} : H^{1,2}_{\mathbb{R}}(\gamma^*\mathcal{D}) \to L^2_{\mathbb{R}}(\gamma^*TM) \\ &F_{\text{loc},\gamma}(\xi)(t) = \nabla_t\xi(t) + g\big(t,\xi(t)\big) \ , \end{aligned}$$

where $g : \overline{\mathbb{R}} \times \gamma^*\mathcal{D} \to \gamma^*TM$ is smooth and fibre respecting, satisfies the identity $g(\pm\infty, 0) = 0$ and is endowed with the asymptotic fibre derivatives $D_2g(\pm\infty, 0)$ which are conjugated linear operators of the Hessians $H^2f(x)$ and $H^2f(y)$, respectively.

Appendix B

The Geometric Boundary Operator

Referring to the brief historical outline in the introduction, we notice that another, geometric definition of the characteristic signs for isolated ∂-trajectories of the time-independent gradient flow was given there:

Definition B.1 *Let f be a smooth Morse function on M together with the critical points $x, y \in \operatorname{Crit} f$. We assume that the Morse-Smale condition is fulfilled, so that*

$$\mathcal{M}^f_{x,y} \approx W^u(x) \pitchfork W^s(y) = \mathcal{M}(x,y) \tag{B.1}$$

represents a $\mu(x) - \mu(y)$-dimensional manifold consisting of parametrized gradient trajectories. Let all unstable manifolds $W^u(x)$ for $x \in \operatorname{Crit} f$ be endowed with arbitrary and noncanonically fixed orientations. Provided with the additional assumption that M is orientable and oriented, we thus obtain immediately induced orientations for the associated, codimensional stable manifolds $W^s(x)$, $x \in \operatorname{Crit} f$. We now consider the short exact sequence associated to a non-critical point $p \in \mathcal{M}(x,y)$

$$0 \longrightarrow T_p\mathcal{M}(x,y) \xrightarrow{i} T_pW^u(x) \oplus T_pW^s(y) \xrightarrow{j} T_pM \longrightarrow 0 \tag{B.2}$$

where

$$\begin{aligned} i(u) &= (u, -u) \\ j(v,w) &= v + w \ . \end{aligned}$$

We note that exactness is essentially due to the transversal intersection as indicated in (B.1). Starting from the given orientations on $T_pW^u(x)$, $T_pW^s(y)$ and T_pM, we are supplied with an orientation on the connected component $\mathcal{M}(x,y)^o_p$ of p, which is induced by this short exact sequence. Thus we obtain *orientations* $o_{\text{geom}}[u]$ *of the operator classes* $[u, D_u] \in \Lambda$ *in a geometrical way for orientable manifolds* M. *These operator classes belong to the trajectories*

$$u = u_{xy} \in \mathcal{M}^f_{x,y}\,,$$
$$(x,y) \in X = \left\{\, (x,y) \in \operatorname{Crit}_k f \times \operatorname{Crit}_l f \,\middle|\, k > l,\ \mathcal{M}(x,y) \neq \emptyset \,\right\}\,.$$

Consequently, the comparison of these geometric orientations with the canonical orientation for isolated trajectories by the free $\mathbb{R}$-action from time-shifting provides the characteristic signs of the geometric boundary operator. Thus the sum of all characteristic signs for fixed endpoints x, y is nothing more than the intersection number yielded by the transversal intersection

$$W^u(x) \pitchfork W^s(y) \pitchfork M^a \tag{B.3}$$

within the level surface M^a associated to a regular value $f(y) < a < f(x)$. This level surface is oriented canonically by the normal field $-\nabla f$.

Theorem 13 *The noncanonical geometric orientation* o_{geom} *of the operator classes* $[u_{xy}]$, $(x,y) \in X$ *can be extended to a coherent orientation* $\sigma \in \mathcal{C}_\Lambda$.

Proof. Let us mention first that it suffices to consider the equivalence classes $\{\sigma\} \in \widetilde{\mathcal{C}}_\Lambda$ in the sense of (4.18) from Section 4.1.4. We further notice that the transformation group $\{\pm 1\}^{\operatorname{Crit} f}$ on the set of these geometric orientations acts in the same way on $\widetilde{\mathcal{C}}_\Lambda$ as it is expressed in (4.19). Therefore it is sufficient to verify the condition of coherence

$$o_{\text{geom}}[u_{xy}] \,\#\, o_{\text{geom}}[u_{yz}] \;=\; o_{\text{geom}}[u_{xy} \#_\rho u_{yz}] \tag{B.4}$$

for the geometric orientation. This will be reduced step by step to a simpler relation for the linear gluing version from Section 3.1.2.

Throughout the following technical discussions we will refer to the notations from Section 2.2 on Fredholm theory.

Definition B.2 *We define* $\Sigma^{\text{diag}}_{\text{triv}}$, *the so-called set of* Σ_{triv}*-operators in ordered diagonal form, as the subset of* Σ_{triv} *comprising the operators* $K = \frac{\partial}{\partial t} + D$, *for which* $D \in \mathcal{A}$ *has the following shape:*

$$D = \operatorname{diag}(\lambda_1, \ldots, \lambda_n),\ \ \lambda_i \in C^0(\overline{\mathbb{R}}, \mathbb{R})$$
$$\textit{where}\quad \lambda_i(\pm\infty) = \lambda_i^{\pm} \begin{cases} > 0, & \textit{for } 1 \leqslant i \leqslant n - \mu(A^{\pm}) \\ < 0, & \textit{for } n - \mu(A^{\pm}) < i \leqslant n\,. \end{cases}$$

Lemma B.3 *Let us assume a* $K_A \in \Sigma_{\text{triv}}$ *with* $A \in \mathcal{A}$ *smooth and asymptotically constant, i.e.* $A(t) = \text{const}^{\pm}$ *for* $|t| \geqslant T$. *Then there exists a conjugation transformation by an asymptotically constant* $C \in C^\infty(\overline{\mathbb{R}}, \operatorname{GL}(n,\mathbb{R}))$, *so that* $\widetilde{K} = C^{-1} K C \in \Sigma^{\text{diag}}_{\text{triv}}$ *has ordered diagonal form.*

Proof. The first step is to construct a $C^{\pm} \in \mathrm{GL}^+(n, \mathbb{R})$, such that we obtain the representation

(B.5) $$C^{\pm^{-1}} A^{\pm} C^{\pm} = \mathrm{diag}(\lambda_1^{\pm}, \ldots, \lambda_n^{\pm})$$

with ordered eigenvalues as specified in Definition B.2. We consequently choose an asymptotically constant curve $C \in C^{\infty}(\overline{\mathbb{R}}, \mathrm{GL}^+(n, \mathbb{R}))$ with $C(\pm\infty) = C^{\pm}$ as in the conclusion of Lemma 2.15. Thus, without loss of generality, A is already endowed with ends in ordered diagonal form,

(B.6) $$A^{\pm} = \mathrm{diag}(\lambda_1^{\pm}, \ldots, \lambda_n^{\pm}) \ .$$

In the following step we consider the ordinary linear differential equation

$$\dot{s}(t) = -A(t) \cdot s(t)$$

and the associated evolution operator

$$U(\cdot) = U(-T, \cdot) \in C^{\infty}(\mathbb{R}, \mathrm{GL}^+(n, \mathbb{R}))$$

with respect to

(B.7) $$\dot{U}(t) = -A(t) \cdot U(t), \ \ U(-T) = \mathbb{1} \ .$$

The formula

$$U(t) = \exp\Big(\int_{-T}^{t} A(\tau) d\tau \Big)$$

together with identity (B.6) yields the representations

(B.8) $$U(t) = \begin{pmatrix} e^{-\lambda_1^{-} \cdot (t+T)} & & \\ & \ddots & \\ & & e^{-\lambda_n^{-} \cdot (t+T)} \end{pmatrix}, \qquad \text{for } t \leqslant -T \ ,$$

(B.9) $$U(t) = U(T) \cdot \begin{pmatrix} e^{-\lambda_1^{+} \cdot (t-T)} & & \\ & \ddots & \\ & & e^{-\lambda_n^{+} \cdot (t-T)} \end{pmatrix}, \qquad \text{for } t \geqslant T \ .$$

Then we define $C \in C^{\infty}(\overline{\mathbb{R}}, \mathrm{GL}^+(n, \mathbb{R}))$ as follows:[1]

$$C(t) = \begin{cases} \begin{matrix} \beta^-(t+T) \cdot \mathbb{1} \\ +(1 - \beta^-(t+T)) \cdot U(t) \end{matrix}, & t \leqslant -T \ , \\ U(t), & -T \leqslant t \leqslant T \ , \\ U(T) \cdot \left[\begin{matrix} \beta^+(t-T) \cdot \mathbb{1} \\ +(1 - \beta^+(t-T)) \cdot U(T)^{-1} U(t) \end{matrix} \right], & t \geqslant T \ . \end{cases}$$

1As to the definition of $\beta^{\pm}$, see Definition 2.47.

Straightforward computation using identities (B.8) and (B.9) together with the formula

$$C^{-1}KC = \frac{\partial}{\partial t} + C^{-1}\dot{C} + C^{-1}A\,C \ ,$$

where A is already given in asymptotically constant diagonal form, shows that conjugation by C yields the diagonalization as described in the assertion. □

Corollary B.4 *Let $K \in \Sigma_{\mathrm{triv}}$ be surjective and asymptotically constant. Due to Lemma B.3 we may assume without loss of generality that K is given in ordered diagonal form. Then the kernel of K is generated by the curves $v_1, \ldots, v_{\mu(K^-)-\mu(K^+)}$ satisfying* [2]

$$\begin{aligned}
&v_i = c_i \cdot e_{i+n-\mu(K^-)},\ c_i \in C^\infty(\overline{\mathbb{R}}, \mathbb{R}_+),\\
&c_i(t) = \begin{cases} c^- \mathrm{e}^{-\lambda_i^- t}, & t \leqslant -T \\ c^+ \mathrm{e}^{-\lambda_i^+ t}, & t \geqslant T \end{cases}, \quad c_i(t) > 0 \ \textit{always},\\
&1 \leqslant i \leqslant \mu(K^-) - \mu(K^+) \ .
\end{aligned}$$

The case where we cannot necessarily assume an asymptotical constant operators can be elucidated by the following

Lemma B.5 *Given $K = \frac{\partial}{\partial t} + A \in \Sigma_{\mathrm{triv}}$, $A \in C^1\big(\overline{\mathbb{R}}, \mathrm{End}(\mathbb{R}^n)\big)$ such that $A(t)$ is self-adjoint for all $|t| > T$ for some $T > 0$, the following asymptotical property holds for the solutions $v \in \ker K \subset H^{1,2}(\mathbb{R}, \mathbb{R}^n)$: $\lim\limits_{t\to\pm\infty} \frac{v(t)}{\|v(t)\|} = \mathrm{e}^{\pm}$ exist and satisfy $A^\pm \mathrm{e}^\pm = \lambda^\pm \mathrm{e}^\pm$ with $\lambda^+ > 0$, $\lambda^- < 0$.*

Proof. Without loss of generality, we may assume that $A(t)$ is self-adjoint for all $t \in \overline{\mathbb{R}}$. Suppose that $v \in \ker K$ is given, i.e.

$$\dot{v}(t) = A(t) \cdot v(t),\ t \in \mathbb{R} \ .$$

We denote by $e(t) = \frac{v(t)}{\|v(t)\|}$, $t \in \mathbb{R}$, the curve $e \in C^\infty(\mathbb{R}, S^{n-1})$. Straightforward computation shows that e satisfies the ordinary differential equation on $S^{n-1} \subset \mathbb{R}^n$:

(B.10)
$$\dot{e}(t) = \mathrm{pr}_{e^\perp(t)} \circ A(t) \cdot e(t) \ ,$$

where $e^\perp(t) = T_{e(t)} S^{n-1} \subset \mathbb{R}^n$ denotes the subspaces perpendicular to $e(t)$ and $\mathrm{pr}_{e^\perp}$ the orthogonal projection onto the tangent space $e^\perp$. Due to the self-adjointness of $A(t)$ for all $t \in \overline{\mathbb{R}}$ this differential equation can be restated as the equation associated to the time-dependent gradient flow of

$$f : \mathbb{R} \times S^{n-1} \to \mathbb{R},\ \ f_t(e) = \langle A(t)e, e\rangle \ .$$

[2] e_i denotes the i-th vector of the canonical standard basis of $\mathbb{R}^n$.

Similar to the calculations in the compactness analysis for time-dependent trajectories we obtain

$$\text{(B.11)}\quad \langle \dot e(s), \dot e(s)\rangle = \langle \nabla f_s(e(s)), \dot e(s)\rangle = \frac{d}{dt}\Big(f_t(e(t))\Big)(s) - \frac{\partial}{\partial t} f_t(e(s))$$

where $\frac{\partial}{\partial t} f_t(e(s)) = \langle \dot A(s)e(s), e(s)\rangle$. Thus, Lemma 2.2 applied to $A \in C^1(\overline{\mathbb{R}}, \mathrm{End}(\mathbb{R}^n))$ implies that

$$\int_{-\infty}^{\infty} \frac{\partial}{\partial t} f_t(e(s))ds < \infty$$

exists, and since S^{n-1} is compact, $t \mapsto f_t(e(t))$ is bounded. Altogether we obtain the finite L^2-norm

$$\text{(B.12)}\qquad \int_{-\infty}^{\infty} \langle \dot e(s), \dot e(s)\rangle ds < \infty$$

and consequently $\lim_{t\to\pm\infty} \dot e(t) = 0$. Due to the compactness of S^{n-1} the assertion now follows. □

Continuation of the proof of Theorem 13. Given $(u_{xy}, u_{yz}) \in \mathcal{M}^f_{x,y} \times \mathcal{M}^f_{y,z}$ and $(x,y),(y,z) \in X$, we choose chart neighbourhoods $\mathcal{U}_{xy}$ and $\mathcal{U}_{yz}$ on the Banach manifolds $\mathcal{P}^{1,2}_{x,y}$ aund $\mathcal{P}^{1,2}_{y,z}$ at asymptotically constant smooth curves h_{xy} and h_{yz}, i.e.

$$u_{xy} = \exp_{h_{xy}} \xi,\ \ \xi \in C^\infty(\overline{\mathbb{R}}, h^*_{xy}TM)\ .$$

In these local coordinates, the focal Fredholm map $\frac{\partial}{\partial t} + \nabla f$ is expressed as

$$F_{u_{\dots}}(\xi) = \Theta(\xi)\cdot \dot h + \nabla_t \xi + \nabla_2 \exp(\xi)^{-1}\cdot (\nabla f \circ \exp_h \xi)\ .$$

We denote the differential of $F_{u_{\dots}}$ at $\xi \in H^{1,2}_{\mathbb{R}}(h^*_{\dots}TM)$ by $D_{u_{\dots}}(\xi)$. In the following analysis we argue that we may assume without loss of generality that $D_{u_{\dots}}(\xi) = \nabla_t + A(\xi)$ is asymptotically constant, which generally is only given if ξ has compact support. Defining a cut-off function $\beta_T \in C^\infty(\overline{\mathbb{R}}, [0,1])$ for $T \in [0,1]$ by

$$\beta_0 \equiv 1,\ \ \beta_T = \begin{cases} 1, & |t| \leqslant \frac{1}{T} - 1 \\ 0, & |t| \geqslant \frac{1}{T} \\ \text{monotone}, & \text{otherwise} \end{cases} \qquad \text{for } T > 0$$

and the continuous path

$$\begin{aligned} \beta_T :\ [0,1] &\to C^\infty(\overline{\mathbb{R}}, h^*_{\dots}TM) \\ T &\mapsto \beta_T \cdot \xi \end{aligned}$$

we obtain the continuous path of Fredholm operators $T \mapsto D_{u_{\dots}}(\xi_T)$ with $D_{u_{\dots}}(\xi) = D_{u_{\dots}}(\xi_0)$. We find a $T_o > 0$ such that $D_{u_{\dots}}(\xi_T)$ is surjective for all $0 \leqslant T \leqslant T_o$. Thus, the geometric orientation $o_{\text{geom}}[u_{\dots}]$ of $\ker D_{u_{\dots}}$ induces a unique orientation of $\ker D_{u_{\dots}}(\xi_{T_o})$, which can be determined by explicit analysis of the asymptotic behaviour of the solutions generating the kernel in the same way as for the original operator which is not asymptotically constant. This follows from the given continuation argument for $T \to 0$ together with Lemma B.5 by using the fact that ξ_T is $\overline{\mathbb{R}}$-differentiable for all $T \in [0, T_o]$. Referring to Lemma 3.13 we notice that, concerning the verification of the coherence condition, it suffices to analyse the local situation at u_{xy}, u_{yz} and $u_{xy}\#^o_\rho u_{yz}$ by means of local coordinates and trivializations. To sum up, we may start without loss of generality from asymptotically constant and surjective operators $D_{u_{xy}}$, $D_{u_{yz}}$.

Now let $T_xW^u(x)$, $T_yW^u(y)$ and $T_zW^u(z)$ be endowed with fixed orientations. Using the trivializations of $u^*_{xy}TM$ and $u^*_{yz}TM$, which coincide at T_yM and transform the operators $D_{u_{xy}}$, $D_{u_{yz}}$ into ordered diagonal form, we are led to the orientations

$$\begin{aligned} &e_{i_1} \wedge \ldots \wedge e_{i_{\mu(x)}} \quad \text{on} \quad \operatorname{span}(e_{n-\mu(x)+1}, \ldots, e_n) \cong T_xW^u(x)\ , \\ &e_{j_1} \wedge \ldots \wedge e_{j_{\mu(y)}} \quad \text{on} \quad \operatorname{span}(e_{n-\mu(y)+1}, \ldots, e_n) \cong T_yW^u(y)\ , \\ &e_{k_1} \wedge \ldots \wedge e_{k_{\mu(z)}} \quad \text{on} \quad \operatorname{span}(e_{n-\mu(z)+1}, \ldots, e_n) \cong T_zW^u(z)\ . \end{aligned} \tag{B.13}$$

According to Corollary B.4 the kernels $\ker D_{u_{xy},\text{triv}}$ and $\ker D_{u_{yz},\text{triv}}$ are generated by

$$v_1, \ldots, v_{\mu(x)-\mu(y)},\ \ v_i = c_i \cdot e_{i+n-\mu(x)},\ \ c_i > 0,\ \ 1 \leqslant i \leqslant \mu(x) - \mu(y)$$

and

$$w_1, \ldots, w_{\mu(y)-\mu(z)},\ \ w_j = d_j \cdot e_{j+n-\mu(y)},\ \ d_j > 0,\ \ 1 \leqslant j \leqslant \mu(y) - \mu(z)\ ,$$

respectively. Then the respective geometric orientations

$$\begin{aligned} o_{\text{geom}}[u_{xy}]_{\text{triv}} &= \epsilon_1 \cdot v_1 \wedge \ldots \wedge v_{\mu(x)-\mu(y)}, \quad \epsilon_1 \in \{\pm 1\} \\ o_{\text{geom}}[u_{yz}]_{\text{triv}} &= \epsilon_2 \cdot w_1 \wedge \ldots \wedge w_{\mu(y)-\mu(z)}, \quad \epsilon_2 \in \{\pm 1\} \end{aligned} \tag{B.14}$$

are determined uniquely by

$$\begin{aligned} e_{i_1} \wedge \ldots \wedge e_{i_{\mu(x)}} &= o_{\text{geom}}[u_{xy}]_{\text{triv}} \wedge e_{j_1} \wedge \ldots \wedge e_{j_{\mu(y)}}, \\ e_{j_1} \wedge \ldots \wedge e_{j_{\mu(y)}} &= o_{\text{geom}}[u_{yz}]_{\text{triv}} \wedge \ldots \wedge e_{k_{\mu(z)}}\ , \end{aligned} \tag{B.15}$$

due to Definition B.1. It likewise holds that

$$e_{i_1} \wedge \ldots \wedge e_{i_{\mu(x)}} = o_{\text{geom}}[u_{xy}\#^o_\rho u_{yz}]_{\text{triv}} \wedge e_{k_1} \wedge \ldots \wedge e_{k_{\mu(z)}}\ . \tag{B.16}$$

Thus, it remains merely to analyse the linear gluing version

$$
\text{(B.17)} \quad \begin{array}{rrcl} \#_\rho: & D_{u_{xy},\mathrm{triv}} \times \ker D_{u_{yz},\mathrm{triv}} & \xrightarrow{\cong} & \ker\left(D_{u_{xy},\mathrm{triv}} \# \rho D_{u_{yz},\mathrm{triv}}\right) \\ & (\xi,\zeta) & \mapsto & \mathrm{Proj}^{L^2}_{\ker\ldots}(\xi_\rho + \zeta_{-\rho}) \end{array}
$$

and to verify that the orientation

$$
\epsilon_1\epsilon_2 \cdot (v_1\#_\rho 0) \wedge \ldots \wedge (v_{\mu(x)-\mu(y)}\#_\rho 0) \wedge (0\#_\rho w_1) \wedge \ldots \wedge (0\#_\rho w_{\mu(y)-\mu(z)})
$$

coincides with the geometric orientation $o_{\mathrm{geom}}[u_{xy}\#^o_\rho u_{yz}]_{\mathrm{triv}}$.

Given the surjective operators $D_{u_{xy},\mathrm{triv}} = \frac{\partial}{\partial t} + A_{xy} \in \Sigma_{\mathrm{triv}}$ and $D_{u_{yz},\mathrm{triv}} = \frac{\partial}{\partial t} + A_{yz} \in \Sigma_{\mathrm{triv}}$ in ordered diagonal form

$$
\begin{array}{c} A_{xy} = \mathrm{diag}\,(\lambda_1,\ldots,\lambda_n), \;\; A_{yz} = \mathrm{diag}\,(\nu_1,\ldots,\nu_n) \\ \text{with} \quad \lambda_i^+ = \nu_i^-,\; i=1,\ldots,n \;, \end{array}
$$

we construct the following homotopies:

$$
\text{(B.18)} \quad \begin{array}{l} [0,1] \ni \tau \mapsto \frac{\partial}{\partial t} + A^\tau_{xy},\; \frac{\partial}{\partial t} + A^\tau_{yz} \in \Sigma_{\mathrm{triv}} \subset \mathcal{F}(H^{1,2},L^2), \\ A^\tau_{xy} = \begin{array}{l} (1-\tau)\cdot \mathrm{diag}\,(\lambda_1,\ldots,\lambda_n) \\ + \;\; \tau \cdot \mathrm{diag}\,(\lambda_1,\ldots,\lambda_{n-\mu(y)},\lambda^+_{n-\mu(y)+1},\ldots,\lambda^+_n) \end{array}, \\ A^\tau_{yz} = \begin{array}{l} (1-\tau)\cdot \mathrm{diag}\,(\nu_1,\ldots,\nu_n) \\ + \;\; \tau \cdot \mathrm{diag}\,(\nu^-_1,\ldots,\nu^-_{n-\mu(y)},\nu_{n-\mu(y)+1},\ldots,\nu_n) \end{array}. \end{array}
$$

Obviously, both homotopies $\frac{\partial}{\partial t} + A^\tau_{xy}$ and $\frac{\partial}{\partial t} + A^\tau_{yz}$ run continuously through surjective operators from Σ_{triv}, which respectively have constant kernels

$$
\begin{array}{rl} & \ker\left(\frac{\partial}{\partial t} + A^\tau_{xy}\right) = \mathrm{span}\,(v_1,\ldots,v_{\mu(x)-\mu(y)}) \\ \text{and} & \ker\left(\frac{\partial}{\partial t} + A^\tau_{yz}\right) = \mathrm{span}\,(w_1,\ldots,w_{\mu(y)-\mu(z)}) \;. \end{array}
$$

Thus, the geometric orientations $o_{\mathrm{geom}}[u_{xy}]_{\mathrm{triv}}$ and $o_{\mathrm{geom}}[u_{yz}]_{\mathrm{triv}}$ remain constant during these homotopies. Moreover these homotopies are compatible with the linear gluing version (B.17). This means that $\tau \mapsto \frac{\partial}{\partial t} + (A^\tau_{xy}\#_\rho A^\tau_{yz})$ describes a continuous homotopy through surjective operators, where $\rho \geqslant \max_{\tau\in[0,1]} \rho_0(D^\tau_{u_{xy},\mathrm{triv}}, D^\tau_{u_{yz},\mathrm{triv}})$ may be chosen as independent of τ. During this homotopy the kernel $\ker\left(\frac{\partial}{\partial t} + A^\tau_{xy}\#_\rho A^\tau_{yz}\right)$ is generally not kept invariant, but the orientations $o_{\mathrm{geom}}[u_{xy}\#^o_\rho u_{yz}]_{\mathrm{triv}}$ and $o_{\mathrm{geom}}[u_{xy}]_{\mathrm{triv}}\# o_{\mathrm{geom}}[u_{yz}]_{\mathrm{triv}}$ remain fixed by reasons of continuity. Due to the construction it holds that

$$
\begin{array}{l} \epsilon_1\epsilon_2 \cdot v_{1,\rho} \wedge \ldots \wedge v_{\mu(x)-\mu(y),\rho} \wedge w_{1,-\rho} \wedge \\ \ldots \wedge w_{\mu(y)-\mu(z),-\rho} \wedge e_{k_1} \wedge \ldots \wedge e_{k_{\mu(z)}} \end{array} \simeq e_{i_1} \wedge \ldots \wedge e_{i_{\mu(x)}} \;.
$$

Thus, the identity

$$o_{\text{geom}}[u_{xy}\#^o_\rho u_{yz}]_{\text{triv}} = o_{\text{geom}}[u_{xy}]_{\text{triv}}\# o_{\text{geom}}[u_{yz}]_{\text{triv}} \tag{B.19}$$

follows from the explicit representation of the linear gluing version $\#_\rho$ in (B.17) for $\tau = 1$:

$$\begin{aligned} v_i\#_\rho 0 &= v_{i,\rho}, \quad i = 1, \ldots, \mu(x) - \mu(y) \\ 0\#_\rho w_j &= w_{j,\rho}, \quad j = 1, \ldots, \mu(y) - \mu(z) \ . \end{aligned}$$

□

Bibliography

[Bo] R. Bott, *Morse theory indomitable*, Publ. Math. I.H.E.S. **68** (1988), 99–114.

[C] C. C. Conley, *Isolated invariant sets and the Morse index*, CBMS Reg. Conf. Series in Math. **38** (1978), A.M.S., Providence, R.I.

[C-Z] C. C. Conley and E. Zehnder, *Morse–type index theory for flows and periodic solutions for Hamiltonian equations*, Comm. Pure Appl. Math. **37** (1984), 207–253.

[Cou] R. Courant, *Dirichlet's principle, conformal mappings and mimimal surfaces*, Interscience, New York, 1950, reprinted: Springer, New York–Heidelberg–Berlin, 1977.

[Don] S. K. Donaldson, *The orientation of Yang–Mills modulispaces and four manifold topology*, J. Diff. Geom. **26** (1987), 397–428.

[Eli] H. I. Eliasson, *Geometry of manifolds of maps*, J. Diff. Geom. **1** (1967), 169–194.

[E-S] S. Eilenberg and N. Steenrod, *Foundations of algebraic topology*, Princeton University Press, 1952.

[F-H] A. Floer and H. Hofer, *Coherent orientations for periodic orbit problems in symplectic geometry*, Math. Z. **212** (1993), 13–38.

[F1] A. Floer, *Morse theory for Lagrangian intersections*, J. Diff. Geom. **28** (1988), 513–547.

[F2] A. Floer, *A relative Morse index for the symplectic action*, Comm. Pure Appl. Math. **41** (1988), 393–407.

[F3] A. Floer, *The unregularized gradient flow of the symplectic action*, Comm. Pure Appl. Math. **41** (1988), 775–813.

[F4] A. Floer, *An instanton–invariant for 3–manifolds*, Comm. Math. Phys. **118** (1988), 215–240.

[F5] A. Floer, *Witten's complex and infinite dimensional Morse theory*, J. Diff. Geom. **30** (1989), 207–221.

[F6] A. Floer, *Symplectic fixed points and holomorphic spheres*, Comm. Math. Phys. **120** (1989), 575–611.

[Fr] J. M. Franks, *Morse–Smale flows and homotopy theory*, Topology **18** (1979), 199–215.

[Hi] M. W. Hirsch, *Differential topology*, Graduate Texts in Math. **33**, Springer, New York, 1976.

[Kli] W. Klingenberg, *Lectures on closed geodesics*, Grundl. der Math. Wiss. **230**, Springer, 1978.

[L-McO] R. B. Lockhart and R. C. McOwen, *Elliptic operators on noncompact manifolds*, Ann. Sci. Norm. Sup. Pisa IV-**12** (1985), 409–446.

[McD] D. McDuff, *Elliptic methods in symplectic geometry*, Bull. A. M. S. **23** (1990), 311–358.

[M1] J. W. Milnor, *Morse theory*, Ann. of Math. Studies **51**, Princeton Univ. Press, 1963.

[M2] J. W. Milnor, *Lectures on the h–cobordism theorem*, Math. Notes, Princeton Univ. Press, Princeton, 1965.

[M-S] J. W. Milnor and J. Stasheff, *Characteristic classes*, Ann. of Math. Studies **76**, Princeton Univ. Press, 1974.

[S] D. Salamon, *Morse theory, the Conley index and Floer homology*, Bull. L. M. S. **22** (1990), 113–140.

[S-Z1] D. Salamon and E. Zehnder, *Floer homology, the Maslov index and periodic orbits of Hamiltonian equations*, in: Analysis et cetera, edited by P. H. Rabinowitz and E. Zehnder, Academic Press, 1990, 573–600.

[S-Z2] D. Salamon and E. Zehnder, *Morse theory for periodic solutions of Hamiltonian systems and the Maslov index*, Comm. Pure Appl. Math. **45** (1992), 1303–1360.

[Sm1] S. Smale, *On gradient dynamical systems*, Ann. of Math. **74** (1961), 199–206.

[Sm2] S. Smale, *Differentiable dynamical systems*, Bull. A. M. S. **73** (1967), 747–817.

[Sm3] S. Smale, *An infinite dimensional version of Sard's theorem*, Ann. of Math. **87** (1973), 213–221.

[Sp] E. Spanier, *Algebraic topology*, McGraw–Hill, New York, 1966.

[T] C. H. Taubes, *Self–dual Yang–Mills connections on non–self–dual 4–manifolds*, J. Diff. Geom. **17** (1982), 139–170.

[Th] R. Thom, *Sur une partition en cellules associés à une fonction sur une variété*, C. R. Acad. Sci. Paris **228** (1949), 973–975.

[W] E. Witten, *Supersymmetry and Morse theory*, J. Diff. Geom. **17** (1982), 661–692.

Index

Progress in Mathematics

Edited by:

J. Oesterlé
Départment de Mathématiques
Université de Paris VI
4, Place Jussieu
75230 Paris Cedex 05, France

A. Weinstein
Department of Mathematics
University of California
Berkeley, CA 94720
U.S.A.

Progress in Mathematics is a series of books intended for professional mathematicians and scientists, encompassing all areas of pure mathematics. This distinguished series, which began in 1979, includes authored monographs, and edited collections of papers on important research developments as well as expositions of particular subject areas.

We encourage preparation of manuscripts in such form of TeX for delivery in camera-ready copy which leads to rapid publication, or in electronic form for interfacing with laser printers or typesetters.

Proposals should be sent directly to the editors or to: Birkhäuser Boston, 675 Massachusetts Avenue, Cambridge, MA 02139, U.S.A.

91 GOLDSTEIN. Séminaire de Théorie des Nombres, Paris 1988-89
92 CONNES/DUFLO/JOSEPH/RENTSCHLER. Operator Algebras, Unitary Representations, Enveloping Algebras, and Invariant Theory. A Collection of Articles in Honor of the 65th Birthday of Jacques Dixmier
93 AUDIN. The Topology of Torus Actions on Symplectic Manifolds
94 MORA/TRAVERSO (eds.) Effective Methods in Algebraic Geometry
95 MICHLER/RINGEL (eds.) Representation Theory of Finite Groups and Finite–Dimensional Algebras
96 MALGRANGE. Equations Différentielles à Coefficients Polynomiaux
97 MUMFORD/NORMAN/NORI. Tata Lectures on Theta III
98 GODBILLON. Feuilletages, Etudes géométriques
99 DONATO/DUVAL/ELHADAD/TUYNMAN. Symplectic Geometry and Mathematical Physics. A Collection of Articles in Honor of J.-M. Souriau
100 TAYLOR. Pseudodifferential Operators and Nonlinear PDE
101 BARKER/SALLY. Harmonic Analysis on Reductive Groups
102 DAVID. Séminaire de Théorie des Nombres, Paris 1989-90
103 ANGER/PORTENIER. Radon Integrals
104 ADAMS/BARBASCH/VOGAN. The Langlands Classification and Irreducible Characters for Real Reductive Groups
105 TIRAO/WALLACH. New Developments in Lie Theory and Their Applications
106 BUSER. Geometry and Spectra of Compact Riemann Surfaces
107 BRYLINSKI. Loop Spaces, Characteristic Classes and Geometric Quantization
108 DAVID. Séminaire de Théorie des Nombres, Paris 1990-91
109 EYSSETTE/GALLIGO. Computational 00..Algebraic Geometry
110 LUSZTIG. Introduction to Quantum Groups
111 SCHWARZ. Morse Homology